NanoScience and Technology

NanoScience and Technology

Series Editors:
P. Avouris B. Bhushan D. Bimberg K. von Klitzing H. Sakaki R. Wiesendanger

Anatolie Sidorenko
Editor

Fundamentals of Superconducting Nanoelectronics

With 496 Figures

Springer

Editor

Prof.Dr. Anatolie Sidorenko

Academy of Science of Moldova
Institute of Electronic Engineering and Nanotechnologies
Academiei Street 3/3, 2028 Chisinau, Moldova
E-mail: anatoli.sidorenko@kit.edu

Series Editors:

Professor Dr. Phaedon Avouris

IBM Research Division
Nanometer Scale Science & Technology
Thomas J. Watson Research Center
P.O. Box 218
Yorktown Heights, NY 10598, USA

Professor Dr. Bharat Bhushan

Ohio State University
Nanotribology Laboratory
for Information Storage
and MEMS/NEMS (NLIM)
Suite 255, Ackerman Road 650
Columbus, Ohio 43210, USA

Professor Dr. Dieter Bimberg

TU Berlin, Fakutät Mathematik/
Naturwissenschaften
Institut für Festkörperphyisk
Hardenbergstr. 36
10623 Berlin, Germany

Professor Dr., Dres. h.c. Klaus von Klitzing

Max-Planck-Institut
für Festkörperforschung
Heisenbergstr. 1
70569 Stuttgart, Germany

Professor Hiroyuki Sakaki

University of Tokyo
Institute of Industrial Science
4-6-1 Komaba, Meguro-ku
Tokyo 153-8505, Japan

Professor Dr. Roland Wiesendanger

Institut für Angewandte Physik
Universität Hamburg
Jungiusstr. 11
20355 Hamburg, Germany

NanoScience and Technology ISSN 1434-4904
ISBN 978-3-642-20157-8 e-ISBN 978-3-642-20158-5
DOI 10.1007/978-3-642-20158-5
Springer Heidelberg Dordrecht London New York

Library of Congress Control Number: 2011929351

Cover design: eStudio Calamar Steinen

Printed on acid-free paper

Springer is part of Springer Science+Business Media (www.springer.com)

Foreword

Over the last decade, we have witnessed a large number of novel developments and rapid progress in superconducting electronics. This has been triggered by mainly two facts:

- The use of novel physical phenomena, which have been theoretically predicted already during the previous decades but detected only recently in superconducting materials.
- The astonishing advances in thin film and nanotechnology, which allows us the reproducible fabrication of the superconducting devices and circuits with nanometer resolution.

This pleasant development is also reflected in a considerable increase in funding and investments related to the field of superconducting electronics. An increase by almost a factor of three has been observed worldwide over the last decade.

The main goal of this book is to give a profound insight into the basic phenomena occurring in superconductors and hybrid systems composed of superconductors and other materials such as ferromagnets on a nanometer scale (e.g., Fulde–Ferrell–Larkin–Ovchinikov state, triplet superconductivity, crossed Andreev reflection, coherent dynamics of superconducting junctions). Moreover, this book provides a comprehensive overview on the application of these phenomena in novel nanoelectronic devices and circuits (e.g., SIS tunnel junction phase detectors, pi-junction-based logic circuits, superconducting spintronic devices). Information on these new developments so far has to be collected from publications in technical journals or presentations at conferences and workshops. Therefore, the collection of review papers presented in this book provides a valuable overview on the Frontiers of Superconducting Electronics, ranging from the discussion of the theoretical background and the experimental study of the phenomena to their application in novel devices. Although it is impossible to cover all aspects of broad and complex field of superconducting electronics, this book addresses the most interesting phenomena and key developments of the last decade in a comprehensive way.

The book is organized as following: Chapter 1 is devoted to one of the fundamental problems of superconductivity – fluctuations, which are also important for numerous technical applications of superconductivity such as ultra-sensitive detectors. Chapter 2 deals with the experimental study of superconducting fluctuations in systems with reduced dimensionality – nanowires. Chapters 3 and 4 present the results of both theoretical (Chap. 3) and experimental (Chap. 4) studies of the crossed Andreev reflection and nonlocal transport phenomena in superconductor–ferromagnet hybrid structures. Since there has been a particularly rapid progress in the new research area of superconducting spintronics, this book puts special emphasis on the advanced theoretical description (Chaps. 5–7) as well as the experimental study of superconductor–ferromagnet hybrid structures (Chap. 8). Chapter 9 shows how point-contact spectroscopy of superconducting materials can be used as a powerful method for the investigation of the energy gap and the electron–phonon interaction in superconductors. In Chaps. 10 and 11, superconducting integrated receivers and cryogenic phase-lock loop systems are presented as prominent examples on how theoretical concepts, novel physics, and advanced fabrication technologies can be combined to achieve smart superconducting devices.

The authors of the chapters and the editor of the book Prof. Anatolie Sidorenko are well-known leading specialists, who have been involved in various research programs dealing with superconducting devices and their applications. All of them were former lecturers at NATO-ARW and NATO-ASI.

This book will be useful for a broad readership including researchers, engineers, lecturers, Ph.D. students, and all others, who would like to gain insight into the Frontiers of Superconductivity at nanoscale.

Garching *Rudolf Gross*
June 2011

Preface

The idea to write this book appeared after a series of workshops devoted to superconductivity of low-dimensional objects, which we organized last decade. In 2004 director of Walther-Meißner-Institut Professor Rudolf Gross and I organized an NATO Advanced Research Workshop "Nanoscale Devices, Fundamentals and Applications" and published the book with the same title, collecting the best of reports, presented on that workshop. As we realized a bit later, the book was in demand by colleagues, who deal with applications of superconductivity. For example, the group of researchers is engaged in development and fabrication of a very sensitive superconducting sensor for infrared radiation, superconducting thin-film bolometer, would like to achieve the highest possible sensitivity. They develop different technological processes for improvement of the quality of the superconducting film, trying to obtain thin films with the narrowest width of superconducting transition. In case, when a member of such group has knowledge in superconducting fluctuations (which are rather noticeable for low-dimensional objects) that there exists a limitation of the smallest possible width of the superconducting transition, ΔT_c, given by the Ginsburg criteria, $\Delta T_c = Gi T_c$, then such group of researchers can save a lot of time and instead of many experimental attempts to improve the quality of the films, just select the most suitable material with the smallest value of the parameter Gi. This is a simple example how the knowledge of the intrinsic phenomena in superconductivity at nanoscale can help the experimentalists to save their resources and time to achieve the desirable result.

Recently, some very interesting effects were first predicted theoretically and then detected experimentally in layered and low-dimensional superconductors – triplet superconductivity, crossed Andreev reflection, and pi-shift. How one can use them for novel devices? What kind of nanostructures should be prepared for detection and application of those effects? In order to highlight some of the rosen questions, well-known experts were invited to write chapters for this book.

We believe that the book can attract attention of researchers, engineers, Ph.D. students and others, who would like to gain knowledge about some intrinsic effects of Superconductivity at nanoscale.

Kishinev, June 2011 *Anatolie Sidorenko*

Contents

Contributors

K.Yu. Arutyunov Nano Science Center, Department of Physics, University of Jyväskylä, PB35, 0014 Jyväskylä, Finland
and
Nuclear Physics Institute, Moscow State University, 119992 Moscow, Russia, konstantin.arutyunov@phys.jyu.fi

Y. Asano Department of Applied Physics, Hokkaido University, Sapporo 060-8628, Japan

D. Beckmann Institut für Nanotechnologie, Karlsruher Institut für Technologie, P.O. Box 3640, 76021 Karlsruhe, Germany, detlef.beckmann@kit.edu

Manfred Birk DLR German Aerospace Centre, Remote Sensing Technology Institute, 82234 Wessling, Germany

Dick Boersma SRON Netherlands Institute for Space Research, 9700 AV Groningen, The Netherlands

Johannes Dercksen SRON Netherlands Institute for Space Research, 9700 AV Groningen, The Netherlands

Pavel Dmitriev Kotel'nikov Institute of Radio Engineering and Electronics, Russian Academy of Science, Mokhovaya st. 11/7, 125009 Moscow, Russia

Andrey B. Ermakov Kotel'nikov Institute of Radio Engineering and Electronics, Russian Academy of Science, Mokhovaya st. 11/7, 125009 Moscow, Russia
and
SRON Netherlands Institute for Space Research, 9700 AV Groningen, The Netherlands

Lyudmila V. Filippenko SRON Netherlands Institute for Space Research, 9700 AV Groningen, The Netherlands

E. Goldobin Physikalisches Institut–Experimental physic II and Center for Collective Quantum Phenomena, Universität Tübingen, Auf der Morgenstelle 14, 72076 Tübingen, Germany, gold@uni-tuebingen.de

Hans Golstein SRON Netherlands Institute for Space Research, 9700 AV Groningen, The Netherlands

A.A. Golubov Faculty of Science and Technology, University of Twente, 7500 AE Enschede, The Netherlands
and
MESA+ Institute of Nanotechnology, University of Twente, 7500 AE Enschede, The Netherlands, A.A.Golubov@tnw.utwente.nl

Ruud W.M. Hoogeveen SRON Netherlands Institute for Space Research, 9700 AV Groningen, The Netherlands

Leo de Jong SRON Netherlands Institute for Space Research, 9700 AV Groningen, The Netherlands

K.V. Kalashnikov Kotel'nikov Institute of Radio Engineering and Electronics, Russian Academy of Science, Mokhovaya st. 11/7, 125009 Moscow, Russia

Mikhail S. Kalenkov I.E. Tamm Department of Theoretical Physics, P.N. Lebedev Physics Institute, 119991 Moscow, Russia

T.Yu. Karminskaya Nuclear Physics Institute, Moscow State University, 119992 Moscow, Russia

Andrey V. Khudchenko Kotel'nikov Institute of Radio Engineering and Electronics, Russian Academy of Science, Mokhovaya st. 11/7, 125009 Moscow, Russia
and
SRON Netherlands Institute for Space Research, 9700 AV Groningen, The Netherlands, Khudchenko@hitech.cplire.ru

Nickolay V. Kinev Kotel'nikov Institute of Radio Engineering and Electronics, Russian Academy of Science, Mokhovaya st. 11/7, 125009 Moscow, Russia
and
SRON Netherlands Institute for Space Research, 9700 AV Groningen, The Netherlands

Oleg S. Kiselev Kotel'nikov Institute of Radio Engineering and Electronics, Russian Academy of Science, Mokhovaya st. 11/7, 125009 Moscow, Russia
and
SRON Netherlands Institute for Space Research, 9700 AV Groningen, The Netherlands

R. Kleiner Physikalisches Institut–Experimental physic II and Center for Collective Quantum Phenomena, Universität Tübingen, Auf der Morgenstelle 14, 72076 Tübingen, Germany, kleiner@uni-tuebingen.de

D. Koelle Physikalisches Institut–Experimental physic II and Center for Collective Quantum Phenomena, Universität Tübingen, Auf der Morgenstelle 14, D-72076 Tübingen, Germany, koelle@uni-tuebingen.de

Valery P. Koshelets Kotel'nikov Institute of Radio Engineering and Electronics, Russian Academy of Science, Mokhovaya st. 11/7, 125009 Moscow, Russia
and
SRON Netherlands Institute for Space Research, 9700 AV Groningen, The Netherlands, valery@hitech.cplire.ru

Pavel V. Kudryashov Kotel'nikov Institute of Radio Engineering and Electronics, Russian Academy of Science, Mokhovaya st. 11/7, 125009 Moscow, Russia

Bart van Kuik SRON Netherlands Institute for Space Research, 9700 AV Groningen, The Netherlands

M.Yu. Kupriyanov Skobeltsyn Institute of Nuclear Physics, M.V. Lomonosov Moscow State University, GSP-1, Leninskie Gory, 119991 Moscow, Russia
and
Nuclear Physics Institute, Moscow State University, 119992 Moscow, Russia, mkupr@pn.sinp.msu.ru

C. Lacroix Institut N'eel, CNRS-UJF, BP166, 38042 Grenoble Cedex 9, France, claudine.lacroix@grenoble.cnrs.fr

Arno de Lange SRON Netherlands Institute for Space Research, 9700 AV Groningen, The Netherlands

Gert de Lange SRON Netherlands Institute for Space Research, 9700 AV Groningen, The Netherlands

Irina L. Lapitsky SRON Netherlands Institute for Space Research, 9700 AV Groningen, The Netherlands

Yu. G. Naidyuk B. Verkin Institute for Low Temperature Physics and Engineering, National Academy of Sciences of Ukraine, 47 Lenin ave., 61103 Kharkiv, Ukraine, naidyuk@ilt.kharkov.ua

Sergey I. Pripolzin Institute for Physics of Microstructure, Russian Academy of Science, Ulyanova 46, GSP-105, Nizhny Novgorod, Russia

N.G. Pugach Faculty of Physics, M.V. Lomonosov Moscow State University, GSP-2, Leninskie Gory, 119991 Moscow, Russia, pugach@magn.ru

Joris van Rantwijk SRON Netherlands Institute for Space Research, 9700 AV Groningen, The Netherlands

V.V. Ryazanov Institute of Solid State Physics, Russian Academy of Sciences, Chernogolovka, Moscow District, 142432 Russia, valery.ryazanov@gmail.com

Avri M. Selig SRON Netherlands Institute for Space Research, 9700 AV Groningen, The Netherlands

A.S. Sidorenko Institute of Electronic Engineering and Nanotechnologies ASM, MD2028 Kishinev, Moldova
and
Institute of Nanotechnology, Karlsruhe Institute of Technology (KIT), D-76021 Karlsruhe, Germany, anatoli.sidorenko@kit.edu

Alexander S. Sobolev Kotel'nikov Institute of Radio Engineering and Electronics, Russian Academy of Science, Mokhovaya st. 11/7, 125009 Moscow, Russia

Y. Tanaka Department of Applied Physics, Nagoya University, Nagoya 464-8603, Japan

Y. Tanuma Institute of Physics, Kanagawa University, 3-7-1, Rokkakubashi, Kanagawa-ku, Yokohama 221-8686, Japan

Mikhail Yu. Torgashin Kotel'nikov Institute of Radio Engineering and Electronics, Russian Academy of Science, Mokhovaya st. 11/7, 125009 Moscow, Russia

Vladimir L. Vaks Institute for Physics of Microstructure, Russian Academy of Science, Ulyanova 46, GSP-105, Nizhny Novgorod, Russia

A.A. Varlamov Institute of Superconductivity and Innovative Materials of National Research Council (SPIN-CNR), Viale del Politecnico 1, I-00133 Rome, Italy, andrei.varlamov@spin.cnr.it

Ed de Vries SRON Netherlands Institute for Space Research, 9700 AV Groningen, The Netherlands

Georg Wagner DLR German Aerospace Centre, Remote Sensing Technology Institute, 82234 Wessling, Germany

Pavel A. Yagoubov SRON Netherlands Institute for Space Research, 9700 AV Groningen, The Netherlands
and
European Organization for Astronomical Research in the Southern Hemisphere (ESO), Karl-Schwarzschild-Strasse 2, 85748 Garching bei München, Germany

I.K. Yanson B. Verkin Institute for Low Temperature Physics and Engineering, National Academy of Sciences of Ukraine, 47 Lenin ave., 61103 Kharkiv, Ukraine, naidyuk@ilt.kharkov.ua

Andrei D. Zaikin Institut für Nanotechnologie, Karlsruher Institut für Tehnologie (KIT), 76021 Karlsruhe, Germany
and
I.E. Tamm Department of Theoretical Physics, P.N. Lebedev Physics Institute, 119991 Moscow, Russia, andrei.zaikin@kit.edu

Chapter 1
"Fluctuoscopy" of Superconductors

A.A. Varlamov

Abstract Study of fluctuation phenomena in superconductors (SCs) is the subject of great fundamental and practical importance. Understanding of their physics allowed to clear up the fundamental properties of SC state. Being predicted in 1968, one of the fluctuation effects, namely paraconductivity, was experimentally observed almost simultaneously. Since this time, fluctuations became a noticeable part of research in the field of superconductivity, and a variety of fluctuation effects have been discovered.

The new wave of interest to fluctuations (FL) in superconductors was generated by the discovery of cuprate oxide superconductors (high-temperature superconductors, HTS), where, due to extremely short coherence length and low effective dimensionality of the electron system, superconductive fluctuations manifest themselves in a wide range of temperatures. Moreover, anomalous properties of the normal state of HTS were attributed by many theorists to strong FL in these systems. Being studied in the framework of the phenomenological Ginzburg–Landau theory and, more extensively, in diagrammatic microscopic approach, SC FLs side by side with other quantum corrections (weak localization, etc.) became a new tool for investigation and characterization of such new systems as HTS, disordered electron systems, granular metals, Josephson structures, artificial super-lattices, etc. The characteristic feature of SC FL is their strong dependence on temperature and magnetic fields in the vicinity of phase transition. This allows one to definitely separate the fluctuation effects from other contributions and to use them as the source of information about the microscopic parameters of a material. By their origin, SC FLs are very sensitive to relaxation processes, which break phase coherence. This allows using them for versatile characterization of SC. Today, one can speak about the "fluctuoscopy" of superconductive systems.

A.A. Varlamov (✉)
Institute of Superconductivity and Innovative Materials of National Research Council
(SPIN-CNR), Viale del Politecnico 1, I-00133, Rome, Italy
e-mail: andrei.varlamov@spin.cnr.it

A. Sidorenko (ed.), *Fundamentals of Superconducting Nanoelectronics*,
NanoScience and Technology, DOI 10.1007/978-3-642-20158-5_1,

In review, we present the qualitative picture both of thermodynamic fluctuations close to critical temperature T_{c0} and quantum fluctuations at zero temperature and in vicinity of the second critical field $H_{c2}(0)$. Then in the frameworks of the Ginzburg–Landau theory, we discuss the characteristic crossovers in fluctuation properties of superconductive nanoparticles and layered superconductors. We present the general expression for fluctuation magneto-conductivity valid through all phase diagram of superconductor and apply it to study of the quantum phase transition close to $H_{c2}(0)$. Fluctuation analysis of this transition allows us to present the scenario of fluctuation defragmentation of the Abrikosov lattice.

1.1 Introduction

"Happy families are all alike; every unhappy family is unhappy in its own way", started Leo Tolstoy his novel "Anna Karenina". A similar statement can be made about the electronic couples in superconductors (SCs): while stable Cooper pairs forming below critical temperature T_{c0} a sort of condensate behave all in the same way, the behavior of the fluctuating Cooper pairs (FCPs) above the transition is complex and involves a lot of interesting physics. Such FCPs affect thermodynamic and transport properties of the metal both directly and through the changes which they cause in normal quasi-particle subsystem [1], and study of superconductive fluctuations (SF) presents the unique tool providing the information about the character of superconductive state formation [1]. Difficulties of such "fluctuoscopy" are caused by the quantity of these quantum corrections, necessity of their separation from unknown background, smallness of their magnitude.

The mechanisms of fluctuations in the vicinity of the superconductive critical temperature T_{c0} were deeply understood in 1970s. SFs are commonly described in terms of three principal contributions: Aslamazov–Larkin (AL) process, corresponding to the opening of the new channel of the charge transfer [2], anomalous Maki–Thompson (MT) process, which is a single-particle quantum interference on impurities in presence of SF [3–5], and the change of the single-particle density of states (DOS) due to their involvement in fluctuation pairings [6, 7]. The first two processes (AL and MT) result in appearance of positive and singular close to the superconductive critical temperature T_{c0} contributions to conductivity, while the third one (DOS) results in decrease of the Drude conductivity due to the lack of single-particle excitations at the Fermi level. The latter contribution is less singular in temperature than the first two and can compete with them only when the AL and MT processes are suppressed by some reasons (e.g., c-axis transport in layered superconductors) or far enough from T_{c0}.

The classical results obtained first in the vicinity of T_{c0} later were generalized to the temperatures far from transition [8–10] and relatively high fields [11]. More recently, quantum fluctuations (QFs), taking place in SC at low temperatures and fields close to the second critical field H_{c2} (0), entered the focus. Their manifestation strikingly differs from that one of thermal fluctuations close to T_{c0}. For instance, the direct contribution of FCPs to transport coefficients here is absent. In [12, 13] was found that in granular SC at very low temperatures and close to $H_{c2}(0)$, the positive AL contribution to magneto-conductivity (MC) decays as T^2 while the fluctuation suppression of the quasiparticle density of states (DOS) by QF results in temperature independent negative contribution to MC logarithmically growing in magnitude when $H \to H_{c2}(0)$. Effects of QF on MC and magnetization of two-dimensional (2D) SC were studied at low temperatures and fields close to $H_{c2}(0)$ in [14]. Fluctuation renormalization of the diffusion coefficient (DCR) results in appearance of a giant Nernst–Ettingshausen signal [15]. Moreover, as it was demonstrated recently [16] namely this contribution governs the behavior of fluctuation conductivity through all periphery of the phase diagram of superconductor and especially in the region of quantum phase transition in the vicinity of $H_{c2}(0)$.

1.2 Thermodynamic Superconductive Fluctuations Close to T_{c0}

1.2.1 Rather Rayleigh–Jeans Fields than Boltzmann Particles

In the BCS theory [17, 18], only the Cooper pairs forming a Bose-condensate are considered. Fluctuation theory deals with the Cooper pairs out of the condensate. In some phenomena, these FCPs behave similarly to quasiparticles but with one important difference. While for the well-defined quasiparticle, the energy has to be much larger than its inverse lifetime, for the FCPs the "binding energy" ΔE turns out to be of the same order. The FCPs lifetime τ_{GL} is determined by its decay into two free electrons. Evidently, at the transition temperature the Cooper pairs start to condense and $\tau_{GL} = \infty$. Above T_{c0} τ_{GL} can be estimated using the uncertainty principle: $\tau_{GL} \sim \hbar/\Delta E$, where ΔE is the difference $k_B(T - T_{c0})$ ensuring that τ_{GL} should become infinite at the point of transition. The microscopic theory confirms this hypothesis and gives the exact coefficient:

$$\tau_{GL} = \frac{\pi\hbar}{8k_B(T - T_c)}. \tag{1.1}$$

Another important difference of the FCPs from quasiparticles lies in their large size $\xi(T)$. This size is determined by the distance by which the electrons forming the FCPs move apart during the pair lifetime τ_{GL}. In the case of an impure superconductor, the electron motion is diffusive with the diffusion coefficient $\mathcal{D} \sim v_F^2\tau$ (τ is the electron scattering time [19]), and $\xi_d(T) = \sqrt{\mathcal{D}\tau_{GL}} \sim v_F\sqrt{\tau\tau_{GL}}$. In the case of a clean superconductor, where $k_BT\tau \gg \hbar$, impurity scattering no longer affects the electron correlations. In this case the time of electron ballistic motion turns out to be less than the electron–impurity scattering time τ and is determined by the uncertainty principle: $\tau_{bal} \sim \hbar/k_BT$. Then this time has to be used in this case for the determination of the effective size instead of τ: $\xi_c(T) \sim$

$v_F\sqrt{\hbar\tau_{\mathrm{GL}}/k_{\mathrm{B}}T}$. In both cases, the coherence length grows with the approach to the critical temperature as $\epsilon^{-1/2}$, where

$$\epsilon = \ln\frac{T}{T_{\mathrm{c}}} \approx \frac{T-T_{\mathrm{c}}}{T_{\mathrm{c}}} \tag{1.2}$$

is the reduced temperature. We will write down coherence length in the unique way

$$\xi_{\mathrm{GL}}(\epsilon) = (\mathcal{D}\tau_{\mathrm{GL}})^{1/2} \sim \xi_{\mathrm{BCS}}/\sqrt{\epsilon}. \tag{1.3}$$

Here, $\xi_{\mathrm{BCS}} = \xi_{c,d}$ is the BCS coherence length. We see that the fluctuating order parameter $\Delta^{(\mathrm{fl})}(\mathbf{r},t)$ varies close to $T_{\mathrm{c}0}$ on the large scale $\xi_{\mathrm{GL}}(\epsilon) \gg \xi_{\mathrm{BCS}}$.

Finally, it is necessary to recognize that FCPs can really be treated as classical objects, but that these objects instead of Boltzmann particles appear as classical fields in the sense of Rayleigh–Jeans. This means that in the general Bose–Einstein distribution function only small energies $\mathcal{E}(p)$ are involved and the exponent can be expanded:

$$n(p) = \frac{1}{\exp(\mathcal{E}(p)/k_{\mathrm{B}}T) - 1} = \frac{k_{\mathrm{B}}T}{\mathcal{E}(p)}. \tag{1.4}$$

That is why the more appropriate tool to study fluctuation phenomena is not the Boltzmann transport equation but the GL equation for classical fields. Nevertheless, at the qualitative level the treatment of fluctuation Cooper pairs as particles with the concentration $N_s^{(D)} = \int n(p)d^Dp/(2\pi\hbar)^D$ often turns out to be useful [20].

In the framework of both the phenomenological GL theory and the microscopic BCS theory was found that in the vicinity of the transition

$$\mathcal{E}(p) = k_{\mathrm{B}}(T-T_{\mathrm{c}}) + \frac{\mathbf{p}^2}{2m^*} = \frac{1}{2m^*}\left[\hbar^2/2\xi^2(T) + \mathbf{p}^2\right]. \tag{1.5}$$

Far from the transition temperature, the dependence $n(p)$ turns out to be more sophisticated than (1.4); nevertheless, one can always write it in the form

$$n(p) = \frac{m^*k_{\mathrm{B}}T}{\hbar^2}\xi^2(T)\, f\left(\frac{\xi(T)p}{\hbar}\right). \tag{1.6}$$

The effective GL energy of the FCPs defined by (1.5) can be understood as the sum of its kinetic energy and the binding energy ΔE, which is nothing else as the chemical potential

$$\mu_{\mathrm{C.p.}}(T) = T_{\mathrm{c}} - T \tag{1.7}$$

of the FCPs taken with the opposite sign:

$$\mathcal{E}(p) = \frac{\mathbf{p}^2}{2m^*} - \mu(T).$$

Let us clarify the issue related to the chemical potential of fluctuating Cooper pairs, $\mu_{\text{C.p.}}$. Indeed, it is known that in the thermodynamic equilibrium, the chemical potential of a system with a variable number of particles is zero, with photon and phonon gases being the textbook examples. A naïve application of this "theorem" to fluctuating Cooper pairs "gas" leads to a wrong conclusion that $\mu_{\text{C.p.}} = 0$. However, a delicate issue concerning Cooper pairs is that they do not form an isolated system but are composed of the fermionic quasi-particles, which constitute another subsystem under consideration. In a multicomponent system, the chemical potential of the i'th component, μ_i, is defined as the derivative of the thermodynamic potential with respect to the number of particles of i-th sort:

$$\mu_i = (\partial\Omega/\partial N_i)_{P,V,N_j}, \tag{1.8}$$

provided the numbers of particles of all other species are fixed, $N_{j\neq i} = \text{const.}$ In deriving the condition for thermodynamic equilibrium, one should now take into account that creation of a Cooper pair must be accompanied by removing two electrons from the fermionic subsystem. This leads to $\mu_{\text{C.p.}} - 2\mu_{\text{q.p.}} = 0$, where $\mu_{\text{q.p.}}$ is the chemical potential of quasi-particles. Therefore, the equilibrium condition does not restrict $\mu_{\text{C.p.}}$ to zero, even though the number of Cooper pairs is not conserved.

1.2.2 Manifestation of SF Close to T_c

In classical field theory, the notions of the particle distribution function $n(p)$ (proportional to $\mathcal{E}^{-1}(p)$ in our case) and Cooper pair mass m^* are poorly determined. At the same time, the characteristic value of the Cooper pair center of mass momentum can be defined and it turns out to be of the order of $p_0 \sim \hbar/\xi(T)$. So for the combination $m^*\mathcal{E}(p_0)$ one can write $m^*\mathcal{E}(p_0) \sim p_0^2 \sim \hbar^2/\xi^2(T)$. The ratio of the FCPs concentration to the corresponding effective mass with the logarithmic accuracy can be expressed in terms of the coherence length:

$$\frac{N_s^{(D)}}{m^*} = \frac{k_B T}{m^*\mathcal{E}(p_0)}\left(\frac{p_0}{\hbar}\right)^D \sim \frac{k_B T}{\hbar^2}\xi_{\text{GL}}^{2-D}(T) \tag{1.9}$$

(p_0^D here estimates the result of momentum integration).

The particles' density enters into many physical values in the combination N/m^*. For example, we can evaluate the direct FCPs contribution to conductivity (Aslamazov–Larkin paraconductivity) by using the Drude formula and noting that the role of scattering time for FCPs plays their lifetime τ_{GL}:

$$\delta\sigma_{(D)}^{\text{AL}} = \frac{N_s^{(D)} e^2 \tau_{\text{GL}}(\epsilon)}{m^*} \Rightarrow \frac{k_B T}{\hbar^2} d^{D-3}\xi_{\text{GL}}^{2-D}(T)(2e)^2\tau_{\text{GL}}(\epsilon) \sim \epsilon^{D/2-2}. \tag{1.10}$$

This contribution to conductivity of the normal phase of superconductor corresponds to opening of the new channel of charge transfer above T_c: due to forming in it FCPs.

Analogously, a qualitative understanding of the increase in the diamagnetic susceptibility above the critical temperature may be obtained from the – known Langevin expression for the atomic susceptibility [21]:

$$\delta\chi^{\mathrm{C.p}} = -\frac{\mathrm{e}^2}{c^2}\frac{n_s^{(D)}}{m^*}\left\langle R^2\right\rangle \Rightarrow -\frac{4\mathrm{e}^2}{c^2}\frac{k_\mathrm{B}T}{\hbar^2}d^{D-3}\xi^{4-D}(T) \sim -\epsilon^{D/2-2}. \tag{1.11}$$

Here, we used the ratio (1.9).

Special attention has been attracted recently by the giant Nernst–Ettingshausen effect observed in the pseudogap state of the underdoped phases of HTSC [22], which motivated speculations [23] about the possibility of existence of some specific vortices and anti-vortices there or the special role of the phase fluctuations [24]. Then, very recently the giant Nernst–Ettingshausen signal (three orders of magnitude more than the value of the Nernst–Ettingshausen coefficient in typical metals) was detected also in the wide range of temperatures in a conventional disordered superconductor Nb_xSi_{1-x} [25]. All these experiments finally have been successfully explained in the frameworks of both phenomenological and microscopic fluctuation theories [15,26,27]. The proposed qualitative consideration of the FCPs allows not only to get in a simple way the correct temperature dependence of the fluctuation NEE coefficient but also to catch the reason of its giant magnitude. Indeed, as it was shown in [15, 28], the Nernst–Ettingshausen coefficient can be related to the temperature derivative of the chemical potential:

$$\delta\mathcal{N}^{\mathrm{C.p}} = \frac{\sigma}{nc\mathrm{e}^2}\left(\frac{\mathrm{d}\mu}{\mathrm{d}T}\right). \tag{1.12}$$

Applying this formula to the subsystem of FCPs close to T_{c0} with $\mu_{\mathrm{C.p.}}(T)$ defined by (1.7) and identifying its conductivity with (1.10), one finds

$$\delta\mathcal{N}^{\mathrm{C.p}} = -\frac{\sigma^{(\mathrm{C.p.})}}{N_s^{(D)}c\mathrm{e}^2} \sim \epsilon^{D/2-2}, \tag{1.13}$$

what fits well the experimental findings obtained in conventional superconductors and optimally doped phases of HTS. The reason of so strong fluctuation effect contains in the extremely strong dependence of the FCPs chemical potential on temperature: $\mathrm{d}\mu_{\mathrm{C.p.}}/\mathrm{d}T = -1$, while for the free electron gas $\mathrm{d}\mu_e/\mathrm{d}T \sim -T/E_F$.

Besides the direct FCPs effect on properties of superconductor in its normal phase, the other, indirect manifestations of SF and their effect on the quasiparticle subsystem take place. These effects, being much more sophisticated, have a purely quantum nature and, in contrast to paraconductivity, require microscopic consideration. First of them is MT contribution [3–5]. It is generated by the coherent scattering of the electrons forming a Cooper pair on the same elastic impurities and can be treated as the result of Andreev reflection of the electron by fluctuation Cooper pairs. This contribution appears only in transport coefficients and often turns out to be important. Its temperature singularity near T_c is similar to that of the paraconductivity, although being extremely sensitive to electron phase-breaking processes and to the type of orbital symmetry of pairing it can be suppressed. Let us evaluate it.

The physical origin of the MT correction consists in the fact that the Cooper interaction of electrons with nearly opposite momenta changes the mean free path (diffusion coefficient) of electrons. The amplitude of the effective BCS interaction increases drastically when $T \to T_c$:

$$g_{\mathrm{eff}} = \frac{g}{1 - \nu g \ln \frac{\omega_D}{2\pi T}} = \frac{1}{\ln \frac{T}{T_c}} \approx \frac{T}{T - T_c} = \frac{1}{\epsilon}.$$

What is the reason for this growth? One can say that the electrons scatter one at another in a resonant way with the virtual Cooper pair formation. Or, it is possible to imagine that the electrons undergo Andreev reflection by fluctuation Cooper pairs, binding in the Cooper pairs themselves. The probability of such induced pair irradiation (let us remember that Cooper pairs are Bose particles) is proportional to their number in the final state that is $n(p)$ (see (1.4)). For small momenta, $n(p) \sim 1/\epsilon$.

One can ask why such an interaction does not manifest itself considerably far from the transition point? This is due to the fact that just a small number of electrons with the total momentum $q \lesssim \xi^{-1}(T)$ interacts so intensively. In accordance with the Heisenberg principle, the minimal distance between such electrons is of the order of $\sim \xi(T)$. On the other hand, such electrons, in order to interact, have to approach one another approximately up to a distance of the Fermi length $\lambda_F \sim 1/p_F$. The probability of such event may be estimated in the spirit of the self-intersecting trajectories contribution evaluation in the weak-localization theory [29].

In the process of diffusion motion, the distance between two electrons increases with time according to the law: $R(t) \sim (\mathcal{D}t)^{1/2}$. Hence, the scattering probability

$$W \sim \int_{t_{\min}}^{t_{\max}} \frac{\lambda_F^{D-1}}{R^D(t)} v_F \mathrm{d}t.$$

The lower limit of the integral can be estimated from the condition $R(t_{\min}) \sim \xi(T)$ (only such electrons interact in the resonant way). The upper limit is determined by the phase-breaking time τ_φ since for larger time intervals the phase coherence, necessary for the pair formation, is broken. As a result, the relative correction to conductivity due to such processes is equal to the product of the scattering probability on the effective interaction constant: $\delta\sigma^{\mathrm{MT}}/\sigma = W\, g_{\mathrm{eff}}$. In the $2D$ case

$$\delta\sigma_{(2)}^{\mathrm{MT(an)}} \sim \frac{\mathrm{e}^2}{8\epsilon} \ln \frac{\mathcal{D}\tau_\varphi}{\xi^2(T)}.$$

However, positive and singular in ϵ close to T_c AL and MT contributions do not capture the complete effect of fluctuations on conductivity. The involvement of quasi-particles in the fluctuation pairing results in their lack at the Fermi level that is in the opening of the pseudo-gap in the one-electron spectrum and consequent decrease of the one-particle Drude-like conductivity. Such an indirect effect of FCPs formation is usually referred as the DOS one. Being proportional to the concentration of the FCPs $N_s^{(D)}$ the DOS contribution formally appears due to the order parameter Fourier-component $\left\langle \left| \Delta^{(\mathrm{fl})}(\mathbf{q}, \omega) \right|^2 \right\rangle$ integrated over all long-wave-length fluctuation modes ($q \lesssim \xi_{\mathrm{BCS}}^{-1} \sqrt{\epsilon}$):

$$\delta\sigma_{(2)}^{\mathrm{DOS}} \sim -\frac{2n_{\mathrm{c.p.}}\mathrm{e}^2\tau}{m_\mathrm{e}} \sim -\mathrm{e}^2 \int \frac{\xi_{\mathrm{BCS}}^2 \mathrm{d}^2\mathbf{q}}{\epsilon + \xi_{\mathrm{BCS}}^2 q^2} \sim -\frac{\mathrm{e}^2}{\hbar} \ln \frac{1}{\epsilon}. \tag{1.14}$$

It is seen that DOS contribution has an opposite sign with respect to the AL and MT contributions, but close to T_{c0} does not compete with them since it turns to be less singular as a function of temperature [1].

Finally, the renormalization of the one-electron diffusion coefficient (DCR) in the presence of fluctuation pairing takes place. Close to T_{c0} this contribution is not singular in ϵ

$$\delta\sigma_{\mathrm{xx}}^{\mathrm{DCR}} \sim \frac{\mathrm{e}^2}{\hbar} \ln \ln \frac{1}{T_{c0}\tau} + O(\epsilon)$$

and was always ignored, but as was found in [15, 16] it becomes of primary importance relatively far from T_{c0}, and at very low temperatures. It is the account for $\delta\sigma_{\mathrm{xx}}^{\mathrm{DCR}}$, which changes the sign of the total contribution of fluctuations to conductivity $\delta\sigma_{(2)}^{(\mathrm{tot})}$ in the wide domain of the phase diagram and especially close to $T = 0$, in the region of quantum fluctuations [16] (see Fig. 1.1, where the regions with the dominating fluctuation contributions to magnetoconductivity are shown).

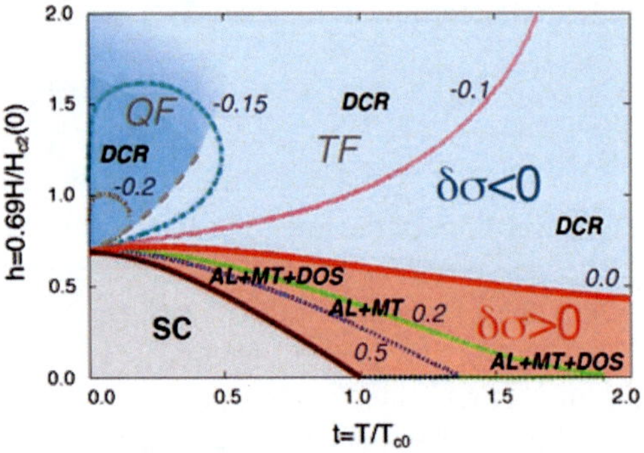

Fig. 1.1 Contours of constant fluctuation conductivity [$\delta\sigma = \delta\sigma_{\mathrm{xx}}^{(\mathrm{tot})}(t, h)$ shown in units of e^2]. The dominant FC contributions are indicated in bold-italic labels. The dashed line separates the domain of quantum fluctuations (QFs) [dark area of $\delta\sigma > 0$] and thermal fluctuations (TFs)

1.3 Ginzburg–Landau Theory

1.3.1 GL Functional

Let us consider the model of metal being close to transition to the superconductive state. The complete description of its thermodynamic properties can be done through the calculation of the partition function [30]:

$$Z = \mathrm{tr}\left\{\exp\left(-\frac{\widehat{\mathcal{H}}}{T}\right)\right\}. \tag{1.15}$$

As discussed above, in the vicinity of the superconductive transition, side by side with the fermionic electron excitations, fluctuation Cooper pairs of a bosonic nature appear in the system. They can be described by means of classical bosonic complex fields $\Psi(\mathbf{r})$, which can be treated as "Cooper pair wave functions". Therefore, the calculation of the trace in (1.15) can be separated into a summation over the "fast" electron degrees of freedom and a further functional integration carried out over all possible configurations of the "steady flow" Cooper pairs wave functions:

$$Z = \int \mathfrak{D}^2\Psi(\mathbf{r})\mathcal{Z}[\Psi(\mathbf{r})], \tag{1.16}$$

where

$$\mathcal{Z}[\Psi(\mathbf{r})] = \exp\left(-\frac{\mathcal{F}[\Psi(\mathbf{r})]}{T}\right) \tag{1.17}$$

is the system partition function in a fixed bosonic field $\Psi(\mathbf{r})$, already summed over the electronic degrees of freedom.

The "steady flow" of wave functions means that they are supposed to vary over a scale much larger than the interatomic distances. The classical part of the Hamiltonian, dependent on bosonic fields, may be chosen in the spirit of the Landau theory of phase transitions. However, in view of the space dependence of wave functions, Ginzburg and Landau included in it additionally the first nonvanishing term of the expansion over the gradient of the fluctuation field. Symmetry analysis shows that it should be quadratic. The weakness of the field coordinate dependence allows us to omit the high order terms of such an expansion. Therefore, the classical part of the Hamiltonian of a metal close to superconductive transition related to the presence of the fluctuation Cooper pairs in it (so-called GL functional) can be written as [31]:

$$\mathcal{F}[\Psi(\mathbf{r})] = F_N + \int \mathrm{d}V\left\{a|\Psi(\mathbf{r})|^2 + \frac{b}{2}|\Psi(\mathbf{r})|^4 + \frac{1}{4m}|\nabla\Psi(\mathbf{r})|^2\right\}. \tag{1.18}$$

Let us discuss the coefficients of this functional. In accordance with the Landau hypothesis, the coefficient a goes to zero at the transition point T_{c0} and depends linearly on $T - T_{c0}$. Then $a = \alpha T_c \epsilon$; all the coefficients α, b, and m are supposed to be positive and temperature independent. Concerning the magnitude of the coefficients, it is necessary to make the following comment. One of these coefficients can always be chosen arbitrarily: this option is related to the arbitrariness of the Cooper pair wave function normalization. Nevertheless, the product of two of them is fixed by dimensional analysis: $ma \sim \xi^{-2}(T)$. Another combination of the coefficients, independent of the wave function normalization and temperature, is α^2/b. One can see that it has the dimensionality of the density of states. Since these coefficients were obtained by a summation over the electronic degrees of freedom, the only reasonable candidate for this value is the one electron DOS ν (for one spin at the Fermi level). One can notice that the arbitrariness of the order parameter amplitude results in the ambiguity in the choice of the Cooper pair mass, introduced in (1.18) as $2m$. Indeed, this value enters in (1.20) as the product with the coefficient α, hence one of these parameters has to be set down.

In the phenomenological GL theory, normalization of the order parameter Ψ is usually chosen in such a way that the coefficient m corresponds to the free electron mass. At that, the coefficient α for D-dimensional clean superconductor is determined by the expression

$$\alpha_{(D)} = \frac{2D\pi^2}{7\zeta(3)} \frac{T_{c0}}{E_F}. \tag{1.19}$$

Yet, the other normalization when the order parameter, denoted as $\Delta(\mathbf{r})$, coincides with the value of the gap in spectrum of one-particle excitations of a homogeneous superconductor turns out to be more convenient. As it will be shown below, in vicinity of T_{c0} the microscopic theory allows to present the free energy of superconductor in the form of the GL expansion namely over the powers of $\Delta(\mathbf{r})$. At that turn out to be defined also the exact values of the coefficients α and b:

$$4m\alpha T_{c0} = \xi^{-2}; \alpha^2/b = \frac{8\pi^2}{7\zeta(3)}\nu, \tag{1.20}$$

where $\zeta(x)$ is the Riemann zeta function, $\zeta(3) = 1.202$.

Let us stress that at such choice of the order parameter normalization the GL parameter $C = 1/4m$ turns out to be dependent on the concentration of impurities.

1.3.2 *Zero Dimensionality: The Exact Solution for the Heat Capacity Jump*

In a system of finite volume, the fluctuations smear out the jump of the heat capacity. Let us demonstrate this on the example of a small superconductive sample with the characteristic size $d \ll \xi(T)$. Due to the small size of the granule with respect to the

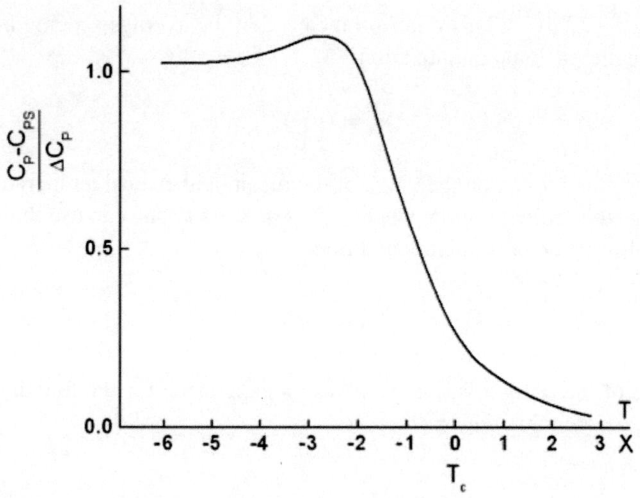

Fig. 1.2 Temperature dependence of the heat capacity of superconductive grains in the region of the critical temperature

GL coherence length, the order parameter Ψ does not depend on the space variables and the free energy can be calculated exactly for all temperatures including the critical region. It the space independent mode $\Psi_0 = \Psi\sqrt{V}$, which defines here the main contribution to the free energy:

$$Z_{(0)} = \int \mathrm{d}^2\Psi_0 \exp\left(-\frac{\mathcal{F}[\Psi_0]}{T}\right) = \pi \int \mathrm{d}|\Psi_0|^2 \exp\left(-\frac{\left(a|\Psi_0|^2 + \frac{b}{2V}|\Psi_0|^4\right)}{T}\right)$$

$$= \sqrt{\frac{\pi^3 VT}{2b}} \exp(x^2)(1 - \mathrm{erf}(x))|_{x=a\sqrt{\frac{V}{2bT}}}. \tag{1.21}$$

By evaluating the second derivative of this exact result [32], one can find the temperature dependence of the heat capacity of the superconductive granule (see Fig. 1.2). One can see that this function is analytic in temperature, therefore fluctuations remove phase transition in the $0D$ system. The smearing of the heat capacity jump takes place in the region of temperatures in the vicinity of T_{c0} where $x \sim 1$, that is

$$\epsilon_{\mathrm{cr}} = Gi_{(0)} = \frac{\sqrt{7\zeta(3)}}{2\pi} \frac{1}{\sqrt{\nu T_{c0} V}} \approx 13.3 \left(\frac{T_{c0}}{E_F}\right) \sqrt{\frac{\xi_{\mathrm{BCS}}^3}{V}}.$$

Here, T_{c0} and ξ_{BCS} are the mean field critical temperature and the zero temperature coherence length of the appropriate bulk material. It is interesting that the width of this smearing does not depend on impurities concentration. From this formula, one can see that the smearing of the transition is very narrow ($\epsilon_{\mathrm{cr}} \ll 1$) when the granule volume $V \gg (\nu T_{c0})^{-1}$. This criterion means that the average spacing between the levels of the dimensional quantization:

$$\delta = (\nu V)^{-1} \tag{1.22}$$

still remains much less than the value of the mean field critical temperature, T_{c0}.

Far above the critical region, where $Gi_{(0)} \ll \epsilon \ll 1$, one can use the asymptotic expression for the erf(x) function and find

$$F_{(0)} = -T \ln Z_{(0)} = -T \ln \frac{\pi}{\alpha \epsilon}. \tag{1.23}$$

Calculation of the second derivative gives an expression for the fluctuation part of the heat capacity in this region:

$$\delta C_{(0)} = \frac{1}{V\epsilon^2}. \tag{1.24}$$

The experimental study of the heat capacity of small Sn particles in the vicinity of the transition was done in [33].

One can estimate the fluctuation contribution to heat capacity for a specimen of an arbitrary effective dimensionality on the basis of the following observation. The volume of the specimen may be divided into regions of size $\xi(T)$, which are weakly correlated with each other. Then the whole free energy can be estimated as the free energy of one such $0D$ specimen (1.23), multiplied by their number $N_{(D)} = V\xi^{-D}(T)$:

$$F_{(D)} = -TV\xi^{-D}(T) \ln \frac{\pi}{\alpha \epsilon}. \tag{1.25}$$

This formula gives the correct temperature dependence of the free energy not too close to T_c for the specimens of the even dimensionalities. As we will demonstrate below, a more accurate treatment removes the $\ln \epsilon$ dependence from it in the case of the odd dimensions.

In the Ginzburg–Landau region, one can omit the fourth-order term in $\Psi(\mathbf{r})$ with respect to the quadratic one and write down the GL functional, expanding the order parameter in a Fourier series:

$$F[\Psi_{\mathbf{k}}] = F_N + \sum_{\mathbf{k}} \left[a + \frac{\mathbf{k}^2}{4m} \right] |\Psi_{\mathbf{k}}|^2 = F_N + \alpha T_c \sum_{\mathbf{k}} \left(\epsilon + \xi^2 \mathbf{k}^2 \right) |\Psi_{\mathbf{k}}|^2. \tag{1.26}$$

Here, $\Psi_{\mathbf{k}} = \frac{1}{\sqrt{V}} \int \Psi(\mathbf{r}) \mathrm{e}^{-i\mathbf{k}\mathbf{r}} \mathrm{d}V$ and the summation is carried out over the wave vectors $\mathbf{k}$ (fluctuation modes). For the specimen of dimensions L_x, L_y, L_z $k_i L_i = 2\pi n_i$. The functional integral for the partition function (1.17) can be factored out to a product of Gaussian-type integrals over these modes:

$$Z = \prod_{\mathbf{k}} \int d^2\Psi_{\mathbf{k}} \exp\left\{-\alpha\left(\epsilon + \frac{\mathbf{k}^2}{4m\alpha T_c}\right)|\Psi_{\mathbf{k}}|^2\right\}. \tag{1.27}$$

Carrying out these integrals, one gets the fluctuation contribution to the free energy:

$$F(\epsilon > 0) = -T\ln Z = -T\sum_{\mathbf{k}} \ln \frac{\pi}{\alpha\left(\epsilon + \frac{\mathbf{k}^2}{4m\alpha T_c}\right)}. \tag{1.28}$$

1.3.3 Zero Dimensionality: The Exact Solution for the Fluctuation Magnetization

For quantitative analysis of the fluctuation diamagnetism, we start from the GL functional for the free energy written down in the presence of the magnetic field. The generalization of the functional (1.18) in the presence of magnetic field requires first of all the gauge invariance; therefore, the momentum operator $-i\nabla$ must be substituted by its gauge invariant form $-i\nabla - 2e\mathbf{A}(\mathbf{r})$ [34]. Moreover, the presence of a magnetic field results in the accumulation of some residual energy of the magnetic field in the volume of superconductor. Finally, the superconductor itself interacts with the external magnetic field $\mathbf{H}$. Taking into account these three observations one can write the generalization of the functional (1.18) in the form

$$\begin{aligned}\mathcal{F}[\Psi(\mathbf{r})] = F_n + \int dV \Big\{ & a|\Psi(\mathbf{r})|^2 + \frac{b}{2}|\Psi(\mathbf{r})|^4 + \frac{1}{4m}|(-i\nabla - 2e\mathbf{A}(\mathbf{r}))\Psi(\mathbf{r})|^2 \\ & + \frac{[\nabla\times\mathbf{A}(\mathbf{r})]^2}{8\pi} - \frac{\nabla\times\mathbf{A}(\mathbf{r})\cdot\mathbf{H}}{4\pi}\Big\}. \end{aligned} \tag{1.29}$$

The fluctuation contribution to the diamagnetic susceptibility in the simplest case of a "zero-dimensional" superconductor (spherical superconductive granule of diameter $d \ll \xi(\epsilon)$) was considered by Shmidt [32]. As above, the smallness $d \ll \xi(T)$ allows us to omit in (1.29) the term $-i\nabla$. Then, due to the smallness of the granule size with respect to the magnetic field penetration depth in superconductor λ, one can assume the equivalence of the average magnetic field in metal $\mathbf{B}$ with the external field $\mathbf{H}$. This allows us to omit also the last two terms in (1.29) since in the assumed approximation they do not depend on fluctuations. It is why formally the effect of a magnetic field in this case is reduced to the renormalization of the coefficient a, or, in other words, to the suppression of the critical temperature:

$$T_c(H) = T_{c0}\left(1 - \frac{4\pi^2\xi^2}{\Phi_0^2}\langle\mathbf{A}^2\rangle\right). \tag{1.30}$$

Here, $\Phi_0 = \pi/e$ is the magnetic flux quantum and $\langle \cdots \rangle$ means the averaging over the sample volume. That is why for the granule in a magnetic field one can use the partition function in the same form (1.21) as in the absence of the field but with the renormalized GL parameter $a(H) = a + \frac{e^2}{m}\langle \mathbf{A}^2 \rangle$:

$$Z_{(0)}(H) = \pi \int d|\Psi_0|^2 \exp\left(-\frac{\left[a + \frac{e^2}{m}\langle \mathbf{A}^2\rangle\right]|\Psi_0|^2 + \frac{b}{2V}|\Psi_0|^4}{T} \right)$$

$$= \sqrt{\frac{\pi^3 V T}{2b}} \exp\left[\frac{a^2(H)V}{2bT}\right]\left\{1 - \operatorname{erf}\left[a(H)\sqrt{\frac{V}{2bT}}\right]\right\}. \quad (1.31)$$

Such a trivial dependence of the properties of $0D$ samples on the magnetic field immediately allows us to understand its effect on the heat capacity of a granular sample. Indeed, with the growth of the field the temperature dependence of the heat capacity presented in Fig. 1.2 just moves in the direction of lower temperatures. Equation (1.31) allows to calculate exactly the fluctuation part of the free energy and corresponding magnetization as the function of temperature and magnetic field, which can be used for the quantitative analysis of the experiments on nanoparticles (see below).

In the GL region $Gi_{(0)} \lesssim \epsilon$, one can easily write the asymptotic expression (1.23) for the free energy:

$$F_{(0)}(\epsilon, H) = -T \ln \frac{\pi}{\alpha\left(\epsilon + \frac{4\pi^2\xi^2}{\Phi_0^2}\langle \mathbf{A}^2\rangle\right)}.$$

In the case of a spherical particle, one has to choose the gauge of the vector-potential $\mathbf{A} = \frac{1}{2}\mathbf{H} \times \mathbf{r}$ yielding $\langle \mathbf{A}^2 \rangle = \frac{1}{40}H^2 d^2$ (calculation of this average value is completely analogous to the calculation of the moment of inertia of a solid sphere). In this way, an expression for the $0D$ fluctuation magnetization valid for all fields $H \ll H_{c2}(0)$ can be found:

$$M_{(0)}(\epsilon, H) = -\frac{1}{V}\frac{\partial F_{(0)}(\epsilon, H)}{\partial H} = -\frac{6\pi T\xi^2}{5\Phi_0^2 d}\frac{H}{\left(\epsilon + \frac{\pi^2\xi^2}{10\Phi_0^2}H^2 d^2\right)}. \quad (1.32)$$

One can see that the fluctuation magnetization turns out to be negative and linear up to some crossover field $H_{up}(\epsilon) \sim \frac{\Phi_0}{d\xi(\epsilon)} \sim \frac{\xi}{d} H_{c2}(0)\sqrt{\epsilon}$ [35] at which it reaches a minimum (this field can be called the temperature dependent upper critical field of the granule). At higher fields, $H_{up}(\epsilon) \lesssim H \ll H_{c2}(0)$ the fluctuation magnetization of the $0D$ granule decreases as $1/H$. In the weak field region $H \ll H_{c2(0)}(\epsilon)$ the diamagnetic susceptibility is:

$$\chi_{(0)}(\epsilon, H) = -\frac{6\pi T\xi_0^2}{5\Phi_0^2 d}\frac{1}{\epsilon} \approx -10^2 \chi_P \left(\frac{\xi}{d}\right)\frac{1}{\epsilon}.$$

Let us underline that the temperature dependence of the $0D$ fluctuation diamagnetic susceptibility turns out to be less singular than the $0D$ heat capacity correction: ϵ^{-1} instead of ϵ^{-2}.

The expression for the fluctuation part of free energy (1.28) is also applicable to the cases of a wire or a film placed in a parallel field: as was already mentioned above all its dependence on the magnetic field is manifested by the shift of the critical temperature (1.30). In the case of a wire in a parallel field the gauge of the vector-potential can be chosen as above what yields $\langle \mathbf{A}^2 \rangle_{(\text{wire},\parallel)} = H^2 d^2/32$. For a wire in a perpendicular field, or a film in a parallel field, the gauge has to be chosen in the form $\mathbf{A} = (0, Hx, 0)$. One can find $\langle \mathbf{A}^2 \rangle_{(\text{wire},\perp)} = H^2 d^2/16$ for a wire and $\langle \mathbf{A}^2 \rangle_{(\text{film},\parallel)} = H^2 d^2/12$ for a film.

Calculating the second derivative of (1.28) with the appropriate magnetic field dependencies of the critical temperature, one can find the following expressions for the diamagnetic susceptibility:

$$\chi_{(D)}(\epsilon) = -2\pi \frac{\xi T}{v_F} \chi_P \begin{cases} \frac{1}{\sqrt{\epsilon}}, & \text{wire in parallel field,} \\ \frac{2}{\sqrt{\epsilon}}, & \text{wire in perpendicular field,} \\ \frac{d}{3\xi} \ln \frac{1}{\epsilon}, & \text{film in parallel field.} \end{cases} \tag{1.33}$$

1.3.4 Fluctuation Diamagnetism in Lead Nanoparticles

Recently, in [36] the $0D$ fluctuating diamagnetism was carefully studied in lead nanoparticles with size $d \ll \xi$ by means of high-resolution superconductive quantum interference device magnetization measurements. In result, the diamagnetic magnetization $M_{\text{dia}}(H, T = \text{const})$ was reported as a function of the applied magnetic field H at constant temperatures in a wide range of temperatures around T_{c0} including the critical region. The magnetization curves were analyzed in the framework of the presented above exact fluctuation theory based on the Ginzburg–Landau functional.

The representative isothermal magnetization curves in the temperature range around T_{c0} are reported in Fig. 1.3. The extraction of the diamagnetic contribution from the magnetization requires a detailed subtraction procedure when the magnetic field is increased to relatively strong values. In fact, in the range $H > H_{\text{up}}$ $|M_{\text{dia}}|$ decreases on increasing the field (see Fig. 1.3), while the paramagnetic contributions due to the Pauli paramagnetism and to a small amount of paramagnetic impurities continue to increase on increasing H. Thus, from the computer-stored raw magnetization data around T_{c0}, the magnetization values measured at a higher temperature (around 8 K) where the SFs are negligible have been subtracted. The slight variation of the paramagnetic contribution with temperature did not prevent reliable estimates of M_{dia} for magnetic field up to about 600 Oe, as indicated by the error bars in Fig. 1.5b.

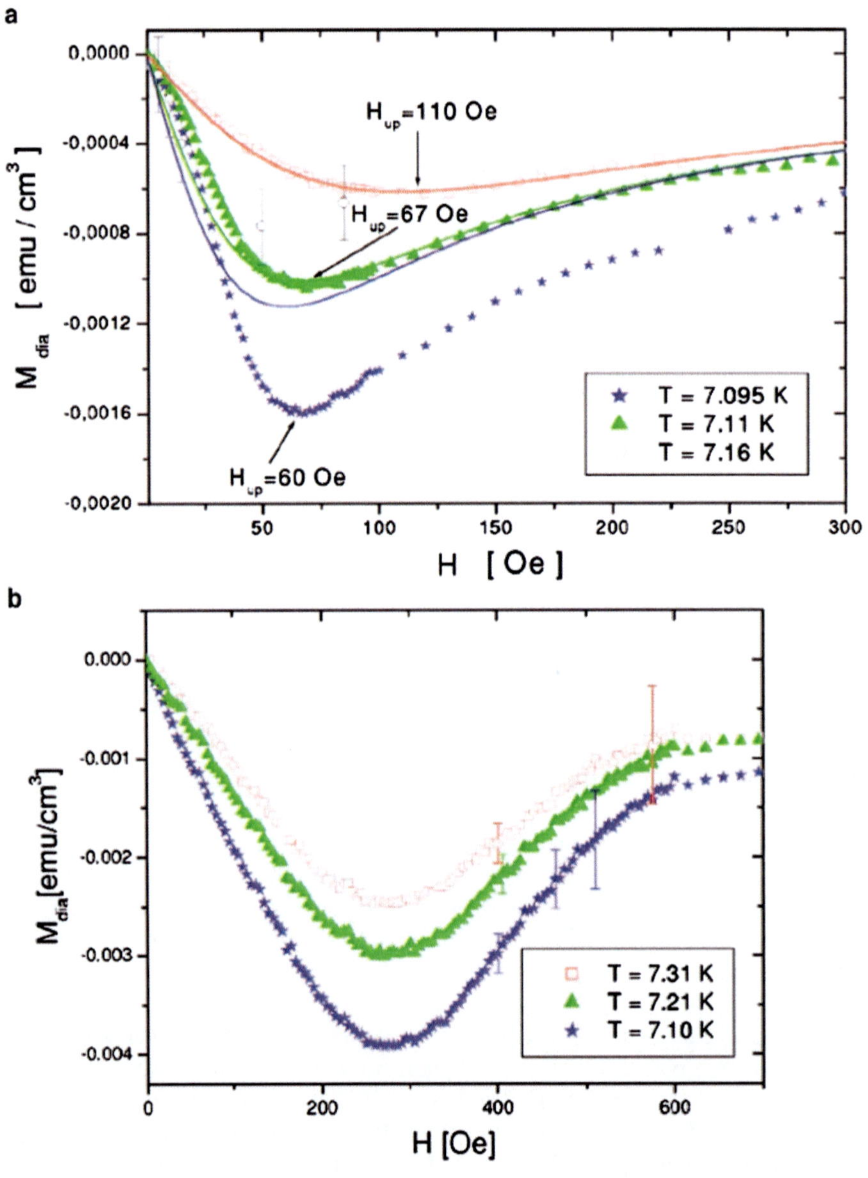

Fig. 1.3 (**a**) Magnetization M_{dia} vs H for the sample with the characteristic diameter of grains $d_3 \simeq 75\,\mathrm{nm}$ at representative temperatures above T_{c0}. The solid lines correspond to (1.32) in the text for critical field of the grain 1,150 Oe . For $\epsilon \lesssim \epsilon_c$, the curves depart from the behavior described by (1.32). (**b**) Magnetization curves for sample with the characteristic size of grains $d_1 \simeq 16\,\mathrm{nm}$, all corresponding to temperature range where $\epsilon \lesssim \epsilon_c$, namely, within the critical region. The open circles in part (a) correspond to the data obtained from the iso-field measurements as a function of temperature, with large experimental errors

The first-order fluctuation correction is found to be valid only outside the critical region $\epsilon \gtrsim \epsilon_c$, where it accurately describes the behavior M_{dia} for magnetic fields $H \lesssim H_{\text{up}}$. Also, the scaling properties of $\mathrm{d}T_c(H)/\mathrm{d}H$ for small fields and of the upturn field H_{up} in the magnetization curves are well described within that approximation.

In the critical region, however, the role of the field and the limits of validity of the first-order fluctuation correction have been analyzed by comparing the experimental findings to the derivation of M_{dia} as a function of the magnetic field starting from the complete form of the GL functional and with the exact expression of the zero-dimensional partition function. The authors found that the role of the $|\Psi(\mathbf{r})|^4$ term in the GL functional is crucial in describing the data in the critical region. For the sample with average grain diameter of 75 nm, the fluctuating diamagnetism can be well described by our extended model even in the critical region, without introducing any adjustable parameters. For the sample with the smallest average diameter of 16 nm, the agreement of the numerically derived M_{dia} with the experimental findings is again good for fields of the order of H_{up}. Poor agreement between the theoretically predicted M_{dia} vs H and the authors data is observed for fields above H_{up}, when the fluctuating diamagnetic contribution is approaching zero and the subtraction procedure of the paramagnetic term introduces large errors.

The temperature dependence of the upturn field and the scaling properties with the grain size are also well described by the exact theory both outside and inside the critical region, with the product $(H_{\text{up}}d)$ vs reduced temperature being approximately size independent and following the predicted temperature dependence, even though the mean field result $H_{\text{up}} \sim \epsilon^{1/2}/d$ evidently breaks down. The relevance of the magnetization curves vs H and of the upturn field H_{up} for the study of the fluctuating diamagnetism above the superconductive transition temperature has been emphasized.

1.4 Fluctuation Thermodynamics of Layered Superconductor in Magnetic Field

1.4.1 Lawrence–Doniach Model

Let us pass now to the quantitative analysis of the temperature and field dependencies of the fluctuation magnetization of a layered superconductor. This system has a great practical importance because of its direct applicability to HTS, where the fluctuation effects are very noticeable. Moreover, the general results obtained will allow us to analyze as limiting cases $3D$ and already familiar $2D$ situations. The effects of a magnetic field are more pronounced for a perpendicular orientation, so let us first consider this case.

The generalization of the GL functional for a layered superconductor (Lawrence–Doniach (LD) functional [37]) in a perpendicular magnetic field can be written as

$$\mathcal{F}_{\mathrm{LD}}[\Psi] = \sum_l \int \mathrm{d}^2 r \left(a\left|\Psi_l\right|^2 + \frac{b}{2}\left|\Psi_l\right|^4 + \frac{1}{4m}\left|\left(\nabla_{\parallel} - 2\mathrm{i}e\mathbf{A}_{\parallel}\right)\Psi_l\right|^2 \right.$$
$$\left. + \mathcal{J}\left|\Psi_{l+1} - \Psi_l\right|^2 \right), \tag{1.34}$$

where Ψ_l is the order parameter of the l-th superconductive layer and the phenomenological constant $\mathcal{J}$ is proportional to the energy of the Josephson coupling between adjacent planes. The gauge with $A_z = 0$ is chosen in (1.34). In the immediate vicinity of T_c, the LD functional is reduced to the GL one with the effective mass $M = (4\mathcal{J}s^2)^{-1}$ along c-direction, where s is the inter-layer spacing. One can relate the value of $\mathcal{J}$ to the coherence length along the z-direction: $\mathcal{J} = 2\alpha T_c \xi_z^2 / s^2$. Since we are dealing with the GL region, the fourth order term in (1.34) can be omitted.

The Landau representation is the most appropriate for solution of the problems related to the motion of a charged particle in a uniform magnetic field. The fluctuation Cooper pair wave function can be written as the product of a plane wave propagating along the magnetic field direction and a Landau state wave function $\phi_n(\mathbf{r})$. Let us expand the order parameter $\Psi_l(\mathbf{r})$ on the basis of these eigenfunctions:

$$\Psi_l(\mathbf{r}) = \sum_{\mathbf{n},k_z} \Psi_{n,k_z} \phi_n(\mathbf{r}) \exp(\mathrm{i}k_z l), \tag{1.35}$$

where $\mathbf{n}$ is the quantum number related to the degenerate Landau state and k_z is the momentum component along the direction of the magnetic field. Substituting this expansion into (1.34), one can find the LD free energy as a functional of the Ψ_{n,k_z} coefficients:

$$\mathcal{F}_{\mathrm{LD}}\left[\Psi_{\{n,k_z\}}\right] = \sum_{n,k_z} \left\{ \alpha T_c \epsilon + \omega_c\left(n + \frac{1}{2}\right) + \mathcal{J}\left[1 - \cos(k_z s)\right] \right\} |\Psi_{n,k_z}|^2. \tag{1.36}$$

In complete analogy with the case of an isotropic spectrum, the functional integral over the order parameter configurations Ψ_{n,k_z} in the partition function can be reduced to a product of ordinary Gaussian integrals, and the fluctuation part of the free energy of a layered superconductor in magnetic field takes the form:

$$F(\epsilon, H) = -\frac{SH}{\Phi_0} T \sum_{n,k_z} \ln \frac{\pi T}{\alpha T_c \epsilon + \omega_c\left(n + \frac{1}{2}\right) + \mathcal{J}\left[1 - \cos(k_z s)\right]} \tag{1.37}$$

(compare this expression with the (1.28)).

In the limit of weak fields, one can carry out the summation over the Landau states by means of the Euler–Maclaurin's transformation and obtain

$$F(\epsilon, H) = F(\epsilon, 0) + \frac{\pi S T H^2}{24 m \Phi_0^2} \int_{-\pi/s}^{\pi/s} \frac{\mathcal{N} s \mathrm{d}k_z}{2\pi} \left\{ \frac{1}{\alpha T_c \epsilon + \mathcal{J}(1 - \cos(k_z s))} \right\}. \tag{1.38}$$

Here $\mathcal{N}$, is the total number of layers. Carrying out the final integration over the transversal momentum, one gets:

$$F(\epsilon, H) = F(\epsilon, 0) + \frac{TV}{24\pi s \xi_{xy}^2} \frac{h^2}{\sqrt{\epsilon(\epsilon + r)}}$$

with the anisotropy parameter defined as

$$r = \frac{2\mathcal{J}}{\alpha T} = \frac{4\xi_z^2(0)}{s^2} \tag{1.39}$$

and $h = 2\pi\xi^2 H/\Phi_0$ as reduced magnetic field. The diamagnetic susceptibility in a weak field turns out [38, 39] to be

$$\chi_{(\text{layer},\perp)} = -\frac{\mathrm{e}^2 T}{3\pi s} \frac{\xi_{xy}^2}{\sqrt{\epsilon(\epsilon + r)}}. \tag{1.40}$$

In the $2D$ and 3D limits, this formula reproduces (1.33). Note that (1.40) predicts a nontrivial increase of diamagnetic susceptibility for clean metals [39]. The usual statement that fluctuations are most important in dirty superconductors with a short electronic mean free path does not hold in the particular case of susceptibility because here ξ turns out to be in the numerator of the fluctuation correction.

1.4.2 General Formula for the Fluctuation Free Energy in Magnetic Field

Now we will demonstrate that, besides the crossovers in its temperature dependence, the fluctuation-induced magnetization and heat capacity are also nonlinear functions of magnetic field. These nonlinearities, different for various dimensionalities, take place at relatively weak fields. This, strong in comparison with the expected scale of $H_{c2}(0)$, manifestation of the nonlinear regime in fluctuation magnetization and hence, field-dependent fluctuation susceptibility was the subject of the intensive debates in early 1970s [40–49] (see also the old but excellent review of Scokpol and Tinkham [50]) and after the discovery of HTS [51–54]. We will mainly follow here the recent essay of Mishonov and Penev [55] and the paper of Buzdin et al. [56], dealing with the fluctuation magnetization of a layered superconductor, which allows observing in a unique way all variety of the crossover phenomena in temperature and magnetic field.

One can evaluate the general expression (1.37) without taking the magnetic field to be small and get

$$F_{\rm LD}(\epsilon,h) = -\frac{TV}{2\pi s\xi_{\rm xy}^2}\left[h\int_0^{2\pi}\frac{d\theta}{2\pi}\ln\frac{\Gamma(1/2+\widetilde{\epsilon}(\theta)/2h)}{\sqrt{2\pi}}\right.$$
$$\left.+\frac{1}{2}\left(\epsilon+\frac{r}{2}\right)\ln h+\text{const}\right]. \tag{1.41}$$

This formula is valid for any anisotropy parameter.

1.4.3 Fluctuation Magnetization of Layered Superconductor and its Crossovers

Direct derivation of (1.41) over magnetic field gives for fluctuation part of magnetization:

$$M_{\rm LD}(\epsilon,h;r) = -\frac{T}{\Phi_0 s}\int_0^{\pi/2}\frac{d\phi}{\pi/2}\left\{\frac{\epsilon+r\sin^2\phi}{2h}\left[\psi\left(\frac{\epsilon+r\sin^2\phi}{2h}+\frac{1}{2}\right)-1\right]\right.$$
$$\left.-\ln\Gamma\left(\frac{\epsilon+r\sin^2\phi}{2h}+\frac{1}{2}\right)+\frac{1}{2}\ln(2\pi)\right\}.$$

Handling with the Hurvitz zeta functions the general formula for an arbitrary magnetic field in $3D$ case ($\epsilon < r$) can be carried out [44, 55]:

$$M_{(3)}(\epsilon\ll r,h) = 3\frac{T}{\Phi_0 s}\left(\frac{2}{r}\right)^{1/2}\sqrt{h}$$
$$\times\left[\zeta\left(-\frac{1}{2},\frac{1}{2}+\frac{\epsilon}{2h}\right)-\zeta\left(\frac{1}{2},\frac{1}{2}+\frac{\epsilon}{2h}\right)\frac{\epsilon}{6h}\right], \tag{1.42}$$

while in the opposite case of extremely high anisotropy $r < |\epsilon|, h \ll 1$ one obtains the $2D$ result.

Let us comment on the different crossovers in the $M(\epsilon, H)$ field dependence. Let us fix the temperature $\epsilon \ll r$. In this case, the c-axis coherence length exceeds the interlayer distance ($\xi_z \gg s$) and in the absence of a magnetic field the fluctuation Cooper pairs motion has a $3D$ character. For weak fields ($h \ll \epsilon$), the magnetization grows linearly with magnetic field, justifying our preliminary qualitative results:

$$M_{(3)}(\epsilon\ll r,h\to 0) = -\frac{e^2TH}{6\pi}\xi_{\rm xy}(\epsilon). \tag{1.43}$$

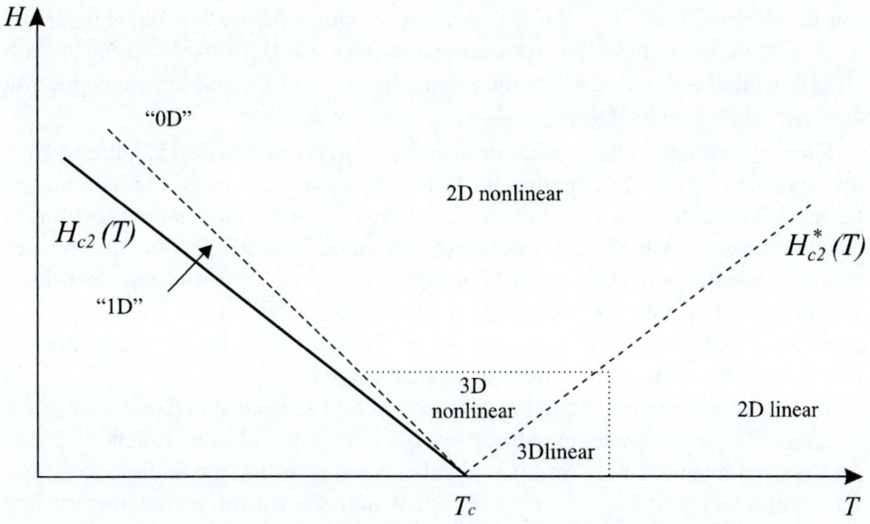

Fig. 1.4 Schematic representation of the different regimes for fluctuation magnetization in the (H, T) diagram. The line $H^*_{c2}(T)$ is mirror-symmetric to the $H_{c2}(T)$ line with respect to a y-axis passing through $T = T_c$. This line defines the crossover between linear and nonlinear behavior of the fluctuation magnetization above T_c [56]

Nevertheless, this linear growth is changed to the nonlinear $3D$ high field regime $M \sim \sqrt{H}$ already in the region of a relatively small fields $H_{c2}(\epsilon) \lesssim H$ $(\epsilon \lesssim h)$ (see Fig. 1.4). The further increase of magnetic field at $h \sim r$ leads to the next $3D \to 2D$ crossover in the magnetization field dependence. In the limit $\epsilon \ll h$, magnetization saturates at the value M_∞.

The substitution of $\epsilon = 0$ gives the result typical of $2D$ superconductors. Therefore, at $h \sim r$ we have a $3D \to 2D$ crossover in $M(H)$ behavior in spite of the fact that all sizes of fluctuation Cooper pair exceed considerably the lattice parameters. Let us stress that this crossover occurs in the region of already strongly nonlinear dependence of $M(H)$ and therefore for a rather strong magnetic field from the experimental point of view in HTS.

Let us mention the particular case of strong magnetic fields $\epsilon \ll h$ (1.42) reproduces the result by Prange [42] with an anisotropy correction multiplier [55] $\xi_{xy}(0)/\xi_z(0)$:

$$M_{(3)}(0, h) = -\frac{0.32T}{\Phi_0^{3/2}} \frac{\xi_{xy}(0)}{\xi_z(0)} \sqrt{H}. \tag{1.44}$$

Near the line of the upper critical field ($h_{c2}(\epsilon) = -\epsilon$), the contribution of the term with $n = 0$ in the sum (1.37) becomes the most important and for the magnetization the expression

$$M(h) = -0.346 \left(\frac{T}{\Phi_0 s}\right) \frac{h}{\sqrt{(h - h_{c2}(\epsilon))(h - h_{c2}(\epsilon) + r)}} \tag{1.45}$$

can be obtained [56]. It contains the already familiar for us "$0D$" regime ($r \ll h-h_{c2} \ll 1$), where the magnetization decreases as $-M(h) \sim \frac{1}{h-h_{c2}}$ (compare with (1.32)), while for $h - h_{c2} << r$ the regime becomes "$1D$" and the magnetization decreases slower, as $-M(h) \sim \frac{1}{\sqrt{h-h_{c2}}}$.

Such an analogy is observed in the next orders in Gi too. In the [57], the analogy was demonstrated for the example of the first eleven terms for the $2D$ case and nine for the $3D$ case. Summation of the series of high-order fluctuation contributions to the heat capacity by the Pade–Borel method resulted in its temperature dependence similar to the $0D$ and $1D$ cases without a magnetic field. Nevertheless, a considerable difference is not to be forgotten: in the $0D$ and $1D$ cases, no phase transition takes place while in the $2D$ and $3D$ cases in a magnetic field a phase transition of first order to the Abrikosov vortex lattice state occurs.

In conclusion, the fluctuation magnetization of a layered superconductor in the vicinity of the transition temperature turns out to be a complicated function of temperature and magnetic field, and it evidently cannot be factorized in these variables. The fit of the experimental data is very sensitive to the anisotropy parameter r and allows determination of the latter with a rather high precision [58, 59]. In Fig. 1.5, the successful application of the described approach to fit the experimental data on $YBa_2Cu_3O_7$ is shown [60].

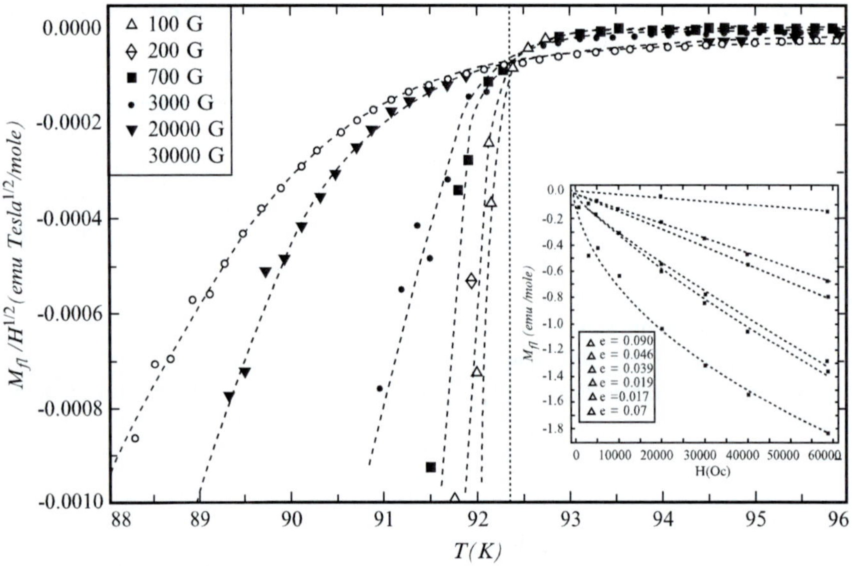

Fig. 1.5 Fluctuation magnetization of a YBaCO123 normalized on $\sqrt{H}$ as the function of temperature in accordance with the described theory shows the crossing of the iso-field curves at $T = T_c(0) = 92.3$ K. The best fit obtained for anisotropy parameter $r = 0.09$. In the inset, the magnetization curves as the function of magnetic field are reported

1.5 Fluctuation Conductivity of Layered Superconductor

The appearance of fluctuating Cooper pairs above T_c leads to the opening of a "new channel" for charge transfer. The fluctuation Cooper pairs were treated above as carriers with charge $2e$, while their lifetime τ_{GL} was chosen to play the role of the scattering time in the Drude formula. Such a qualitative consideration results in the Aslamazov–Larkin (AL) pair contribution to conductivity (1.10) (the so-called paraconductivity [61]). Below we will present the generalization of the phenomenological GL functional approach to transport phenomena. Dealing with the fluctuation order parameter, it is possible to describe correctly the paraconductivity-type fluctuation contributions to the normal resistance and magnetoconductivity, Hall effect, thermoelectric power, and thermal conductivity at the edge of the transition. Unfortunately, the indirect fluctuation contributions are beyond the possibilities of the description by time-dependent GL (TDGL) approach, and they can be calculated only in the framework of the microscopic theory (see below).

1.5.1 Time-Dependent GL Equation

In previous sections, we have demonstrated how the GL functional formalism allows one to accounting for fluctuation corrections to thermodynamic quantities. Let us discuss the effect of fluctuations on the transport properties of a superconductor above the critical temperature.

To find the value of paraconductivity, some time-dependent generalization of the GL equations is required. Indeed, the conductivity characterizes the response of the system to the applied electric field. It can be defined as $\mathbf{E} = -\partial \mathbf{A}/\partial t$ but, in contrast to the previous section, $\mathbf{A}$ has to be regarded as being time dependent. The general nonstationary BCS equations are very complicated, even in the limit of slow time and space variations of the field and the order parameter. For our purposes, it will be sufficient, following [62–70], to write a model equation in the vicinity of T_c, which in general correctly reflects the qualitative aspects of the order parameter dynamics and in some cases is exact.

Let us revise the GL functional formalism introduced above. One can see that the derived above stationary GL equations do not describe correctly the superconductive properties when a deviation from equilibrium is assumed. Indeed, in the absence of equilibrium, the order parameter Ψ becomes time dependent and this in no way was included in the scheme. Nevertheless, the scheme can be improved. For small deviations from the equilibrium, it is natural to assume that in the process of order parameter relaxation its time derivative $\partial\Psi/\partial t$ is proportional to the variational derivative of the free energy $\delta\mathcal{F}/\delta\Psi^*$, which is equal to zero at the equilibrium. But this is not all: side by side with the normal relaxation of the order parameter the effect of thermodynamic fluctuations on it has to be taken into account. This can be done by the introduction, the Langevin forces $\zeta(\mathbf{r}, t)$ in the right-hand side of the equation describing the order parameter dynamics. Finally, gauge invariance requires that $\partial\Psi/\partial t$ should be included in the equation in the combination $\partial\Psi/\partial t + 2ie\varphi\Psi$, where φ is the scalar potential of the electric field. By including all these considerations, one can write the model time-dependent GL equation in the form

$$-\gamma_{\mathrm{GL}}\left(\frac{\partial}{\partial t}+2ie\varphi\right)\Psi=\frac{\delta\mathcal{F}}{\delta\Psi^{*}}+\zeta(\mathbf{r},t) \tag{1.46}$$

with the GL functional $\mathcal{F}$ determined by (1.18), (1.29), (1.34) [71]. The dimensionless coefficient γ_{GL} in the left-hand-side of the equation can be related to pair lifetime τ_{GL} (1.1): $\gamma_{\mathrm{GL}}=\alpha T_{\mathrm{c}}\epsilon\tau_{\mathrm{GL}}=\pi\alpha/8$ by the substitution in (1.46) of the first term of (1.18) only [72].

Neglecting the fourth-order term in the GL functional, (1.46) can be rewritten in operator form as

$$[\widehat{L}^{-1}-2ie\gamma_{\mathrm{GL}}\varphi(r,t)]\Psi(\mathbf{r},t)=\zeta(\mathbf{r},t) \tag{1.47}$$

with the TDGL operator $\widehat{L}$ and Hamiltonian $\widehat{\mathcal{H}}$ defined as

$$\widehat{L}=\left[\gamma_{\mathrm{GL}}\frac{\partial}{\partial t}+\widehat{\mathcal{H}}\right]^{-1},\widehat{\mathcal{H}}=\alpha T_{\mathrm{c}}\left[\epsilon-\widehat{\xi}^{2}(\widehat{\nabla}-2ie\mathbf{A})^{2}\right]. \tag{1.48}$$

We have introduced here the formal operator of the coherence length $\widehat{\xi}$ to have the possibility to deal with an arbitrary type of spectrum. For example, in the most interesting case for our applications to layered superconductors, the action of this operator is defined by (1.34).

In the absence of an electric field, one can write the formal solution of (1.47) as

$$\Psi^{(0)}(\mathbf{r},t)=\widehat{L}\zeta(\mathbf{r},t). \tag{1.49}$$

The correlator of the Langevin forces introduced above must satisfy the fluctuation–dissipation theorem. This requirement is fulfilled if the Langevin forces $\zeta(\mathbf{r},t)$ and $\zeta^{*}(\mathbf{r},t)$ are correlated by the Gaussian white-noise law

$$\langle\zeta^{*}(\mathbf{r},t)\zeta(\mathbf{r}',t')\rangle=2T\,\mathrm{Re}\,\gamma_{\mathrm{GL}}\delta(\mathbf{r}-\mathbf{r}')\delta(t-t'). \tag{1.50}$$

The fundamental solution $L(\mathbf{p},\Omega)$ can be found by making a Fourier transform of (1.48), what gives:

$$L(\mathbf{p},\Omega)=(-i\gamma_{\mathrm{GL}}\Omega+\varepsilon_{\mathbf{p}})^{-1}. \tag{1.51}$$

with

$$\varepsilon_{\mathbf{p}}=\alpha T_{\mathrm{c}}(\epsilon+\widehat{\xi}^{2}\mathbf{p}^{2}) \tag{1.52}$$

as the fluctuation Cooper pair energy spectrum.

1.5.2 General Expression for Paraconductivity

By means of the qualitative consideration based on the Drude formula, we obtained in the Introduction the expression for paraconductivity, which correctly reflects its temperature singularity in any dimension. Following this way, one could write down some kind of master equation for fluctuation Cooper pairs and obtain indeed the precise expression for paraconductivity (see [1]). Unfortunately, the applicability of the derived master equation is restricted to relatively weak electric and magnetic fields. For stronger fields $H_{c2}(\epsilon) \lesssim H \ll H_{c2}(0)$, the density matrix has to be introduced and the master equation loses its attractive simplicity. At the same time, as we already know, these fields, quantizing the fluctuation Cooper pair motion, present special interest. That is why to include in the scheme the magnetic field and frequency dependencies of the paraconductivity, we return to the analysis of the general TDGL equation (1.46) without the objective to reduce it to a Boltzmann-type transport equation.

Let us solve it in the case, when the applied electric field can be considered as a perturbation. The method will much resemble an exercise from a course on quantum mechanics. To impose the necessary generality side by side with a formal simplicity of expressions, we will introduce a subscript of the kind $\{i\}$, which includes the complete set of quantum numbers and time. By a repeated subscript, a summation over a discrete and integration over continuous variables (time in particular) is implied.

We will look for the response of the order parameter to a weak electric field applied in the form

$$\Psi_{k_z}(\mathbf{r},t) = \Psi^{(0)}_{\{i\}} + \Psi^{(1)}_{\{i\}}, \tag{1.53}$$

where $\Psi^{(0)}_{\{i\}}$ is determined by (1.49). Substituting this expression into (1.47) and restricting our consideration to linear terms in the electric field, we can write

$$(\widehat{L}^{-1})_{\{ik\}}\Psi^{(1)}_{\{k\}} = 2\mathrm{i}e\gamma_{\mathrm{GL}}\varphi_{\{il\}}\Psi^{(0)}_{\{l\}} \tag{1.54}$$

with the solution in the form

$$\Psi^{(1)}_{\{i\}} = 2\mathrm{i}e\gamma_{\mathrm{GL}}\widehat{L}_{\{ik\}}\varphi_{\{kl\}}\widehat{L}_{\{lm\}}\zeta_{\{m\}}. \tag{1.55}$$

Let us substitute the order parameter (1.53) in the quantum mechanical expression for current:

$$\mathbf{j} = 2e\,\mathrm{Re}\left[\Psi^{(0)*}_{\{i\}}\widehat{\mathbf{v}}_{\{ik\}}\Psi^{(1)}_{\{k\}} + \Psi^{(1)*}_{\{i\}}\widehat{\mathbf{v}}_{\{ik\}}\Psi^{(0)}_{\{k\}}\right], \tag{1.56}$$

where $\widehat{\mathbf{v}}_{\{ik\}}$ is the velocity operator, which can be expressed by means of the commutator of $\mathbf{r}$ with Hamiltonian (1.48):

$$\widehat{\mathbf{v}}_{\{ik\}} = \mathrm{i}\{\widehat{\mathcal{H}}, \mathbf{r}\}_{\{ik\}} \tag{1.57}$$

and average now (1.56) over the Langevin forces. Moving the operator $\widehat{L}^*_{\{ki\}}$ from the beginning to the end of the trace, one finds

$$\mathbf{j} = -16Te^2\,\mathrm{Re}(\gamma_{\mathrm{GL}})\,\mathrm{Im}\{\gamma_{\mathrm{GL}}\widehat{\mathbf{v}}_{\{il\}}\widehat{L}_{\{lm\}}\varphi_{\{mn\}}\widehat{L}_{\{np\}}\widehat{L}^*_{\{pi\}}\}. \tag{1.58}$$

Now we choose the representation where the $\widehat{L}_{\{lm\}}$ operator is diagonal (it is evidently given by the eigenfunctions of the Hamiltonian (1.48)):

$$L_{\{m\}}(\Omega) = \frac{1}{-\mathrm{i}\Omega\gamma_{\mathrm{GL}} + \varepsilon_{\{m\}}}, \tag{1.59}$$

where $\varepsilon_{\{m\}}$ are the appropriate energy eigenvalues. Then we assume that the electric field is coordinate independent but is a monochromatic periodic function of time:

$$\varphi(r,t) = -E^\beta r^\beta \exp(-\mathrm{i}\omega t). \tag{1.60}$$

In doing the Fourier transform in (1.58), one has to remember that the time dependence of the matrix elements $\varphi_{\{mn\}}$ results in a shift of the frequency variable of integration $\Omega \to \Omega - \omega$ in both L-operators placed after $\varphi_{\{mn\}}$ or, what is the same, to a shift of the argument of the previous $\widehat{L}_{\{lm\}}$ for ω:

$$\mathbf{j}^\alpha_\omega = 16Te^2\,\mathrm{Re}(\gamma_{\mathrm{GL}})\int\frac{\mathrm{d}\Omega}{2\pi}\Re\{\gamma_{\mathrm{GL}}\widehat{\mathbf{v}}^\alpha_{\{il\}}\widehat{L}_{\{l\}}(\Omega+\omega)[-\mathrm{i}r^\beta_{\{li\}}]\widehat{L}_{\{i\}}(\Omega)\widehat{L}^*_{\{i\}}(\Omega)\}\mathbf{E}^\beta, \tag{1.61}$$

where $\Re f(\omega) \equiv [f(\omega) + f^*(-\omega)]/2$.

Let us express the matrix element $\mathbf{r}_{\{li\}}$ by means of $\widehat{\mathbf{v}}_{\{li\}}$ using the commutation relation (1.57). One can see that in the representation chosen

$$\widehat{\mathbf{r}}^\beta_{\{li\}} = \mathrm{i}\frac{\widehat{\mathbf{v}}^\beta_{\{li\}}}{\varepsilon_{\{i\}} - \varepsilon_{\{l\}}} \tag{1.62}$$

and, carrying out the frequency integration in (1.61), finally write for the fluctuation conductivity tensor ($\mathbf{j}^\alpha_\omega = \sigma^{\alpha\beta}(\omega)\mathbf{E}^\beta$):

$$\sigma^{\alpha\beta}(\epsilon, H, \omega) = 8e^2T\,\mathrm{Re}(\gamma_{\mathrm{GL}})\sum_{\{i,l\}=0}^{\infty}\Re\left[\gamma_{\mathrm{GL}}\frac{\widehat{\mathbf{v}}^\alpha_{\{il\}}\widehat{\mathbf{v}}^\beta_{\{li\}}}{\varepsilon_{\{i\}}(\gamma_{\mathrm{GL}}\varepsilon_{\{i\}} + \gamma^*_{\mathrm{GL}}\varepsilon_{\{l\}} - \mathrm{i}|\gamma_{\mathrm{GL}}|^2\omega)(\varepsilon_{\{l\}} - \varepsilon_{\{i\}})}\right]. \tag{1.63}$$

This is the most general expression which describes the d.c., galvanomagnetic and high frequency paraconductivity contributions.

The microscopic analysis of the coefficient γ_{GL} demonstrates that its imaginary part $\mathrm{Im}\,\gamma_{GL}$ usually is much smaller than $\mathrm{Re}\,\gamma_{GL}$. Its origin can be related to the electron–hole asymmetry or other peculiarities of the electron spectrum. In the case when one is interested in the diagonal effects only it is enough to accept γ_{GL} as real: ($\gamma_{GL} = \mathrm{Re}\,\gamma_{GL} = \pi\alpha/8$). In this way, (1.64) can be simplified and after symmetrization of the summation variables the d.c. contribution of fluctuation Cooper pairs to magnetoconductivity takes the form:

$$\sigma^{\alpha\alpha}(\epsilon, H) = \frac{\pi}{2}\alpha \mathrm{e}^2 T \sum_{\{i,l\}=0}^{\infty} \Re\left[\frac{\widehat{\mathbf{v}}^{\alpha}_{\{il\}}\widehat{\mathbf{v}}^{\alpha}_{\{li\}}}{\varepsilon_{\{i\}}\varepsilon_{\{l\}}(\varepsilon_{\{i\}}+\varepsilon_{\{l\}})}\right]. \tag{1.64}$$

Let us demonstrate the calculation of the d.c. paraconductivity in the simplest case of a metal with an isotropic spectrum. In this case, we choose a plane wave representation. By using $\varepsilon_{\mathbf{p}}$ defined by (1.52), one has

$$\widehat{\mathbf{v}}_{\{\mathbf{pp}'\}} = \mathbf{v}_{\mathbf{p}}\delta_{\mathbf{pp}'},\ \mathbf{v}_{\mathbf{p}} = \frac{\partial \varepsilon_{\mathbf{p}}}{\partial \mathbf{p}} = 2\alpha T_{\mathrm{c}}\xi^2\mathbf{p}. \tag{1.65}$$

We do not need to keep here the imaginary part of γ_{GL}, which is necessary to calculate particle-hole asymmetric effects only. As a result, one reproduces the AL formula:

$$\sigma^{\alpha\beta}_{(D)}=2\mathrm{e}^2 T\,\mathrm{Re}\,\gamma_{GL}\sum_{\mathbf{p}}\frac{\mathbf{v}^{\alpha}_{\mathbf{p}}\mathbf{v}^{\beta}_{\mathbf{p}}}{\varepsilon^3_{\mathbf{p}}}=\delta^{\alpha\beta}\begin{cases}\dfrac{\mathrm{e}^2}{32\xi}\dfrac{1}{\sqrt{\epsilon}}, & \text{3D case,}\\ \dfrac{\mathrm{e}^2}{16d}\dfrac{1}{\epsilon}, & \text{2D film, thickness}: d \ll \xi,\\ \dfrac{\pi \mathrm{e}^2\xi}{16S}\dfrac{1}{\epsilon^{3/2}}, & \text{1D wire, cross}-\text{section}: S \ll \xi^2.\end{cases}$$

1.5.3 *Paraconductivity of a Layered Superconductor*

Let us return to the discussion of our general formula (1.64) for the fluctuation conductivity tensor. A magnetic field directed along the c-axis still allows separation of variables even in the case of a layered superconductor. The Hamiltonian in this case can be written as in (1.36), (1.48):

$$\widehat{\mathcal{H}} = \alpha T_{\mathrm{c}}\left(\epsilon - \xi^2_{\mathrm{xy}}(\nabla_{\mathrm{xy}} - 2\mathrm{i}e\mathbf{A}_{\mathrm{xy}})^2 - \frac{r}{2}(1-\cos(k_z s))\right). \tag{1.66}$$

It is convenient to work in the Landau representation, where the summation over $\{i\}$ is reduced to one over the ladder of Landau levels $i = 0, 1, 2 \ldots$. (each is degenerate with a density H/Φ_0 per unit square) and integration over the c-axis momentum in the limits of the Brillouin zone. The eigenvalues of the Hamiltonian (1.66) can be written in the form

$$\varepsilon_{\{n\}} = \alpha T_{\rm c}\left[\epsilon + \frac{r}{2}(1-\cos(k_z s)) + h(2n+1)\right] = \varepsilon_{k_z} + \alpha T_{\rm c} h(2n+1), \quad (1.67)$$

where $h = eH/2m\alpha T_{\rm c}$. For the velocity operators one can write

$$\widehat{\mathbf{v}}^{x,y} = \frac{1}{2m}(-\mathrm{i}\nabla - 2\mathrm{i}e\mathbf{A})^{x,y};\ \widehat{\mathbf{v}}^{z} = -\frac{\alpha r s}{2}T_{\rm c}\sin(k_z s). \quad (1.68)$$

1.5.4 In-Plane Conductivity

Let us start from the calculation of the in-plane components. The calculation of the velocity operator matrix elements requires some special consideration. First of all, let us stress that the required matrix elements have to be calculated for the eigenstates of a quantum oscillator whose motion is equivalent to the motion of a charged particle in a magnetic field. The commutation relation for the velocity components follows from (1.68) (see [73]):

$$[\widehat{\mathbf{v}}^{x}, \widehat{\mathbf{v}}^{y}] = \mathrm{i}\frac{eH_z}{2m^2} = \frac{\mathrm{i}\alpha T_{\rm c}}{m}h. \quad (1.69)$$

To calculate the necessary matrix elements, let us present the velocity operator components in the form of boson-type creation and annihilation operators $\widehat{a}^{+}, \widehat{a}$:

$$\langle l|\widehat{a}|n\rangle = \langle n|\widehat{a}^{+}|l\rangle = \sqrt{n}\delta_{n,l+1},$$

which satisfy the commutation relation $[\widehat{a}, \widehat{a}^{+}] = 1$. We obtain

$$\widehat{\mathbf{v}}^{x,y} = \sqrt{\frac{\alpha T_{\rm c} h}{2m}}\begin{pmatrix} \widehat{a}^{+} + \widehat{a} \\ \mathrm{i}\widehat{a}^{+} - \mathrm{i}\widehat{a} \end{pmatrix}.$$

One can check that the correct commutation relation (1.69) is fulfilled and see that the only nonzero matrix elements of the velocity operator are

$$\langle l|\widehat{\mathbf{v}}^{x,y}|n\rangle = \sqrt{\frac{\alpha T_{\rm c} h}{2m}}\begin{pmatrix} \sqrt{l}\delta_{l,n+1} + \sqrt{n}\delta_{n,l+1} \\ \mathrm{i}\sqrt{l}\delta_{l,n+1} - \mathrm{i}\sqrt{n}\delta_{n,l+1} \end{pmatrix}. \quad (1.70)$$

Using these relations, the necessary product of matrix elements can be calculated:

$$\langle l|\widehat{\mathbf{v}}^{x}|n\rangle\langle n|\widehat{\mathbf{v}}^{x}|l\rangle = \frac{\alpha T_{\rm c} h}{2m}(l\delta_{l,n+1} + n\delta_{n,l+1}). \quad (1.71)$$

Its substitution to the expression (1.64) gives for the diagonal in-plane component of the paraconductivity tensor

$$\sigma^{xx}(\epsilon, h) = \frac{\pi\alpha^2 T_c^2 e^2}{4m} h \sum_{\{n,l\}=0}^{\infty} \Re \frac{(l\delta_{l,n+1} + n\delta_{n,l+1})}{\varepsilon_{\{l\}}\varepsilon_{\{n\}}\left[\varepsilon_{\{l\}} + \varepsilon_{\{n\}}\right]}.$$

Summation over the subscript $\{l\}$ and accounting of the degeneracy of the Landau levels $H/\Phi_0 = 2m\alpha T_c h/\pi$ (the layer area we assume to be equal one) gives for the diagonal component of the in-plane paraconductivity tensor:

$$\sigma^{xx}(\epsilon, H) = \frac{e^2 (\alpha T_c)^3 h^2}{2} \int_{-\frac{\pi}{s}}^{\frac{\pi}{s}} \frac{dk_z}{2\pi} \sum_{n=0}^{\infty} \Re \frac{n+1}{\varepsilon_{n+1}\varepsilon_n(\varepsilon_{n+1} + \varepsilon_n)}. \quad (1.72)$$

1.5.5 *Out-of Plane Conductivity*

The situation with the out-of plane component of paraconductivity turns out to be even simpler because of the diagonal structure of the

$$\widehat{\mathbf{v}}^z_{\{in\}} = -\frac{\alpha r s}{2} T_c \sin(k_z s) \times \delta_{\text{in}} \times \delta(k_z - k_{z'}).$$

Taking into account that the Landau state degeneracy, we write

$$\sigma^{zz}(\epsilon, H) = \frac{\pi e^2 (\alpha T_c)^3}{32} \left(\frac{sr}{\xi_{xy}}\right)^2 h \sum_{n=0}^{\infty} \int_{-\frac{\pi}{s}}^{\frac{\pi}{s}} \frac{dk_z}{2\pi} \Re \left[\frac{\sin^2(k_z s)}{\varepsilon_n^2(k_z)[\varepsilon_n(k_z)]}\right].$$

1.5.6 *Analysis of the Limiting Cases*

In principle, the expressions derived above give an exact solution for the d.c. paraconductivity tensor of a layered superconductor in a perpendicular magnetic field $H \ll H_{c2}$ ($h \ll 1$) in the vicinity of the critical temperature ($\epsilon \ll 1$). The interplay of the parameters r, ϵ, h, as we have seen in the example of fluctuation magnetization yields a variety of crossover phenomena.

The simplest and most important results which can be derived are the components of the d.c. paraconductivity of a layered superconductor in the absence of magnetic field. Setting $h \to 0$ one can change the summations over Landau levels into integration and find

$$\sigma^{xx}(\epsilon, h \to 0, \omega = 0) = \frac{e^2}{16s} \frac{1}{\sqrt{[\epsilon(r+\epsilon)]}}, \quad (1.73)$$

$$\sigma^{zz}(\epsilon, h \to 0, \omega = 0) = \frac{e^2 s}{32\xi_{xy}^2}\left(\frac{\epsilon + r/2}{[\epsilon(\epsilon + r)]^{1/2}} - 1\right). \tag{1.74}$$

For $2D$ case, the sum can be calculated exactly in terms of the ψ-functions:and one finds the expression for the 2D magnetoconductivity:

$$\sigma_{(2)}^{xx}(\epsilon, h) = \frac{e^2}{2s}\frac{1}{\epsilon}F\left(\frac{\epsilon}{2h}\right) = \frac{e^2}{16s}\begin{cases} 1/\epsilon, & h \ll \epsilon \\ 2/h, & \epsilon \ll h \\ 4/(\epsilon + h), & \epsilon + h \to 0 \end{cases}, \tag{1.75}$$

where

$$F(x) = x^2\left[\psi\left(\frac{1}{2} + x\right) - \psi(x) - \frac{1}{2x}\right]. \tag{1.76}$$

The $2D$ AL theory was extended [16, 74] to the high temperature region by taking into account the short-wavelength and dynamic fluctuations. The following universal formula for paraconductivity of a $2D$ superconductor as a function of the generalized reduced temperature $\epsilon = \ln T/T_c$ and magnetic field was obtained [16]:

$$\delta\sigma_{xx}^{AL}(t, h) = \frac{e^2}{\pi}\sum_{m=0}^{\infty}(m+1)\int_{-\infty}^{\infty}\frac{dx}{\sinh^2 \pi x}$$
$$\times\left\{\frac{\mathrm{Im}^2\mathcal{E}_m}{|\mathcal{E}_m|^2} + \frac{\mathrm{Im}^2\mathcal{E}_{m+1}}{|\mathcal{E}_{m+1}|^2} + \frac{\mathrm{Im}^2\mathcal{E}_{m+1} - \mathrm{Im}^2\mathcal{E}_m}{|\mathcal{E}_m|^2|\mathcal{E}_{m+1}|^2}\mathrm{Re}\left[\mathcal{E}_m\mathcal{E}_{m+1}\right]\right\} \tag{1.77}$$

with

$$\mathcal{E}_m \equiv \mathcal{E}_m(\epsilon, h, iz) = \epsilon + \psi\left[\frac{1 + iz}{2} + \frac{2h}{t}\frac{(2m+1)}{\pi^2}\right] - \psi\left(\frac{1}{2}\right), \tag{1.78}$$

and

$$h = \frac{\pi^2}{8\gamma_E}\frac{H}{H_{c2}(0)}. \tag{1.79}$$

In the limit of zero fields, one can find [74]:

$$\sigma_{xx}^{(fl)} = \frac{e^2}{16}\begin{cases} \dfrac{1}{\epsilon} & \epsilon \ll 1 \\ \dfrac{0.11}{\epsilon^3} & \epsilon \gtrsim 1 \end{cases}.$$

Here, it is worth making an important comment. The proportionality of the fluctuation magnetoconductivity to h^2 is valid when using the parametrization $\epsilon = (T - T_{c0})/T_{c0}$ only. Often the analysis of the experimental data is carried out by choosing as the reduced temperature parameter $\epsilon_h = (T - T_c(H))/T_c(H)$. At that point, it is important to recognize that the effect of a weak magnetic field on the fluctuation conductivity cannot be reduced to a simple replacement of T_{c0} by $T_c(H)$ in the appropriate formula without the field. In this parametrization, one can get a term in the magnetoconductivity linear in h. Point is that besides the cases of the special specimen geometry, a weak magnetic field shifts the critical temperature linearly. Such linear correction is exactly compensated by the change in the functional dependence of the paraconductivity in magnetic field, and finally it contains the negative quadratic contribution only.

1.5.7 Comparison with the Experiment

Although the in-plane and out-of-plane components of the fluctuation conductivity tensor of a layered superconductor contain the same fluctuation contributions, their temperature behavior may be qualitatively different. In fact, for $\sigma_{xx}^{(fl)}$, the negative contributions are considerably less than the positive ones in the entire experimentally accessible temperature range above the transition, and it is a positive monotonic function of the temperature. Moreover, for HTS compounds, where the pair-breaking is strong and the anomalous MT contribution is in the saturated regime, it is almost always enough to take into account only the paraconductivity to fit experimental data. Some examples of the experimental findings for in-plane fluctuation conductivity of HTS materials can be seen in [75–82].

In Fig. 1.6, the fluctuation part of in-plane conductivity $\sigma_{xx}^{(fl)}$ is plotted as a function of $\epsilon = \ln T/T_c$ on a double logarithmic scale for three HTS samples (the solid line represents the $2D$ AL behavior $(1/\epsilon)$, the dotted line represents the $3D$ one: $3.2/\sqrt{\epsilon}$) [83]. One can see that paraconductivity of the less anisotropic YBCO compound asymptotically tends to the $3D$ behavior $(1/\epsilon^{1/2})$ for $\epsilon < 0.1$, showing the LD crossover at $\epsilon \approx 0.07$; the curve for more anisotropic 2223 phase of BSCCO starts to bend for $\epsilon < 0.03$ while the most anisotropic 2212 phase of BSCCO shows a $2D$ behavior in the whole temperature range investigated. All three compounds show a universal $2D$ temperature behavior above the LD crossover up to the limits of the GL region. It is interesting that around $\epsilon \approx 0.24$ all the curves bend down and in accordance to [74], follow the same asymptotic $1/\epsilon^3$ behavior (dashed line). Finally at the value $\epsilon \approx 0.45$, all the curves fall down indicating the end of the observable fluctuation regime.

In the case of the out-of-plane conductivity, the situation is quite different. Both positive contributions (AL and anomalous MT) are suppressed here by the necessity of the interlayer tunneling, what results in a competition between positive and negative terms. Such concurrence can lead to formation of a maximum in the temperature dependence of the c-axis resistivity. This nontrivial effect of fluctuations on the transverse resistance of a layered superconductor allows a successful fit to the data observed on optimally doped and overdoped HTS samples (see, e.g., Fig. 1.7), where the growth of the resistance still can be treated as a correction. The fluctuation mechanism of the growth of the transverse resistance

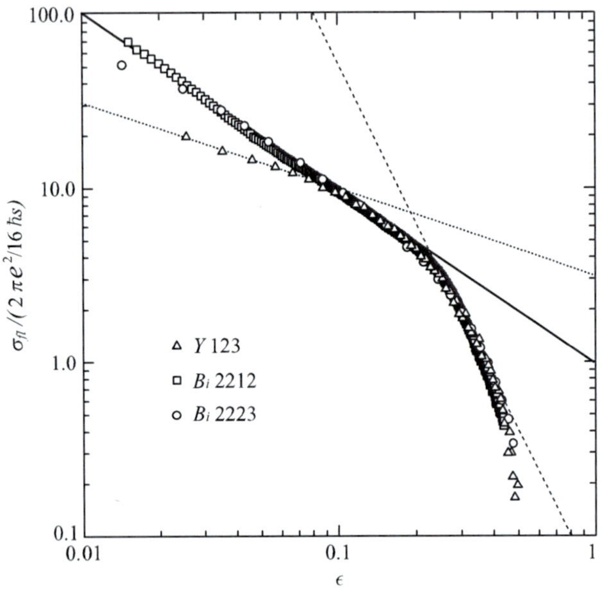

Fig. 1.6 The normalized excess conductivity for samples of YBCO-123 (*triangles*), BSSCO-2212 (*squares*) and BSSCO-2223 (*circles*) plotted against $\epsilon = \ln T/T_c$ on a ln-ln plot as described in [83]. The *dotted and solid lines* are the AL theory in $3D$ and $2D$ respectively. The *dashed line* is the extended theory of [74]

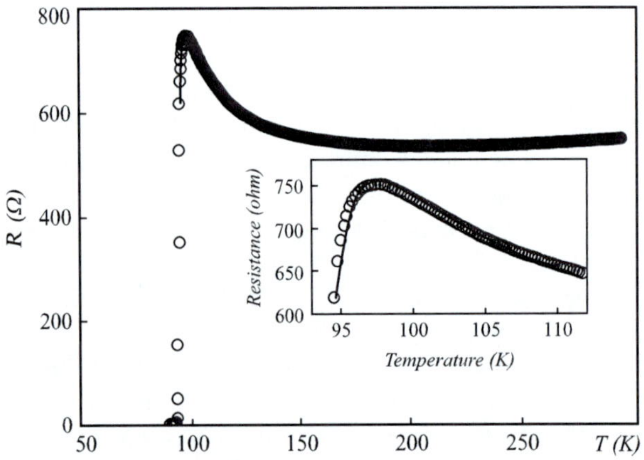

Fig. 1.7 Fit of the temperature dependence of the transverse resistance of a slightly underdoped BSCCO c-axis oriented film with the results of the fluctuation theory [84]. The inset shows the details of the fit in the temperature range between T_c and 110 K

can be easily understood in a qualitative manner. Indeed, to modify the in-plane result for the case of c-axis paraconductivity, one has to take into account the hopping character of the electronic motion in this direction. If the probability of one-electron interlayer hopping is $\mathcal{P}_1$, then the probability of coherent hopping for two electrons during the fluctuation Cooper pair lifetime τ_{GL} is the conditional probability of these two events: $\mathcal{P}_2 = \mathcal{P}_1(\mathcal{P}_1\tau_{\mathrm{GL}})$. The transverse paraconductivity may thus be estimated as $\sigma_{\perp}^{\mathrm{AL}} \sim \mathcal{P}_2\sigma_{\parallel}^{\mathrm{AL}} \sim \mathcal{P}_1^2\frac{1}{\epsilon^2}$, in complete accordance with the result of microscopic theory. We see that the temperature singularity of $\sigma_{\perp}^{\mathrm{AL}}$ turns out to be stronger than that in $\sigma_{\parallel}^{\mathrm{AL}}$, however, for a strongly anisotropic layered superconductor $\sigma_{\perp}^{\mathrm{AL}}$ is considerably suppressed by the square of the small probability of interlayer electron hopping, which enters in the prefactor. It is this suppression which leads to the necessity of taking into account the DOS contribution to the transverse conductivity. The latter is less singular in temperature but, in contrast to the paraconductivity, manifests itself in the first, not the second, order in the interlayer transparency $\sigma_{\perp}^{\mathrm{DOS}} \sim -\mathcal{P}_1 \ln(1/\epsilon)$. The DOS fluctuation correction to the one-electron transverse conductivity is negative and, being proportional to the first order of $\mathcal{P}_1$, can completely change the traditional picture of fluctuations just rounding the resistivity temperature dependence around transition. The shape of the temperature dependence of the transverse resistance mainly is determined by the competition between the opposite sign contributions: the paraconductivity and MT term, which are strongly temperature dependent but are suppressed by the square of the barrier transparency and the DOS contribution, which has a weaker temperature dependence but depends only linearly on the barrier transparency.

1.6 Quantum Superconductive Fluctuations Above $H_{c2}(0)$

1.6.1 Dynamic Clustering of FCPs

The qualitative picture for SF in the quantum region at very low temperatures and close to $H_{c2}(0)$ drastically differs from the Ginzburg–Landau one, valid close to T_{c0}. As we saw above, the latter can be described in terms of the set of long-wavelength fluctuation modes (with $\lambda \gtrsim \xi_{\mathrm{GL}}(T) \gg \xi_{\mathrm{BCS}}$) of the order parameter, with the characteristic lifetime $\tau_{\mathrm{GL}} = \pi\hbar/8k_{\mathrm{B}}(T - T_{c0})$. In the former, the order parameter oscillates in much smaller scale, the fluctuation modes with the wave-lengths up ξ_{BCS} are excited. One can imagine that FCPs here rotate in magnetic field with the Larmor radius $\sim \xi_{\mathrm{BCS}}$ and cyclotron frequency $\omega_c \sim \Delta_{\mathrm{BCS}}^{-1}$. The microscopic theory shows below that close to $H_{c2}(0)$ these FCPs form some kind of quantum liquid with the long coherence length $\xi_{\mathrm{QF}} \sim \xi_{\mathrm{BCS}}/\widetilde{h}^{1/2}$ and slow relaxation with the characteristic time

$$\tau_{\mathrm{QF}} \sim \hbar\left(\Delta_{\mathrm{BCS}}\widetilde{h}\right)^{-1}, \widetilde{h} = (H - H_{c2}(0))/H_{c2}(0) \tag{1.80}$$

One sees that the functional form of τ_{QF} is completely analogous to that of τ_{GL}: $\Delta_{BCS} \sim T_{c0}$ and the reduced field $\widetilde{h}$ plays the role of reduced temperature ϵ. Equation (1.80) can also be obtained also from the uncertainty principle. Indeed, the energy, characterizing the proximity to the quantum phase transition is $\Delta E = \hbar\omega_c(H) - \hbar\omega_c(H_{c2}(0)) \sim \Delta_{BCS}\widetilde{h}$ and namely this value should be used in the Heisenberg relation instead of $k_B(T - T_{c0})$, as was done in the vicinity of T_{c0}. The spatial coherence scale $\xi_{QF}\left(\widetilde{h}\right)$ can be estimated from the value of τ_{QF} analogously to consideration near T_{c0}. Namely, two electrons with the coherent phase starting from the same point after the time τ_{QF} get separated by the distance

$$\xi_{QF}\left(\widetilde{h}\right) \sim (\mathcal{D}\tau_{QF})^{1/2} \sim \xi_{BCS}/\sqrt{\widetilde{h}}.$$

To clarify the physical meaning of τ_{QF} and ξ_{QF}, note that near the quantum phase transition at zero temperature, where $H \to H_{c2}(0)$, the fluctuations of the order parameter $\Delta^{(fl)}(\mathbf{r}, t)$ become highly inhomogeneous, contrary to the situation near T_{c0}. Indeed, below $H_{c2}(0)$, the spatial distribution of the order parameter at finite magnetic field reflects the existence of Abrikosov vortices with average spacing (close to $H_{c2}(0)$ but in the region where the notion of vortices is still adequate) equal to

$$a(H) = \xi_{BCS}/\sqrt{H/H_{c2}(0)} \to \xi_{BCS}.$$

Therefore, one expects that close to and above $H_{c2}(0)$ the fluctuation order parameter $\Delta^{(fl)}(\mathbf{r}, t)$ also has "vortex-like" spatial structure and varies over the scale of ξ_{BCS} being preserved over the time scale τ_{QF}. In the language of FCPs, one describes this situation in the following way. The FCPs at zero temperature and in magnetic field close to $H_{c2}(0)$ rotates with the Larmor radius $r_L \sim v_F/\omega_c(H_{c2}(0)) \sim v_F/\Delta_{BCS} \sim \xi_{BCS}$, which presents their effective size. During the time τ_{QF}, two initially selected electrons participate in the multiple fluctuating Cooper pairings maintaining their coherence. The coherence length $\xi_{QF}\left(\widetilde{h}\right) \gg \xi_{BCS}$ is thus a characteristic size of a cluster of such coherently rotating FCPs, and τ_{QF} estimates the lifetime of such flickering cluster. One can view the whole system as an ensemble of flickering domains of coherently rotating FCPs, precursors of vortices (see Fig. 1.8).

In view of described qualitative picture of SF in the regime of QPT, let us resume the scenario of Abrikosov lattice defragmentation. Approaching to $H_{c2}(0)$ from below, the paddles of fluctuating vortices, which are nothing else as rotating in magnetic field FCPs, are formed. Their characteristic size is $\xi_{QF}\left(|\widetilde{h}|\right)$, and they flicker with the characteristic time $\tau_{QF}\left(|\widetilde{h}|\right)$. At this stage, the supercurrent still can flow through the sample until these paddles do not break the last percolating superconductive channel. Corresponding field determines the value of the renormalized by QF second critical field: $H_{c2}^{*}(0) = H_{c2}(0)\left[1 - 2\mathrm{Gi}\ln(1/\mathrm{Gi})\right]$ (see [1]). Above this field, no supercurrent can flow through the sample more, that is it is the normal state. Nevertheless, as demonstrate our above estimations its properties are strongly

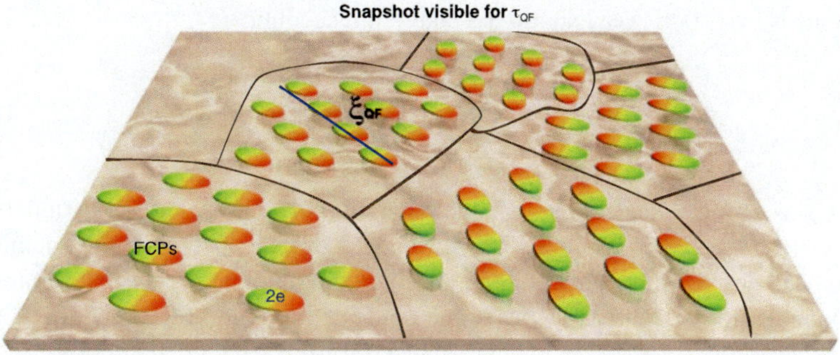

Fig. 1.8 The sketch of cluster structure of fluctuation Cooper pairs above the upper critical field

affected by the QF. The fragments of Abrikosov lattice can be still observed here by the following gedanken experiment. The clusters of rotating FCPs (ex-vortices) of size ξ_{QF} with some kind of the superconductive order should be found at the background of normal metal when one takes the picture with the exposure time shorter than τ_{QF}. When the exposure time is chosen longer than τ_{QF} the picture is smeared out and no traces of Abrikosov vortex state can be found. What kind of the order can be detected is still unclear. It would be attractive to identify these clusters with the splinters of Abrikosov lattice, but more probably this is some kind of quantum FCPs liquid. Indeed, presence of the structural disorder can result in formation close to $H_{c2}^{*}(0)$ of the hexatic phase, where the translational invariance no longer exists, although it still conserves the oriental order in the vortex positioning.

1.6.2 Manifestation of QF Above $H_{c2}(0)$

At zero temperature and fields above $H_{c2}(0)$, the systematics of the fluctuation contributions to the conductivity considerably changes with respect to that close to T_{c0}. The collisionless rotation of FCPs (they do not "feel" the presence of elastic impurities, all information concerning electron scattering is already included in the effective mass of the Cooper pair) results in the lack of their direct contribution to the longitudinal (along the applied electric field) electric transport (analogously to the suppression of one-electron conductivity in strong magnetic fields $(\omega_c \tau \gg 1)$: $\delta\sigma_{xx}^{(e)} \sim (\omega_c \tau)^{-2}$, see [29]) and the AL contribution to $\delta\sigma_{(2)}^{(tot)}$ becomes zero. The anomalous MT and DOS contributions turn zero as well but due to different reasons. Namely, the former vanishes since magnetic fields as large as $H_{c2}(0)$ completely destroy the phase coherence, whereas the latter disappears since magnetic field suppresses the fluctuation gap in the one-electron spectrum. Therefore, the effect of fluctuations on the conductivity at zero temperature is reduced to the renormalization of the one-electron diffusion coefficient. FCPs here occupy the lowest Landau level, but all the dynamic fluctuations in the interval of frequencies from 0 to Δ_{BCS} should be taken into account:

$$\delta\sigma_{\mathrm{xx}}^{\mathrm{DCR}} \sim -\frac{\mathrm{e}^2}{\Delta_{\mathrm{BCS}}}\int_0^{\Delta_{\mathrm{BCS}}}\frac{\mathrm{d}\omega}{\widetilde{h}+\frac{\omega}{\Delta_{\mathrm{BCS}}}} \sim -\frac{\mathrm{e}^2}{\hbar}\ln\frac{1}{\widetilde{h}}. \tag{1.81}$$

In terms of introduced above QF characteristics τ_{QF} and ξ_{QF} one can understand the meaning of QF contributions to different physical values in the vicinity of $H_{c2}(0)$ and derive others which are required. For example, the physical meaning of (1.86) can be understood as follows: one could estimate the FCPs conductivity by mere replacing $\tau_{\mathrm{GL}} \to \tau_{\mathrm{QF}}$ in the classical AL formula, which would give $\delta\widetilde{\sigma}^{\mathrm{AL}} \sim \mathrm{e}^2\tau_{\mathrm{QF}}$. Nevertheless, as we already noticed, the FCPs at zero temperature cannot drift along the electric field but only rotate around the fixed centers. As temperature deviates from zero, the FCPs can change their state due to the interaction with the thermal bath, that is their hopping to an adjacent rotation trajectory along the applied electric field becomes possible. This means that FCPs now can participate in longitudinal charge transfer. This process can be mapped onto the paraconductivity of a granular superconductors [85] at temperatures above T_{c0}, where the FCPs tunneling between grains occurs in two steps: first one electron jumps, then the second follows. The probability of each hopping event is proportional to the inter-grain tunneling rate Γ. To conserve the superconductive coherence between both events, the latter should occur during the FCPs lifetime τ_{GL}. The probability of FCPs tunneling between two grains is determined as the conditional probability of two one-electron hopping events and is proportional to $W_{\Gamma} = \Gamma^2\,\tau_{\mathrm{GL}}$. Coming back to the situation of FCPs above $H_{c2}(0)$, one can identify the tunneling rate with temperature T, while τ_{GL} corresponds to τ_{QF}. Therefore, to get a final expression, $\delta\widetilde{\sigma}^{\mathrm{AL}}$ should be multiplied by the probability factor $W_{\mathrm{QF}} = t^2\tau_{\mathrm{QF}}$ of the FCPs hopping to the neighboring trajectory:

$$\delta\sigma_{\mathrm{xx}}^{\mathrm{AL}} \sim \delta\widetilde{\sigma}^{\mathrm{AL}}W_{\mathrm{QF}} \sim \mathrm{e}^2t^2/\widetilde{h}^2,$$

which corresponds to the asymptotic (1.86).

To estimate the contribution of QF to the fluctuation magnetic susceptibility of the SC in the vicinity of $H_{c2}(0)$, one can apply the Langevin formula to a coherent cluster of FCPs and identifying its average size with the rotator radius to find

$$\chi^{\mathrm{AL}} = \frac{\mathrm{e}^2 n_{\mathrm{c.p.}}}{m_{\mathrm{c.p.}}c}\left\langle\xi_{\mathrm{QF}}^2\left(\widetilde{h}\right)\right\rangle \sim \xi_{\mathrm{BCS}}^2/c\widetilde{h}$$

in complete agreement with the result of [14].

One further reproduces the contribution of QF to the Nernst coefficient. Close to $H_{c2}(0)$ the chemical potential of FCPs can be identified as $\mu_{\mathrm{FCPs}} = \hbar\omega_c(H_{c2}(0)) - \hbar\omega_c(H)$ (as in [15], close to T_{c0}, $\mu_{\mathrm{FCPs}} = k_{\mathrm{B}}(T_{c0} - T)$). Corresponding derivative $\mathrm{d}\mu_{\mathrm{FCPs}}/\mathrm{d}T \sim \mathrm{d}H_{c2}(T)/\mathrm{d}T \sim -T/\Delta_{\mathrm{BCS}}$. Using the relation between the latter and the Nernst coefficient it is possible to reproduce one of the results of [15]:

$$\nu^{\mathrm{AL}} \sim \left[\tau_{\mathrm{QF}}/m_{\mathrm{c.p.}}\right] \mathrm{d}\mu_{\mathrm{FCPs}}/\mathrm{d}T \sim \xi_{\mathrm{BCS}}^2 t/\widetilde{h}.$$

1.7 Fluctuation Conductivity of 2D Superconductor in Magnetic Field: A Complete Picture

The complete expression for the total fluctuation correction to conductivity $\delta\sigma_{\mathrm{xx}}^{(\mathrm{tot})}(T,H)$ of a disordered 2D SC in a perpendicular magnetic field that holds through all T-H phase diagram above the line $H_{c2}(T)$ is given by the sum [16]:

$$\delta\sigma_{\mathrm{xx}}^{(\mathrm{tot})}(t,h) = \delta\sigma_{\mathrm{xx}}^{\mathrm{AL}} + \delta\sigma_{\mathrm{xx}}^{\mathrm{MT}} + \delta\sigma_{\mathrm{xx}}^{\mathrm{DOS}} + \delta\sigma_{\mathrm{xx}}^{\mathrm{DCR}} \tag{1.82}$$

with $\delta\sigma_{\mathrm{xx}}^{\mathrm{AL}}$ defined by (1.77) and

$$\begin{aligned}\delta\sigma_{\mathrm{xx}}^{\mathrm{MT(an)}} + \delta\sigma_{\mathrm{xx}}^{\mathrm{MT(reg2)}} &= \frac{\mathrm{e}^2}{\pi}\left(\frac{h}{t}\right)\sum_{m=0}^{M}\frac{1}{\gamma_\phi + \frac{2h}{t}(m+1/2)}\int_{-\infty}^{\infty}\frac{\mathrm{d}x}{\sinh^2\pi x}\frac{\mathrm{Im}^2\mathcal{E}_m}{|\mathcal{E}_m|^2}\\ \delta\sigma_{\mathrm{xx}}^{\mathrm{MT(reg1)}} &= \frac{\mathrm{e}^2}{\pi^4}\left(\frac{h}{t}\right)\sum_{m=0}^{M}\sum_{k=-\infty}^{\infty}\frac{4\mathcal{E}_m''(t,h,|k|)}{\mathcal{E}_m(t,h,|k|)}\end{aligned} \tag{1.83}$$

$$\delta\sigma_{\mathrm{xx}}^{\mathrm{DOS}} = \frac{4\mathrm{e}^2}{\pi^3}\left(\frac{h}{t}\right)\sum_{m=0}^{M}\int_{-\infty}^{\infty}\frac{\mathrm{d}x}{\sinh^2\pi x}\frac{\mathrm{Im}\mathcal{E}_m\mathrm{Im}\mathcal{E}_m'}{|\mathcal{E}_m|^2} \tag{1.84}$$

$$\delta\sigma_{\mathrm{xx}}^{\mathrm{DCR}} = \frac{4\mathrm{e}^2}{3\pi^6}\left(\frac{h}{t}\right)^2\sum_{m=0}^{M}\left(m+\frac{1}{2}\right)\sum_{k=-\infty}^{\infty}\frac{8\mathcal{E}_m'''(t,h,|k|)}{\mathcal{E}_m(t,h,|k|)}. \tag{1.85}$$

Here, $t = T/T_{\mathrm{c0}}$,

$$h = \frac{\pi^2}{8\gamma_E}\frac{H}{H_{c2}(0)} = 0.69\frac{H}{H_{c2}(0)},$$

$\gamma_E = \mathrm{e}^{\gamma_e}$ (γ_e is the Euler constant), $M = (tT_{\mathrm{c0}}\tau)^{-1}, \gamma_\phi = \pi/(8T_{\mathrm{c0}}\tau_\phi)$, τ_ϕ is the phase-breaking time, $\mathcal{E}_m(t,h,z)$ is defined by (1.78) and its derivatives $\mathcal{E}_m^{(p)}(t,h,z) \equiv \partial_z^p\mathcal{E}_m(t,h,z)$. All of them, side by side with the asymptotic expressions for $\delta\sigma_{\mathrm{xx}}^{(\mathrm{tot})}$ are shown in Table 1.1.

Let us start its discussion from the first line, corresponding to the Ginzburg–Landau region of fluctuations close to T_{c0} and in zero magnetic field (domain I). One can see our general expression naturally reproduces the well-known AL, MT, and DOS contributions. The only news here is the written in the explicit form contribution $\delta\sigma^{(\mathrm{DCR})}$, which was usually ignored in view of the lack of its divergency close to T_{c0}. Nevertheless, one can see that its constant contribution $\sim\ln\ln(T_{\mathrm{c0}}\tau)^{-1}$ is necessary for matching of the GL results with the neighbor

Table 1.1 Asymptotic expressions for different fluctuation contributions overall phase diagram

	$\delta\sigma_{xx}^{AL}$	$\delta\sigma_{xx}^{MT}$	$\delta\sigma_{xx}^{DOS}$	$\delta\sigma_{xx}^{DCR}$	$\delta\sigma_{xx}^{tot}$
I	$\frac{e^2}{16\epsilon} - \frac{7\zeta(3)e^2}{8\pi^4}\ln\frac{1}{\epsilon}$	$\frac{e^2}{8(\epsilon-\gamma_\phi)}\ln\frac{\epsilon}{\gamma_\phi} - \frac{14\zeta(3)e^2}{\pi^4}\ln\frac{1}{\epsilon}$	$-\frac{14\zeta(3)e^2}{\pi^4}\ln\frac{1}{\epsilon}$	$\frac{e^2}{3\pi^2}\ln\ln\frac{1}{T_{c0}\tau} + O(\epsilon)$	$\frac{e^2}{16\epsilon} + \frac{e^2}{8(\epsilon-\gamma_\phi)}\ln\frac{\epsilon}{\gamma_\phi} + \frac{e^2}{3\pi^2}\ln\ln\frac{1}{T_{c0}\tau}$
$I-III$	$\frac{e^2}{2\epsilon}\left(\frac{\epsilon}{2h}\right)^2\left[\psi\left(\frac{1}{2}+\frac{\epsilon}{2h}\right) - \psi\left(\frac{\epsilon}{2h}\right) - \frac{h}{\epsilon}\right]$	$\frac{e^2}{8}\frac{1}{\epsilon-\gamma_\phi}\left[\psi\left(\frac{1}{2}+\frac{t\epsilon}{2h}\right) - \psi\left(\frac{1}{2}+\frac{t\gamma_\phi}{2h}\right)\right] - \frac{14\zeta(3)e^2}{\pi^4}\left[\ln\left(\frac{t}{2h}\right) - \psi\left(\frac{1}{2}+\frac{t\epsilon}{2h}\right)\right]$	$-\frac{14\zeta(3)e^2}{\pi^4}\left[\ln\left(\frac{t}{2h}\right) - \psi\left(\frac{1}{2}+\frac{t\epsilon}{2h}\right)\right]$	$\frac{e^2}{3\pi^2}\ln\ln\frac{1}{T_{c0}\tau} + O\left(\max\left[\epsilon, h^2\right]\right)$	
IV	$\frac{4e^2\gamma_E^2t^2}{3\pi^2\tilde{h}^2}$	$-\frac{2e^2}{\pi^2}\ln\frac{1}{\tilde{h}} - \frac{2\gamma_E e^2}{\pi^2}\left(\frac{t}{\tilde{h}}\right)$	$-\frac{4e^2\gamma_E^2t^2}{3\pi^2\tilde{h}^2}$	$\frac{4e^2}{3\pi^2}\ln\frac{1}{\tilde{h}}$	$-\frac{2e^2}{3\pi^2}\left(\ln\frac{1}{\tilde{h}} + \frac{3t}{\tilde{h}}\right)$
V	$\frac{2\gamma_E e^2}{\pi^2}\left(\frac{t}{\tilde{h}}\right)$	$-\frac{2e^2}{3\pi^2}\ln\frac{1}{4\gamma_E t}$	$-\frac{2\gamma_E e^2}{\pi^2}\left(\frac{t}{\tilde{h}}\right)$	$\frac{4e^2}{3\pi^2}\ln\frac{1}{4\gamma_E t}$	$-\frac{2e^2}{3\pi^2}\ln\frac{1}{4\gamma_E t}$
$VI-VII$	$\frac{e^2}{4}\frac{t}{h-h_{c2}(t)}$	$-\frac{2e^2}{3\pi^2}\ln\frac{2h}{\pi^2 t}$	$-\frac{e^2}{4}\frac{t}{h-h_{c2}(t)}$	$\frac{4e^2}{3\pi^2}\ln\frac{2h}{\pi^2 t}$	$-\frac{2e^2}{3\pi^2}\ln\frac{h_{c2}(t)}{h-h_{c2}(t)}$
VIII	$\frac{e^2}{6\pi^2}\frac{C_1}{\ln^3 t}$	$-\frac{e^2}{\pi^2}\ln\frac{\ln\frac{1}{T_{c0}\tau}}{\ln t} + \frac{\pi^2e^2}{192}\frac{\ln\frac{\pi^2}{2\gamma_\phi}}{\ln^2 t}$	$-\frac{\pi^2e^2}{192}\frac{1}{\ln^2 t}$	$\frac{e^2}{3\pi^2}\ln\frac{\ln\frac{1}{T_{c0}\tau}}{\ln t}$	$-\frac{2e^2}{3\pi^2}\ln\frac{\ln\frac{1}{T_{c0}\tau}}{\ln t}$
IX	$\frac{\pi^2e^2}{192}\left(\frac{t}{h}\right)^2\frac{C_2}{\ln^3\frac{2h}{\pi^2}}$	$-\frac{e^2}{\pi^2}\ln\frac{\ln\frac{1}{T_{c0}\tau}}{\ln\frac{2h}{\pi^2}} + \frac{7\zeta(3)\pi^2e^2}{768}\left(\frac{t}{h}\right)^2\frac{1}{\ln^2\frac{2h}{\pi^2}}$	$-\frac{7\zeta(3)\pi^2e^2}{384}\left(\frac{t}{h}\right)^2\frac{1}{\ln^2\frac{2h}{\pi^2}}$	$\frac{e^2}{3\pi^2}\ln\frac{\ln\frac{1}{T_{c0}\tau}}{\ln\frac{2h}{\pi^2}}$	$-\frac{2e^2}{3\pi^2}\ln\frac{\ln\frac{1}{T_{c0}\tau}}{\ln\frac{2h}{\pi^2}} - \frac{7\zeta(3)\pi^2e^2}{768}\left(\frac{t}{h}\right)^2\frac{1}{\ln^2\frac{2h}{\pi^2}}$

domains YIII, IX. The domains II-III are still described by the GL theory in weak magnetic fields and $\delta\sigma_{\rm xx}^{(\rm tot)}(t,h)$ reproduces all available in literature asymptotic expressions.

What is really surprising in the Table 1.1 is the domain IV, the region of quantum fluctuations (see Fig. 1.1). Looking at the third line, one finds that the positive AL (anomalous MT contributions here is equal to the AL one) decays with the decrease of temperature as T^2. Moreover, it is exactly canceled by the negative contribution of the four DOS-like diagrams 3–6:

$$\delta\sigma_{\rm xx}^{\rm AL} = \delta\sigma_{\rm xx}^{\rm MT(an)} = -\delta\sigma_{\rm xx}^{\rm DOS} = \frac{4e^2\gamma_E^2 t^2}{3\pi^2\tilde{h}^2}. \tag{1.86}$$

The total fluctuation contribution to conductivity $\delta\sigma_{\rm xx}^{(\rm tot)}$ in this important region ($t \ll \tilde{h}$) is *completely determined by the renormalization of the diffusion coefficient.* It turns out to be negative and at zero temperature diverges logarithmically when the magnetic field approaches $H_{c2}(0)$. The nontrivial fact following from (1.82) is that an increase of temperature at a fixed value of the magnetic field in this domain first results in the further decrease of conductivity

$$\delta\sigma_{\rm xx}^{(\rm tot)} = -\frac{2e^2}{3\pi^2}\ln\frac{1}{\tilde{h}} - \frac{6\gamma_E e^2}{\pi^2}\frac{t}{\tilde{h}} + O\left[\left(\frac{t}{\tilde{h}}\right)^2\right]. \tag{1.87}$$

and only at the confine with the domain V, when $t \sim \tilde{h}$, the total fluctuation contribution $\delta\sigma_{\rm xx}^{(\rm tot)}$ pass through the minimum and starts to grow. Such nonmonotonic behavior of the of the conductivity close to $H_{c2}(0)$ was multiply observed in experiments [86,87] (see Fig. 1.9).

The domain Y describes the transition regime between quantum and classical fluctuations, while in the domains YI-YII, extended along the line $H_{c2}(T)$, superconductive fluctuations have already classical (but non-Ginzburg–Landau) character. In all these three regions, one observes the same exact cancellation of the AL and DOS contributions as in the domain IY and $\delta\sigma_{\rm xx}^{(\rm tot)}$ is determined here by the negative DCR contribution.

Finally, in the peripheric domains YIII-IX the direct positive contribution of fluctuation Cooper pairs (AL) to conductivity decays faster than all other: $\sim \ln^{-3}(T/T_{\rm c0})$. Let us stress that this exact result is in complete agreement with the high temperature asymptotical expression for the paraconductivity of the clean 2D superconductor. Such agreement seems natural: fluctuation Cooper pairs transport is insensitive to the impurity scattering. Anomalous MT contribution in complete accordance [8,9] decays as $\sim \ln\gamma_\phi^{-1}/\ln^{-2}(T/T_{\rm c0})$. Contribution of the diagrams 3–6 also decays as $\ln^{-2}(T/T_{\rm c0})$, but without the large factor $\ln\gamma_\phi^{-1}$. Finally, the regular MT contribution and that one of the diagrams 7–10 decay extremely slow, double logarithmically:

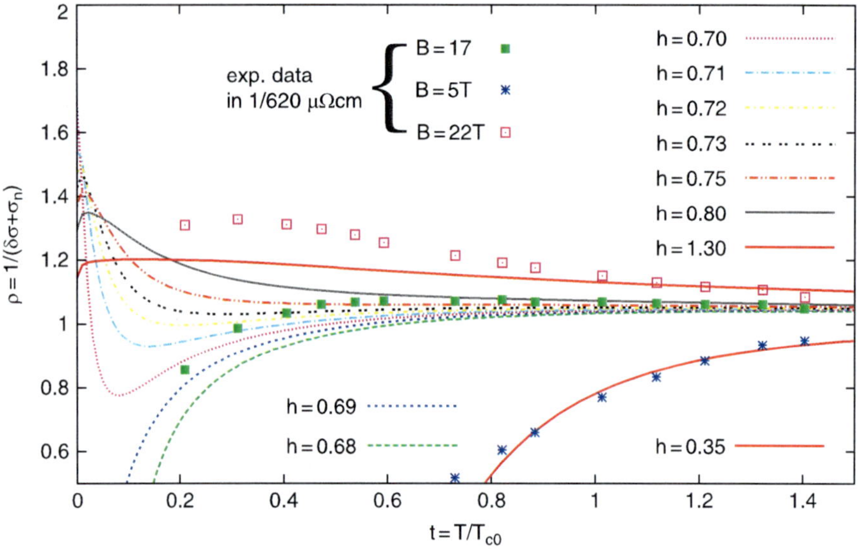

Fig. 1.9 Temperature dependence of the FC at different fields close to $H_{c2}(0)$ and comparison to experimental data for thin films of $La_{2-x}Sr_xCuO_4$ with $T_{c0} \approx 19K$ and $B_{c2}(0) \approx 15T$ (Data is courtesy of B. Leridon, unpublished). Note that for the theoretical curves a fixed $T_{c0}\tau_\phi = 10$ is used, which does not necessarily agree with the experimental value. Nevertheless, the overall behavior can be captured by this rough comparison. All curves are numerically calculated with $T_{c0}\tau = 0.01$

$$\delta\sigma_{\mathrm{xx}}^{(\mathrm{DCR})} = -\frac{2e^2}{3\pi^2}\left(\ln\ln\frac{1}{T_{\mathrm{c0}}\tau} - \ln\ln\frac{T}{T_{\mathrm{c0}}}\right). \tag{1.88}$$

Up to the numerical factor, this expression coincides with the results [8, 10].

Equation (1.82) gives the background for the "fluctuoscopy" of superconductors, that is extraction of its microscopic parameters from the analysis of fluctuation corrections. Indeed, one can see that $\delta\sigma_{\mathrm{xx}}^{(\mathrm{tot})}$ depends on two superconductive parameters: $T_{\mathrm{c0}}, H_{c2}(0)$, the elastic scattering time τ, and magnetic field and temperature dependent phase-breaking time $\tau_\phi(T, H)$. The elastic scattering time can be obtained from the normal state properties of superconductor, while (1.82) can become the instrument of precise definition of the critical temperature T_{c0} (instead of the often "half width of transition") and $H_{c2}(0)$. Moreover, it can be invaluable tool for the study of the temperature and magnetic field dependencies of the phase-breaking time $\tau_\phi(T, H)$.

The characteristic example of the surface $\delta\sigma_{\mathrm{xx}}^{(\mathrm{tot})}(T, H)$ for $T_{\mathrm{c0}}\tau = 0.1$ and $T_{\mathrm{c0}}\tau_\phi = 0.01$ is presented in Fig. 1.10. The value of τ_ϕ determines the behavior of fluctuation corrections only in the region of low fields. Figure 1.10 is convenient to analyze together with Fig. 1.1 where the lines $\delta\sigma_{\mathrm{xx}}^{(\mathrm{tot})}(T, H) = \mathrm{const}$ through all phase diagram are shown. One sees that FC is positive only in the domain

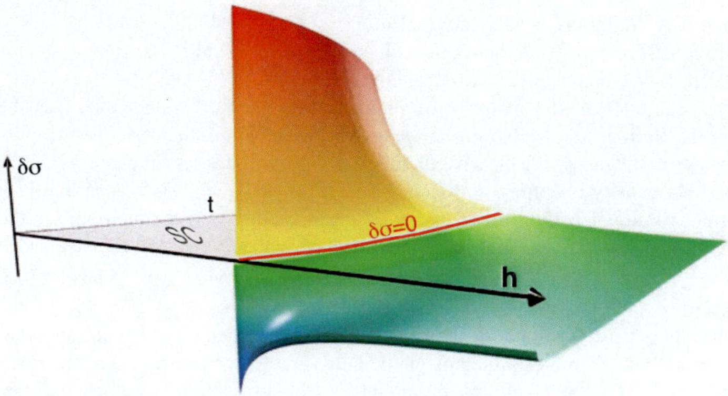

Fig. 1.10 FC as the function of the reduced temperature T/T_{c0} and magnetic field $H/H_{c2}(0)$

restricted by the lines $H_{c2}(T)$ and $\delta\sigma_{xx}^{(tot)}(T, H) = 0$ and is negative through all other parts of the phase diagram. With the growth of the magnetic field, the width of the domain where $\delta\sigma_{xx}^{(tot)}(T, H) > 0$ shrinks and turns zero close to $H_{c2}(0)$. The behavior of FC at low temperatures, in accordance with our asymptotic analysis, becomes nonmonotonic, the surface $\delta\sigma_{xx}^{(tot)}(T, H)$ here has trough-shaped form. It is interesting to note that the numerical analysis of (1.82) shows that the logarithmic asymptotic (1.87) is valid only within the extremely narrow field range $\tilde{h} \lesssim 10^{-6}$.

Acknowledgements The author acknowledges valuable discussions and collaborations with G. Balestrino, A. Glatz, B. Leridon, A. Rigamonti, V. Vinokur and also support of the MIUR under the project PRIN 2008.

References

1. A.I. Larkin, A.A. Varlamov, *Theory of Fluctuations in Superconductors*, 2nd edn (Oxford University Press, USA, 2009)
2. L.G. Aslamazov, A.I. Larkin, Soviet Solid State Phys. **10**, 875 (1968)
3. K. Maki, Prog. Theor. Phys. **39**, 897
4. K. Maki, Prog. Theor. Phys. **40**, 193 (1968)
5. R.S. Thompson, Phys. Rev. B **1**, 327 (1970)
6. L.B. Ioffe, A.I. Larkin, A.A. Varlamov, Yu. Lu, Phys. Rev. B **47**, 8936 (1993)
7. V.V. Dorin, R.A. Klemm, A.A. Varlamov, A.I. Buzdin, D.V. Livanov, Phys. Rev. B **48**, 12591 (1993)
8. L.G. Aslamazov, A.A. Varlamov, J. Low Temp. Phys. **38**, 223 (1980)
9. A.I. Larkin, JETP Lett. **31**, 219 (1980)
10. B.L. Altshuler, M.Yu. Reyzer, A.A. Varlamov, Soviet JETP **57**, 1329 (1983)
11. J.M.B. Lopes dos Santos, E. Abrahams, Phys. Rev. B **31**, 172 (1985)
12. I.S. Beloborodov, K.B. Efetov, Phys. Rev. Lett. **82**, 3332 (1999)
13. I.S. Beloborodov, K.B. Efetov, A.I. Larkin, Phys. Rev. B **61**, 9145 (2000)
14. V.M. Galitski, A.I. Larkin, Phys. Rev. Lett. **87**, 087001 (2001)
15. M.A. Skvortsov, M. Serbin, A.A. Varlamov, V. Galitski, Phys. Rev. Lett. **102**, 067001, (2009)

16. A. Glatz, A.A. Varlamov, V.M. Vinokur, Europhys. Lett. **94**, 47005 (2011)
17. J. Bardeen, L.N. Cooper, J.R. Schrieffer, Phys. Rev. **106**, 162 (1957)
18. J. Bardeen, L.N. Cooper, J.R. Schrieffer, Phys. Rev. **108**, 1175 (1957)
19. Strictly speaking τ in the majority of future results should be understood as the electron transport scattering time τ_{tr}. Nevertheless, as is well known, in the case of isotropic scattering these values coincide; so for the sake of simplicity we will use hereafter the symbol τ
20. This particle density is defined in the (D)-dimensional space. This means that it determines the normal volume density of pairs in the $3D$ case, the density per square unit in the $2D$ case and the number of pairs per unit length in $1D$. The real $3D$ concentration N can be defined too: $N = N_s^{(2)}/d$, where d is the thickness of the film and $N = N_s^{(1)}/S$, where S is the wire cross-section
21. This formula is valid for the dimensionalities $D = 2, 3$, when the fluctuation Cooper pair has the ability to "rotate" in the applied magnetic field and the average square of the rotation radius is $< R^2 >\sim \xi^2(T)$. "Size" effects, important for low-dimensional samples, will be discussed later on
22. Z.A. Xu, et al., Nature **406**, 486 (2000)
23. P.W. Anderson, arXiv:cond-mat/0603726.
24. D. Podolsky, S. Raghu, A. Vishwanath, Phys. Rev. Lett. **99**, 117004 (2007)
25. A. Pouret, et al., Phys. Rev. Lett. **96**, 176402 (2006)
26. I. Ussishkin, S.L. Sondhi, D.A. Huse, Phys. Rev. Lett. **89**, 287001 (2002)
27. A. Levchenko, M. Norman, A.A. Varlamov, Phys. Rev. B **83**, 020506 (R) (2011)
28. A.A. Varlamov, A. Kavokin, Europhys. Lett. **87**, 47007 (2009)
29. A.A. Abrikosov, *Fundamentals of the Theory of Metals* (North Holland, 1988)
30. Hereafter $\hbar = k_{\text{B}} = c = 1$
31. For simplicity in this subsection the magnetic field is assumed to be zero
32. V.V. Schmidt, in *Proceedings of the 10th International Conference on Low Temperature Physics*, **C2**, p. 205, VINITI, Moscow (1967)
33. T. Tsuboi, T. Suzuki, J. Phys. Soc. Jpn **42**, 654 (1977)
34. The precise value of the effective charge $e^* = 2e$ could not be determined in the framework of the GL phenomenology. It was found in the Gor'kov's microscopic rederivation of their equations
35. For a spherical particle $H_{c2(0)}^{\text{sph}}(\epsilon) = \frac{\Phi_0}{\pi d \xi}\sqrt{10\epsilon}$
36. E. Bernardi, et al., Phys. Rev. B **74**, 134509 (2006)
37. W.E. Lawrence, S. Doniach, in *Proceedings* of the 12th International Conference on Low Temperature *Physics*, ed. by E. Kanda, p.361 (Academic Press, Japan, Kyoto, 1971)
38. K. Yamaji, Phys. Lett. A **38**, 43 (1972)
39. L.G. Aslamazov and A.I. Larkin, Zhurnal Eksperimentalnoi i Teoreticheskoi Fisiki **67**, 647 (1973) [Soviet Phys. JETP **40**, 321 (1974)]
40. A. Schmid, Phys. Rev. **180**, 527 (1969)
41. H. Schmidt, Zeitschrift für Physik B **216**, 336 (1968)
42. R.E. Prange, Phys. Rev. B **1**, 2349 (1970)
43. B.R. Patton, V. Ambegaokar, J.W. Wilkins, Solid State Commun. **7**, 1287 (1969)
44. J. Kurkijarvi, V. Ambegaokar, G. Eilenberger, Phys. Rev., B **5**, 868 (1972)
45. J.P. Gollub, M.R. Beasley, R.S. Newbower, M. Tinkham, *Phys. Rev. Lett.* **22**, 1288 (1969)
46. J.P. Gollub, M.R. Beasley, M. Tinkham, Phys. Rev. Lett. **25**, 1646 (1970)
47. J.P. Gollub, M.R. Beasley, R. Callarotti, M. Tinkham, Phys. Rev. B **7**, 3039 (1973)
48. P.A. Lee, S.R. Shenoy, Phys. Rev. Lett. **28**, 1025 (1972)
49. S.P. Farrant, C.E. Gough, Phys. Rev. Lett. **34**, 943 (1975)
50. W.J. Skocpol, M. Tinkham, Rep. Progress Phys. **38**, 1094 (1975)
51. R.S. Thompson, V.Z. Kresin, Modern Phys. Lett. B **2**, 1159 (1988)
52. K.F. Quader, E. Abrahams, Phys. Rev. B **38**, 11977 (1988)
53. W.C. Lee, R.A. Klemm, D.C. Johnson, Phys. Rev. Lett. **63**, 1012 (1989)

54. P. Carretta, A. Lascialfari, A. Rigamonti, A. Rosso, A.A. Varlamov, Phys. Rev. B **61**, 12420 (2000)
55. T.M. Mishonov, E.S. Penev, Int. J. Modern Phys. **14**, 3831 (2000)
56. A.I. Buzdin, V.V. Dorin, in *Fluctuation* phenomena in high temperature *superconductors*, ed. by M. Ausloos, A.A. Varlamov, NATO-ASI Series (Kluwer, Dordrecht, 1997)
57. S. Hikami, A. Fujita, A.I. Larkin, Phys. Rev. B **44**, 10400 (1991)
58. C. Baraduc, A.I. Buzdin, J-Y. Henry, J.P. Brison, L. Puech, Phys. C **248**, 138 (1995)
59. A. Junod, J-Y. Genoud, G. Triscone, Phys. C **294**, 115 (1998)
60. A. Lascialfari, A. Rigamonti, P. Tedesco, A.A. Varlamov, Phys. Rev. B **65**, 144523 (2002)
61. This term may have different origins. First of all, evidently, paraconductivity is analogous to paramagnetism and means excess conductivity. Another possible origin is an incorrect onomatopoeic translation from the Russian "paroprovodimost' " that means pair conductivity
62. A. Schmid, Physik Kondensierter Materie **5**, 302 (1966)
63. C. Caroli, K. Maki, Phys. Rev. **159**, 306 (1967)
64. C. Caroli, K. Maki, Phys. Rev. **159**, 316 (1967)
65. E. Abrahams, T. Tsuneto, Phys. Rev. **152**, 416 (1966)
66. J.W.F. Woo, E. Abrahams, Phys. Rev. **169**, 407 (1968)
67. C. Di Castro, W. Young, Il Nuovo Cimento B **62**, 273 (1969)
68. S. Ullah, A.T. Dorsey, Phys. Rev. B **44**, 262 (1991)
69. S. Ullah, A.T. Dorsey, Phys. Rev. Lett. **65**, 2066 (1990)
70. L.P. Gor'kov, G.M. Eliashberg, Zhurnal Eksperimentalnoi i Teoreticheskoi Fiziki **54**, 612 (1968) [Soviet Phys. JETP **27**, 328 (1968)]
71. An equation of this type was considered by Landau and Khalatnikov in connection with the study of superfluid helium dynamics in early 1950s
72. Account for electron- hole asymmetry leads to the appearance of the imaginary part of γ_{GL} proportional to the derivative $\partial \ln(\rho v^2 \tau)/\partial E|_{E_F} \sim \mathcal{O}(1/E_F)$. This is important for such phenomena as fluctuation Hall effect or fluctuation thermopower and, having in mind the writing of the most general formula, we will suppose $\gamma_{\mathrm{GL}} = \pi\alpha/8 + i\ \mathrm{Im}\,\gamma_{\mathrm{GL}}$, where necessary
73. L.D. Landau, E.M. Lifshitz, *Quantum Mechanics. Course of Theoretical Physics,* vol. 3 (Pergamon Press, Oxford, 1978)
74. L. Reggiani, R. Vaglio, A.A. Varlamov, Phys. Rev. B **44**, 9541 (1991)
75. M. Ausloos, Ch. Laurent, Phys. Rev. B **37**, 611 (1988)
76. P.P. Frietas, C.C. Tsuei, T.S. Plaskett, Phys. Rev. B **36**, 833 (1987)
77. M. Hikita, M. Suzuki, Phys. Rev. B **41**, 834 (1990)
78. M. Akinaga, D. Abukay, L. Rinderer, Modern Phys. Lett. **2**, 891 (1988)
79. A. Poddar, P. Mandal, A.N. Das, B. Ghosh, P. Choudhury, Phys. C **159**, 231 (1989)
80. D.H. Kim, A.M. Goldman, J.H. Kang, K.E. Gray, R.T. Kampwirth, Phys. Rev. B **39**, 12275 (1989)
81. G. Balestrino, A. Nigro, R. Vaglio, Phys. Rev. B **39**, 12264 (1989)
82. G. Kumm, K. Winzer, Phys. B **165-166**, 1361 (1990)
83. M.R. Cimberle, C. Ferdeghini, D. Marrè, M. Putti, S. Siri, F. Federici, A.A. Varlamov, Phys. Rev. B **55**, R14745 (1997)
84. G. Balestrino, E. Milani, A.A. Varlamov, Phys. C **210**, 386 (1993)
85. I.V. Lerner, A.A. Varlamov, V.M. Vinokur, Phys. Rev. Lett. **100**, 117003, (2008)
86. B. Leridon, J. Vanacken, T. Wambecq, V. Moshchalkov, Phys. Rev. B **76**, 012503 (2007)
87. V.F. Gantmakher, S.N. Ermolov, G.E. Tsydynzhapov, A.A. Zhukov, T.I. Baturina, JETP Lett. **77**, 498 (2003)

Chapter 2
Experimental Study of the Fluctuation-Governed Resistive State in Quasi-One-Dimensional Superconductors

K.Yu. Arutyunov

Abstract Physical properties of quasi-one-dimensional superconducting channels (nanowires) can differ significantly from those of bulk superconductors. The reason behind is the impact of thermal and quantum fluctuations. In the particular case of resistive measurements the fluctuations may significantly broaden the superconducting transition $R(T)$. Here we review the activities in the field with main emphasis on experiment, while brief theoretical background is also presented.

2.1 Introduction

The phenomenon of superconductivity was discovered as a sudden drop of resistance to immeasurably small value. With the development of the topic, it was realized that the superconducting phase transition is frequently not at all "sudden" and the measured dependence of the sample resistance $R(T)$ in the vicinity of the critical temperature T_C may have a finite width. One possible reason for this behavior – and frequently the dominating factor – is the sample inhomogeneity, that is the sample might simply consist of regions with different local critical temperatures. However, with improving fabrication technologies it became clear that even for highly homogeneous samples the superconducting phase transition may remain broadened. This effect is usually very small in bulk samples and becomes more pronounced in systems with reduced dimensions. A fundamental physical reason behind such smearing of the transition is *superconducting fluctuations*.

K.Yu. Arutyunov (✉)
NanoScience Center, Department of Physics, University of Jyväskylä, PB 35, FI-40014 Jyväskylä, Finland
and
Nuclear Physics Institute, Moscow State University, 119992 Moscow, Russia
e-mail: konstantin.arutyunov@phys.jyu.fi

A. Sidorenko (ed.), *Fundamentals of Superconducting Nanoelectronics*, NanoScience and Technology, DOI 10.1007/978-3-642-20158-5_2,

An important role of fluctuations in superconductors with reduced dimension is well known [1, 2]. Above T_C such fluctuations yield an enhanced conductivity of metallic systems [3–5]. For instance, the so-called Aslamazov–Larkin fluctuation correction to conductivity $\delta G_{AL} \sim (T - T_C)^{-(2-D/2)}$ becomes large in the vicinity of T_C, and this effect increases with decreasing dimensionality D. In the opposite limit – at temperatures below T_C – it is quite obvious that in $3D$ (bulk) and $2D$ (thin films) systems fluctuations cannot provide any measurable resistive response as the superconducting condensate shunts the dissipative electron current. On the contrary, in quasi-$1D$ systems with transverse dimensions smaller than the superconducting coherence length ξ, the single supercurrent channel can be easily blocked by a fluctuation. This limit has both the significant fundamental and practical importance. Rapidly progressing miniaturization of nanodevices requires better understanding of fundamental limitations for the phenomenon of superconductivity in reduced dimension. Of particular interest are the nanowires, which can be considered as basic elements of any superconducting nanocircuit. A detailed review of the present status of experiments in the field is the main subject of this chapter.

2.2 Theory Background

The detailed overview of the theory of fluctuations in superconductors can be found in specialized reviews [1, 2] and in a more compact way – in Chap. 1. Here, we will very briefly outline the main conclusions of the theory, necessary for understanding the interesting experimental phenomenon – finite resistivity of very narrow superconducting wires below the critical temperature T_C.

It was first pointed out by Little [6] that quasi-one-dimensional wires made of a superconducting material can acquire a finite resistance below T_C of a bulk material due to the mechanism of thermally activated phase slips (TAPS). With the Ginzburg–Landau theory, one can describe a superconducting wire by means of a complex order parameter $\Psi(x) = |\Psi(x)|e^{i\varphi(x)}$. Thermal fluctuations cause deviations of both the modulus and the phase of this order parameter from their equilibrium values. A nontrivial fluctuation corresponds to temporal suppression of $|\Psi(x)|$ down to zero in some point (e.g., $x = 0$) inside the wire. As soon as the modulus of the order parameter $|\Psi(0)|$ vanishes, the phase $\varphi(0)$ becomes unrestricted and can jump by the value $2\pi n$, where n is any integer number. After this process, the modulus $|\Psi(0)|$ gets restored, the phase again becomes single valued and the system returns to its initial state accumulating the net phase shift $2\pi n$. Provided such phase slip events are sufficiently rare, one can restrict n by $n = \pm 1$ and totally disregard fluctuations with $|n| \geq 2$. According to the Josephson relation $V = \hbar\dot{\varphi}/2e$ each such phase slip event causes a nonzero voltage drop V across the wire. In the absence of any bias current, the net average numbers of "positive" ($n = +1$) and "negative" ($n = -1$) phase slips are equal, thus the net voltage drop remains zero. Applying the current $I \propto |\Psi|^2 \nabla\varphi$, one creates nonzero phase gradient along the wire and makes "positive" phase slips more likely than "negative" ones. Hence, the net voltage drop V due to TAPS differs from zero, that is thermal fluctuations cause nonzero resistance $R = V/I$ of superconducting wires even below T_C. We would also like to emphasize that, in contrast to the so-called phase slip centers [7, 8] produced by a large current above the critical one $I > I_C$, here we are dealing with *fluctuation-induced phase slips*, which can occur at arbitrarily small values I.

A quantitative theory of the TAPS phenomenon was first proposed by Langer and Ambegaokar [9] and then completed by McCumber and Halperin [10]. This LAMH theory predicts that below T_C the effective resistance R of a superconducting wire with cross section σ is determined by

$$R(T) = \Omega(T)\exp(-U/k_B T), \quad U \sim H_c^2(T)\sigma\xi(T) \sim I_C, \tag{2.1}$$

where $U(T)$ is the effective potential barrier for TAPS determined simply as the superconducting condensation energy for a part of the wire of a minimal statistically independent volume $\sigma\xi$, and $H_c(T)$ and I_C are the critical magnetic field and current, respectively. Prefactor $\Omega(T)$ has much weaker temperature dependence compared to the exponent. Formally, both the potential barrier $U(T)$ and the prefactor $\Omega(T)$ depend on the bias current I. In a typical experimental configuration $I \ll I_C$, this dependence is not essential and can be neglected.

Following the standard quantum mechanical arguments, one can expect that there should exist an alternative (to TAPS) mechanism, where the order parameter field $\Psi(x)$ tunnels under the barrier U rather than overcomes it by thermal activation. In this case, the corresponding probability of such tunneling should be controlled by the exponent $\sim\exp(-U/\hbar\omega_0)$, that is instead of the thermal energy $k_B T$ in the activation exponent one should just substitute $\hbar\omega_0$, where ω_0 is an effective attempt frequency of quantum fluctuations – *quantum phase slips* (QPS). Since such tunneling process should obviously persist down to infinitely small temperatures, one arrives at a fundamentally important conclusion that *in nanowires superconductivity can be destroyed by quantum fluctuations at any temperature including* $T = 0$.

There have been several attempts to develop a quantitative theory of quantum fluctuations in quasi-1D superconducting channels (for details, see Chap. 1). The microscopic theory of QPS processes in superconducting nanowires was developed with the aid of the imaginary time effective action technique [11–13]. This theory remains applicable down to $T = 0$ and properly accounts for nonequilibrium, dissipative and electromagnetic effects during a QPS event. One of the main conclusions of the theory is that in sufficiently dirty superconducting nanowires with diameters in the ~10 nm range QPS probability can already be large enough to yield experimentally observable resistance at temperatures $T \ll T_c$:

$$R(T) = b\frac{\Delta_0(T)S_{\text{core}}^2 L}{\xi(T)}\exp(-2S_{\text{core}}), \qquad S_{\text{core}} = a\frac{R_q}{R_N}\frac{L}{\xi(T)}. \tag{2.2}$$

Here, R_N is the total normal state resistance of a wire with length L, $R_q = \pi\hbar/2e^2 = 6.453$ kΩ is the "superconduting" resistance quantum, ξ and Δ_0 are the temperature-dependent superconducting coherence length and gap, respectively; $a \simeq 1$ and $b \simeq 1$ are numeric prefactors. It should be noted that the above expression is strictly valid only in the limit of relatively rare QPS events when the effective wire resistance $R(T) \ll R_N$.

2.3 Sample Fabrication

As it follows from (2.1) and (2.2), effective resistance of a quasi-*1D* superconducting channel exponentially depends on its cross section σ. Hence, a reliable experimental observation of the impact of superconducting fluctuations requires fabrication of very narrow superconducting wires [14].

The earliest experimental activity with quasi-1D superconductors utilized, so-called, whiskers: single crystals with characteristic cross-section ~1 μm and length up to 1 mm (Fig. 2.1).

Growth and basic properties of these highly anisotropic objects have been described in literature [15]. In particular, case of widely studied superconducting tin whiskers "squeeze" method was typically applied for their growing [16]. Then the crystals were literally hand picked from the ingots and positioned on substrates. The electrodes were made either by conducting paste or epoxy [17], soldered by Wood's metal [7, 18] or Sn-Pb alloy [19], or squeezed by a soft metal (e.g., indium) [20]. With further development of micro- and nanotechnology hybrid whisker-based nanostructures were fabricated [21]. The 'body' of the samples consisted of a superconducting (tin) whisker embedded in a *Spin-On-Glas* (SOG) polymer on the surface of Si/SiO_x substrate, while the contacts were fabricated utilizing lift-off e-beam lithography (Fig. 2.2) enabling true 4-terminal measuring configuration.

Recently an alternative method – molecular template – has been developed. A single molecule, carbon nanotube or a DNA double helix, is placed over a

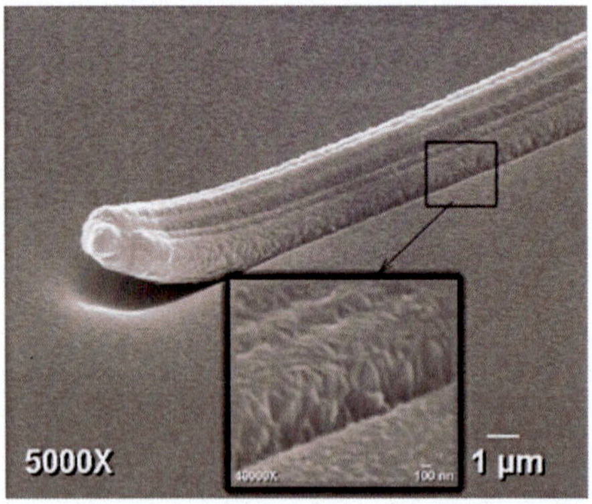

Fig. 2.1 SEM image of a typical tin whisker. Inset shows the magnified view of the crystal surface [14]

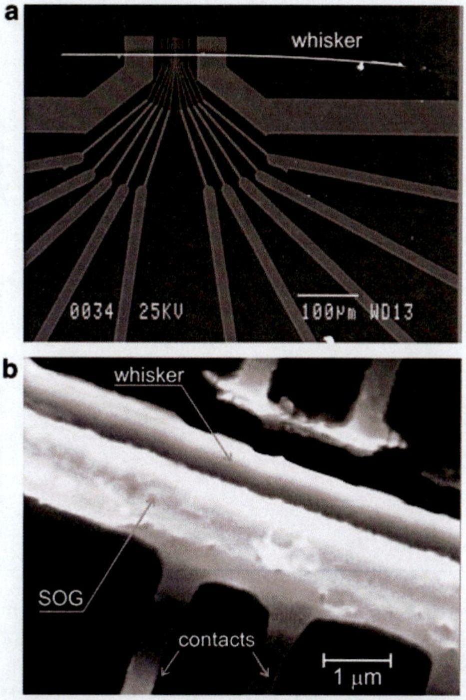

Fig. 2.2 (**a**) SEM image of hybrid microstructure. Bright horizontal line is tin whisker. (**b**) Enlared SEM image of the sample with lithographically fabricated nanocontacts [21]

substrate with a narrow (100 nm to 1 μm) trench and a thin layer of metal is deposited on top of the molecule by sputtering (Fig. 2.3) [22–24]. The approach enables fabrication of superconducting nanowires (MoGe and Nb) with effective diameter down to ∼3 nm and the length up to ∼1 μm. Only pseudo-four-probe configuration is available for measurements with the deposited metal film from both sides of the trench serving as electrodes. Similar technique has been applied using insulating WS_2 nanotube as a molecular substrate covered with disordered superconductor: amorphous α : *InO* [25]. Samples fabricated with molecular template technology enable pseudo-4-terminal measurements.

With rapid progress of nanolithography, fabrication of sub-100 nm nanowires became a routine task. In a number of experimental works, quasi-1D superconducting nanostructures of different complicity have been fabricated using conventional e-beam lithography followed by a lift-off process [14]. The big advantage of the method is the ability to fabricate nanostructures with complicated shape and in a rather reproducible way. Typically, the metal is deposited through the PMMA mask using thermal or e-beam evaporation, or sputtering. For some applications, the minimum line width of e-beam lithographically fabricated superconducting quasi-1D nanostructures might be an issue. Although state-of-the-art e-beam writers should provide ∼15 nm resolution, reproducible fabrication in this limit is still far from being a routine. Several 'tricks' were suggested to obtain ∼10 nm nanowires. "Step lithography" utilizes the shadow effect produced by extremely shallow steps

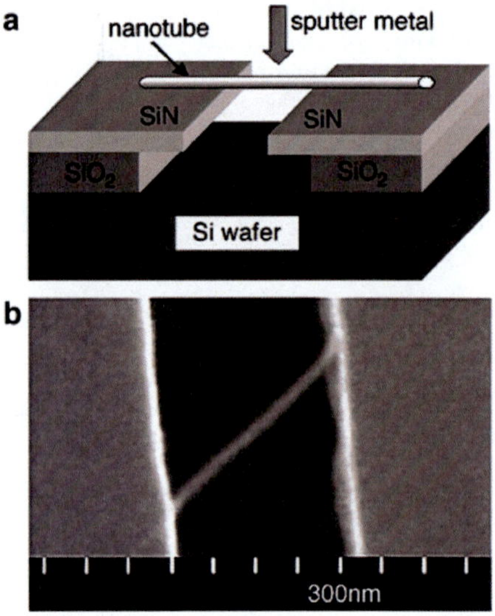

Fig. 2.3 (**a**) Schematics of the molecular template method. (**b**) SEM image of a carbon nanotube-based structure [22, 24]

in substrates and the corresponding tilted ion beam milling of the deposited material. The method has been applied for fabrication of superconducting nanowires (In and Pb-In) with triangular cross-section and the effective diameter down to ~15 nm [26, 27]. Yet another method utilizing non-conventional mask – suspended stencil technique – has been developed enabling fabrication of nanowires with effective diameter down to ~15 nm from various materials including several superconductors: Pb,Sn,Pb-Bi [28,29]. A remarkable feature of this method is the ability to vary the cross-section of the wires in situ without warming up the cryostat in between the sessions of measurements.

An advanced method of nanowire fabrication uses MBE grown InP layer on a cleaved $In_xGa_{1-x}As/InP$ substrate as a support for thermally evaporated metal. The method allows fabrication of very long (up to 100 μm) superconducting (Al) nanowires with effective cross-section down to 7 nm [30] enabling pseudo-4-terminal measurements of the sample.

An alternative and to some extend complementary approach to the described methods has been proposed recently. The idea is to use low energetic ion beam sputtering for progressive reduction of pre-fabricated nanostructures [31,32]. Gentle erosion of a nanostructure surface by inert ions (e.g. Ar^+) enables ~1 nm accurate reduction of the sample dimensions. The method can be applied to a large variety of materials and nanostructures, including superconducting nanowires (Fig. 2.4). The advantage of the approach is the ability to trace evolution of size phenomena eliminating uncertainty related to artifacts of individual nanostructures by performing measurements on a same sample between the sessions of sputtering. Evolution of size-dependent properties of the same Al nanowire starting from 100 nm down to 8 nm has been demonstrated [33, 34].

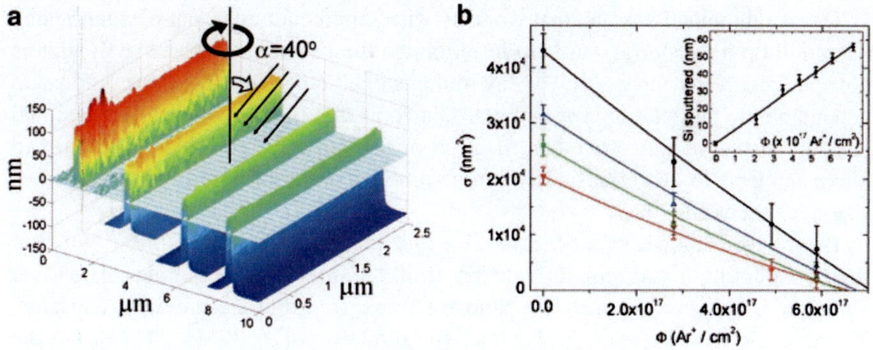

Fig. 2.4 (**a**) SPM images of the same section of *Al* nanowire after series of Ar^+ sputtering. Horizontal plane corresponds to the initial level of *Si* substrate. Bright color above – metal (*Al*), dark under – sputtered substrate (*Si*). (**b**) Variation of the nanowire cross-section with the ion dose. Inset: the corresponding dependence of *Si* substrate erosion [32]

2.4 Experiments

It follows from the theoretical analysis that the impact of fluctuations can be observed in sufficiently thin superconducting wires. On top of that, it is highly desirable to deal with structurally (and chemically) homogeneous samples in order to eliminate spatial variations of T_C along the wire and rule out the effect of (possibly existing) constrictions and tunnel barriers. It has been shown that structural imperfections of real quasi-1D structures (nonuniform cross section, existence of probes, and finite length effects) might effectively mask the phenomena related to thermal or quantum fluctuations [35].

Structural and geometrical homogeneity of quasi-1D samples is the central question in interpretation of experimental data related to contribution of superconducting fluctuations. Theoretically, one usually assumes that (a) the critical temperature T_C of quasi-1D wires under consideration remains spatially constant, that is it does not vary along the wire, (b) the cross-section σ does not vary along the wire either and (c) the measuring probes are noninvasive. Unfortunately, in realistic samples none of these conditions is usually well satisfied. To which extent can these imperfections be neglected while interpreting the experimental data using models developed under the assumptions (a)–(c)?

If *structural* imperfections (e.g., non-uniform chemical composition) can alter the critical temperature, then the shape of the experimentally observed $R(T)$ dependence is determined by the sequence of transitions of various parts of the wire with different local critical temperature $T_C(x)$. If the degree of such inhomogeneity is not too strong, a step-like $R(T)$ transition may not be observed as the variations of the critical temperature are averaged on the scale of the coherence length ξ resulting in a relatively wide "smooth-looking" $R(T)$ dependence to some extent simulating the impact of superconducting fluctuations.

One might naively expect that working with superconducting samples fabricated of initially pure material can help to eliminate the problem. Unfortunately, this is not the case. First, properties of low-dimensional superconductors are known to depend on the fabrication process details, such as thin film deposition rate, residual pressure in the vacuum chamber, material of the substrate, etc. Second, even if to make an effort to keep the fabrication parameters constant, it is hard to get rid of size-dependent effects.

It is a well-known experimental fact that the critical temperature T_C of thin superconducting films frequently differs from the one for bulk samples. The same tendency is observed in metallic nanostructures. In indium, aluminum, and zinc, T_C increases with decreasing characteristic dimension [26, 30, 33, 34, 36]. On the contrary, in lead, niobium, and MoGe, an opposite tendency is observed [28, 37–40]. No noticeable variations of T_C were detected in tin nanowires. The origin of the T_C size dependence phenomenon is not clear. There exist models predicting both suppression [41] and enhancement [42] of the mean field critical temperature for low-dimensional superconductors. One can take the empirical fact of variation of T_C with effective diameter of a superconducting nanowire as granted. It has been shown that size-dependent effects are extremely important for interpretation of fluctuation phenomena in quasi-1D systems [35]. The conclusion of a quantitative analysis [35] is rather disappointing (Fig. 2.5): even chemically pure nanostructures cannot be considered as sufficiently homogeneous as soon as the size dependence of the critical temperature T_C comes into play. One should study atomically homogeneous systems as single crystalline whiskers [18,19]. Unfortunately, modern nanotechnology does not enable growth of high-quality quasi-1D single crystals of arbitrary diameter made of all materials of interest. The lithographic processes result in much lower quality samples. The only exception are materials, such as *Al*,

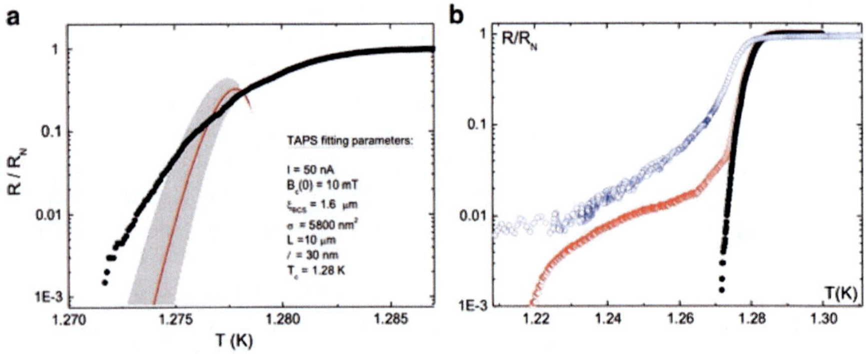

Fig. 2.5 Resitive transition of a typical *Al* nanowire. (**a**) Simulated $R(T)$ dependence. *Bullets* (•) correspond to simulated $R(T)$ transition with size dependent local critical temperature [42], but without taking into consideration finite size effects. Solid line is TAPS calculation using realistic fitting parameters. Shaded region corresponds to TAPS simulation allowing 10% variation of the fitting parameters. (**b**) *Bullets* (•) – same as in *Diamonds* (◇) represent the results of similar calculations with contribution of the node regions. *Open circles* (○) denote the experimentally measured $R(T)$ dependence [35]

Zn, and *In*, where the critical temperature decreases with reduction of the effective dimension (e.g., nanowire diameter). Under certain conditions, experiments on lift-off fabricated nanowires made of these materials can be interpreted using fluctuation models developed for perfect 1D channels. Provided broadening of the $R(T)$ dependencies is detected below the bulk critical temperature T_C^{bulk}, the size effects cannot account for the phenomenon, and presumably "real" physics is observed.

Although measuring the temperature dependence of the system resistance $R(T)$ in the vicinity of the transition point might seem a routine experimental task, in the case of nanostructures more care is required. Typically, $R(T)$ dependencies are measured in the current-biased regime. A standard requirement is to keep the bias current I much smaller than the critical (depairing) current I_C to avoid hysteresis effects due to overheating. Additionally, to stay within the linear response regime, the measuring current I should remain smaller than the characteristic scale $I_0 = k_B T_C / \Phi_0$ [9,10]. Employing room-temperature electronics, it is preferable to use ac lock-in technique to increase the signal-to-noise ratio. This can be associated with some hidden problems. One of them is that even a tiny fraction of dc component (e.g., from the ground loop) adds a parasitic signal $\sim dV/dI(T)$ to the "valuable" one $R(T) \equiv V(T)/I$. For this reason, it is advisable to decouple the ac current source from the sample using a low-noise transformer.

Another problem might originate from the presence of rf filters, which are mandatory to protect nano-sized samples from noisy electromagnetic environments. Very often such filters are just *RLC* circuits shunting the rf component to the ground through a capacitor as high as few μF. Such configuration might provide good results with high-Ohmic systems (e.g., tunnel structures). In a superconducting nanowire measured in 4-probe configuration, each electrode contacting the "body" of the sample is typically made of the same material. Hence, the resistance of these probes also drops down to zero over the same temperature range within the transition. Depending on particular configuration, at a certain T the ac sample impedance might become comparable to that of the current leads through the ground, causing the current re-arrangement throughout the sample. Even at rather low measuring frequencies $\sim$10 Hz, the parasitic effect might manifest itself as a nonmonotonous $R(T)$ dependence with unusual "bumps" or "foot" at the bottom of the transition. Additional complication might arise from the nonnegligible dependence of the gain of the nanovolt pre-amplifier on the total impedance of the load.

Concluding this part, an experimentalist should be extremely cautious in designing the measuring set-up and in selecting samples for unambiguous interpretation of the data on superconducting nanowires.

2.5 Thermally Activated Phase Slips (TAPS)

Very quickly after development of the theory [9, 10], two experimental groups reported experiments aimed at verification of the model [18,19]. At 1970s microfabrication technique was not much developed. In these early studies of 1D superconductivity [18, 19] metallic whiskers (Fig. 2.1) with characteristic diameter $\sim 1\,\mu$m and lengths up to 1 mm were utilized. An example of experimental $R(T)$ dependence measured on tin whisker is shown in Fig. 2.6. One observes a very good quantitative agreement between the experiment and the TAPS model [9, 10]. The superconducting transition $R(T)$ is very steep: resistance of the sample drops five orders of magnitude within the temperature range $\delta T \sim 1\,\text{mK}$ below T_C. This is a consequence of rather large effective diameter values $\sqrt{\sigma} \simeq 0.5\,\mu$m and extremely high homogeneity of single crystals. In the case of thinner single-crystalline structures, the width of the $R(T)$ dependence would be larger, cf. (2.1). Unfortunately, the dimensions of natural grown whiskers made of superconducting materials do not vary much. Manual manipulation of sub-1 μm objects is extremely time consuming and results in a low yield of suitable samples, while the dimensions of whiskers are still too large to use scanning probe (SPM) technique well developed for manipulation of nano-sized objects, such as carbon nanotubes.

There were several reports on experimental studies of 1D superconductivity in lift-off-fabricated superconducting nanostructures (wires and loops) with the linewidth much smaller than of whiskers. The quality of the structures is far from perfect: the majority of superconducting thin-film structures fabricated at room temperatures are polycrystalline with the grain size of about few tens of nm. Deposition of metal (in particular low melting ones, such as tin or indium) on a cryogenically cooled substrate might reduce the grain size. In contrast to whiskers, the lift-off-fabricated superconducting nanostructures studied so far were all in the dirty limit $\ell \ll \xi$. Attempts to quantitatively describe the shape of superconducting transition of these quasi-1D systems using the TAPS model failed: Experimental curves for $R(T)$ were always significantly broader than theoretical predictions. Nevertheless, it was believed that with a certain adjustment of fit parameters (e.g., reduction of the effective size of the phase slip center), a reasonable agreement between the experiments and the TAPS model predictions could be achieved. Later it was shown that inevitable inhomogeneity of lift-off-fabricated nanostructures, the presence of node regions and finite size effects can dramatically broaden the experimentally observed dependencies $R(T)$ making any comparison with the TAPS theory inconclusive or even impossible (Fig. 2.5) [35].

Even thinner quasi-1D structures, obtained with the molecular template decoration method [22–24], were used to study TAPS. Transmission electron microscopy (TEM) study of *Nb* nanowires, fabricated using the approach, revealed polycrystalline structure with an average grain size about few nm. To fit the data obtained in these Nb structures, it has been proposed to represent the measured wire conductance $1/R(T)$ as a sum of contributions from normal electrons above the gap and from TAPS, $1/R(T) = 1/R_N + 1/R_{\text{TAPS}}$. The experimental curves $R(T)$ could be reasonably well fitted [37, 38] to this phenomenological model (Fig. 2.7). It is quite surprising that the inhomogeneities in granular *Nb* nanowires do not seem to broaden the $R(T)$ curves, which can be fitted by the TAPS model developed for homogeneous quasi-1D systems.

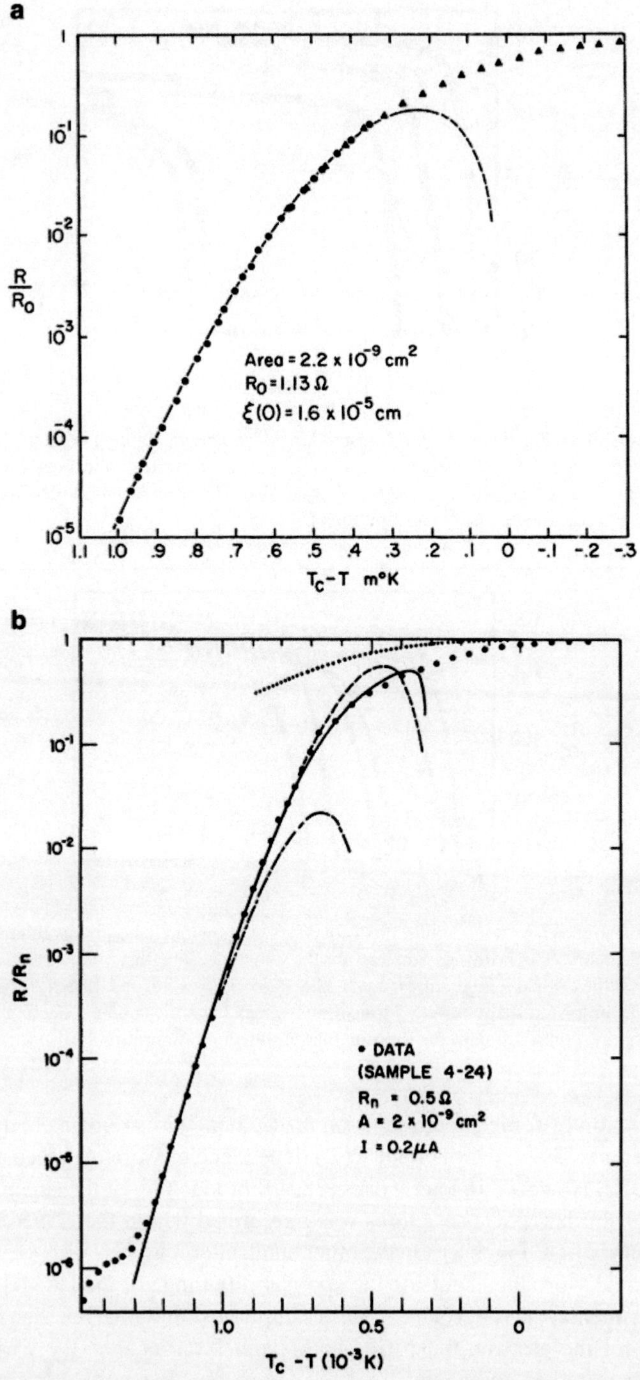

Fig. 2.6 Resistance vs. temperature for tin whiskers. (**a**) Solid line stands for LAMH model [18]. (**b**) Dashed line represents LAMH model, while other lines – some other models [19]

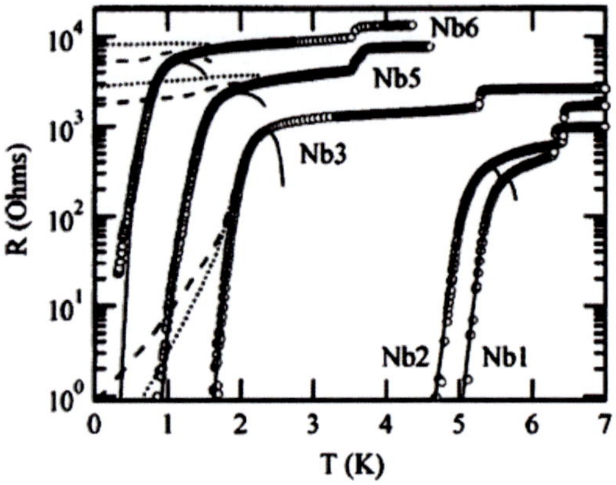

Fig. 2.7 Temperature dependence of the resistance of superconducting *Nb* nanowires obtained by template decoration method using carbon nanotube as a substrate. Solid lines show fits to the phenomenological model $1/R(T) = 1/R_N + 1/R_{\mathrm{TAPS}}$. The dashed and dotted lines are some theoretical curves that include QPS contributions [37]

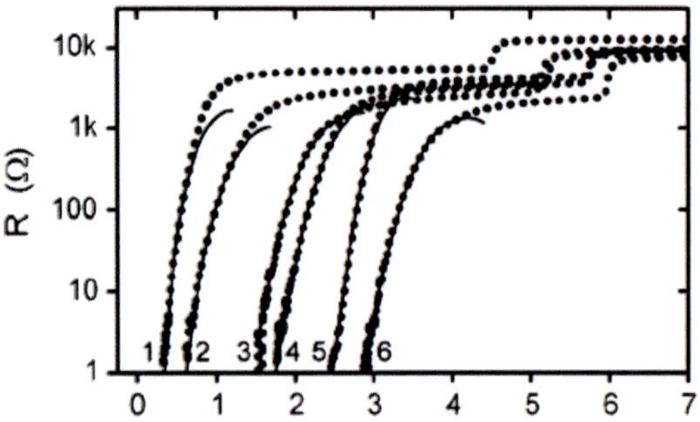

Fig. 2.8 Resistance versus temperature plots for *MoGe* nanowires:solid curves indicate fits to the LAMH-TAPS theory. The fitting parameters are the coherence lengths, 70.0, 19.0, 11.5, 9.4, 5.6, and 6.7 nm, and the critical temperatures, 1.72, 2.28, 3.75, 3.86, 3.80, and 4.80 K, for samples 1–6, respectively. Double-step shape of $R(T)$ transitions comes from the superconducting transition of the contact regions contributing to the 2-probe measurement configuration [40]

Interpretation of some experiments on *MoGe* nanowires remains under debates. In earlier reports [39,43] very broad $R(T)$ dependencies were observed and associated with QPS. However, in later works [38,40] broadening of the superconducting transition in similar MoGe structures was interpreted within the TAPS model with no QPS contribution (Fig. 2.8). On the other hand, later it was argued [44,45] that the fits were actually produced outside the applicability range of the LAMH theory. To fit the experimental curves $R(T)$ for these samples using LAMH model, one should use values for the electron mean free path ℓ much larger than the wire diameter [40], which is rather unrealistic for such structures. We will return to possible interpretation of these experiments below in connection with QPS effects.

2.6 Quantum Phase Slips

Technological and experimental considerations for observation of QPS are even more demanding than for TAPS. As it follows from (2.2), the effective resistance $R(T)$ exponentially depends on the QPS core action S_{core}. In a typical case of dirty-limit $\ell \ll \xi$ superconducting nanostructures $S_{\text{core}} \sim T_C^{1/2}\sigma/\rho_N$, where ρ_N is the wire resistivity in the normal state. Hence, additionally to making the structure cross-section σ as narrow as possible, it is desirable to select superconducting materials with smaller values of T_C and perform experiments on sufficiently dirty nanowires with higher values of ρ_N. The latter requirement is in line with the well-known general observation that fluctuation effects are more pronounced in dirtier systems.

Comparing (2.1) and (2.2) one comes to the conclusion that the crossover between TAPS and QPS regimes can be expected at temperatures $T \sim \Delta(T)/k_B$. For superconductors with typical material parameters and $T_C \sim 1$ K, this condition implies that QPS effects may become important already at $| T - T_C | \geq 100$ mK. To deal with experimentally measurable resistances, it is necessary to fabricate nanowires with effective diameters in the range $\sqrt{\sigma} \sim 10$ nm. The wires should be sufficiently uniform and homogeneous in order not to override fluctuation effects by trivial broadening of $R(T)$ dependencies due to wire imperfections and inhomogeneities [35]. Thus, proper fabrication technology is vitally important for experimental studies of QPS in quasi-1D superconductors.

Perhaps, the first experimental indication of the effect of quantum fluctuations was obtained in amorphous *MoGe* nanowires with effective diameter down to $\sqrt{\sigma} \sim 30$ nm [46]. Although the paper [46] was mainly focused on the effect of disorder in low-dimensional superconductors, it was clearly stated that for the narrowest samples significantly broader curves $R(T)$ were observed than it is predicted by the TAPS model (see Fig. 2.9). The effect of quantum fluctuations was pointed out as a possible reason for this disagreement.

Detailed experimental studies of transport properties of superconducting nanowires have been carried out by Giordano and co-workers [26, 27, 47–49]. In these experiments, *In* and *In* – *Pb* wires were fabricated with effective diameters $\sqrt{\sigma}$ in the range from 15 nm to 100 nm with the grain size from 10 to 20 nm. For wider structures, a reasonable agreement with the TAPS model was observed, while for thinner wires with $\sqrt{\sigma} < 50$ nm clear deviations from TAPSpredictions were demonstrated (see Fig. 2.10). The discrepancy was interpreted as a manifestation of quantum phase slippage. To explain their observations, Giordano and co-workers proposed phenomenological description based on the Caldeira–Leggett model for macroscopic quantum tunneling with dissipation [50]. Although qualitative agreement with this simple phenomenological model has been obtained (Fig. 2.10), quantitative interpretation of the data is problematic due to poor uniformity of the thinnest samples. It appears that strong granularity of the wires was most likely a very important factor in the experiments by Giordano and co-workers. For instance, estimating S_{core} for the thinnest wires [48] with diameters 16 and 25 nm,

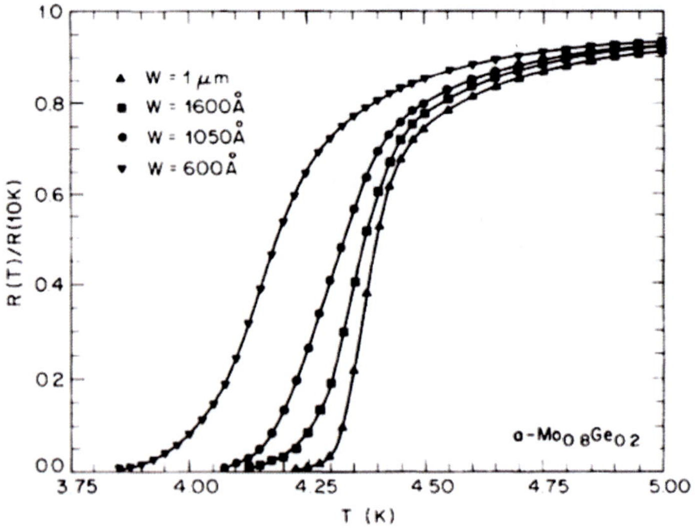

Fig. 2.9 Resistive transitions for a set of *MoGe* wires with different width w upon a single film of thickness 5 nm [46]

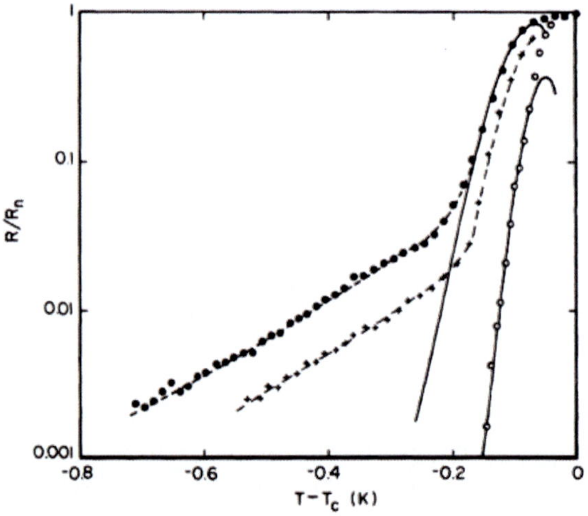

Fig. 2.10 Resistance (normalized by its normal state value) as a function of temperature for three *In* wires; the sample diameters were 41 nm (•), 50.5 nm (+) and 72 nm (o). The solid curves are fits to the TAPS model, while the dashed curves indicate fits to the phenomenological model [26]

for the experimental parameters we obtain, respectively, $S_{\text{core}} \sim 300$ and ~ 700. For uniform wires such huge values of the QPS action would totally prohibit any signature of quantum phase slips. Since QPS effects were very clearly observed, it appears inevitable that these samples contained constrictions with diameters significantly smaller than the quoted average thickness [48].

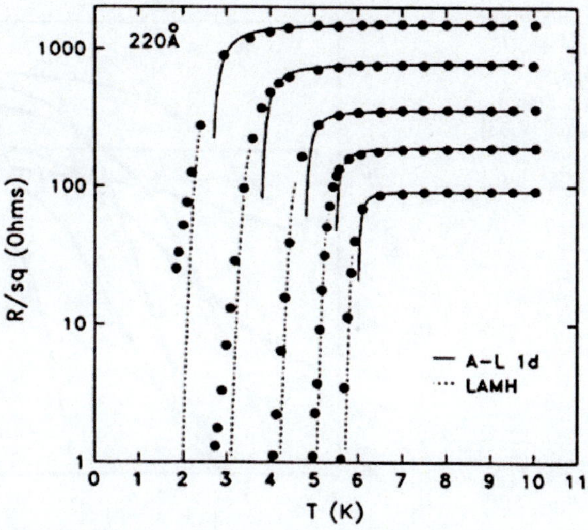

Fig. 2.11 $R(T)$ dependencies for the same Pb nanowire with width $w = 22$ nm and various thicknesses. Solid lines at the top of the transitions are fits to the Aslamazov–Larkin theory. Dashed lines are fits to the TAPS model, which clearly yields steeper $R(T)$ dependencies compared to the experimental data for the thinnest wires [28]

Another set of experiments was performed by Dynes and co-workers [28,29,51]. Suspended stencil technique has been developed enabling fabrication of quench-condensed granular nanowires with cross-section area σ down to 15 nm^2 and length L ranging from 1 to 2 μm from various materials including several superconductors, such as $Pb, Sn, Pb - Bi$. The samples edge roughness was claimed to be about 3 nm. A remarkable feature of the method is the ability to vary the wire thickness and, hence, its cross-section σ at a constant width w in situ inside the cryostat in-between the sessions of truly 4-probe $R(T)$ measurements. The experiments clearly indicated systematic deviations of the experimental data points $R(T)$ from the TAPS model predictions (Fig. 2.11). This discrepancy increases as the wires become narrower. The width of the superconducting transition was found to scale with the normal state resistance R_N. It should be noted that in lead nanowires as narrow as 15 nm and as thin as 10 nm no low temperature resistance tails were observed. Instead, a less dramatic but systematic broadening of the superconducting transition beyond the TAPS limit was noted. On the other hand, long resistance tails were always present in tin structures fabricated and measured using similar technique. Very probably, the discrepancy comes from the difference in material parameters exponentially strongly contributing to the fluctuation-governed $R(T)$ dependence (2.1) and (2.2).

It has been already mentioned in the previous section, the results of experiments on $MoGe$ nanowires, fabricated using molecular template method, are still under debates. In earlier works [39, 43], it was claimed that the resistive state of these structures is governed by quantum fluctuations, while in later experiments on similar

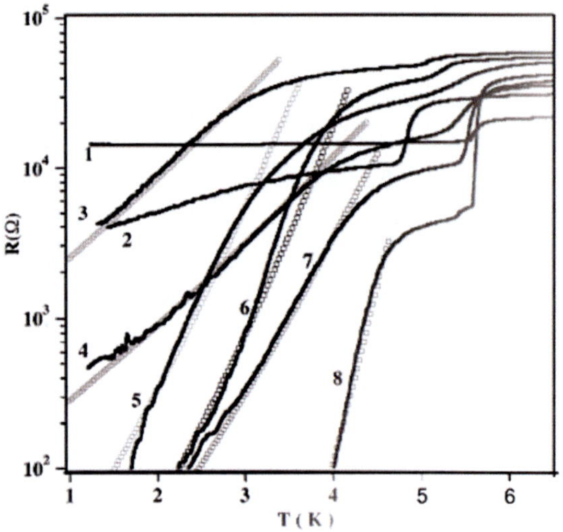

Fig. 2.12 Superconducting transitions of "long" MoGe nanowires on top of insulating carbon nanotube used as the substrate [39,43]. Symbols stand for calculations using the QPS mechanism (2.2)

samples it was argued that the shape of $R(T)$ transitions can be described by alternative mechanisms: for example, TAPS [38] or Coulomb effects [52–55]. A bulk of recent experimental data accumulated on *MoGe* nanowires was recently reviewed in detail by Bezryadin [56]. The main observations can be summarized as follows: (a) "shorter" nanowires ($L < 200\,\text{nm}$) demonstrate either "weakly insulating" behavior with clear features of weak Coulomb blockade [52–55], or relatively steep superconducting transition $R(T)$ with virtually no samples showing an intermediate regime, (b) "longer" samples ($200\,\text{nm} < L < 1\,\mu\text{m}$) typically showed the behavior which – similarly to the earlier data [43] – can reasonably well be interpreted in terms of a crossover between the regimes of small and large QPS rates (corresponding to respectively superconducting and normal behavior). The $R(T)$ curves of longer wires in the regime (b) showed a decrease of the resistance with cooling no matter whether their normal state resistance R_N was smaller or bigger than R_q (contrary to the earlier observation [39]). The crossover between normal and superconducting behavior of these long samples was found to be controlled by the wire cross-section σ or, equivalently, by the ratio R_N/L with the overall picture consistent with the QPS scenario [11–13].

Recently, the experiments on *MoGe* nanowires were analyzed in detail in a comprehensive review [2]. The main conclusion of this analysis is that only the longest *MoGe* nanowires, fabricated using molecular template method, are suitable for experimental study of fluctuation-governed mechanism (Fig. 2.12). In shorter structures, the Josephson physics or/and proximity effects might provide the dominating contribution.

Arrays of metallic nanowires inside nanopore membranes can be fabricated using electrochemical deposition. $R(T)$ transitions on bundles of multiple Sn nanowires showed clear broadening (Fig. 2.13), which was associated with QPS effects [57].

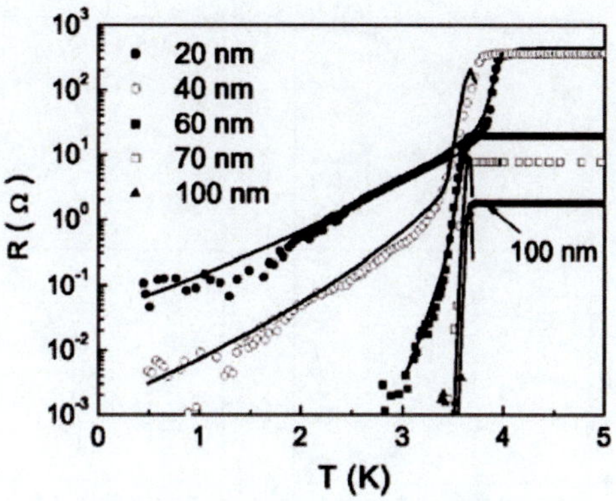

Fig. 2.13 $R(T)$ curves for 20, 40, 60, 70, and 100 nm wide and 6 μm long *Sn* nanowire arrays containing, respectively, 18, 1, 8, 15 and 53 wires in the bundle. The solid lines for 20, 40, and 60 nm wires are the results based on the TAPS model near T_C and QPS model below T_C [57]

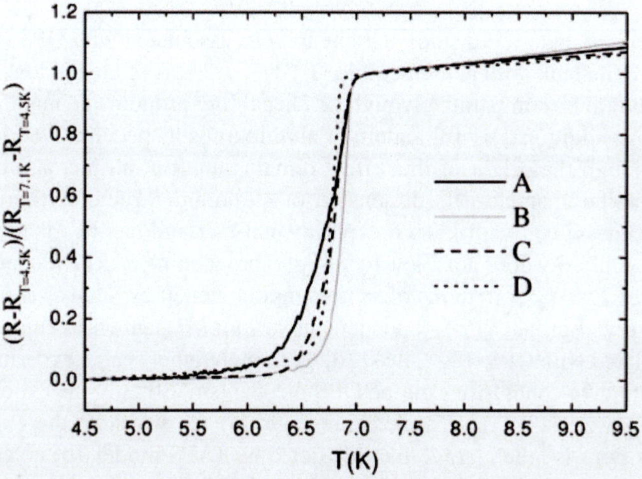

Fig. 2.14 $R(T)$ dependence for four *Pb* nanowires (A) – (D) with diameters 40, 55, 55, and 70 nm, respectively. The transition is broader for narrower wires [58]

As the number of wires in the measured bundle is not known precisely, quantitative comparison with theoretical predictions appears complicated.

Experiments with individual *Pb* nanowires grown using similar technique – nanopore electrochemical deposition – show clear superconducting transitions, which get broadened with reduction of the sample diameter (Fig. 2.14). Unfortunately no theory fits was provided by the authors, while the linear scale of the reported $R(T)$ dependencies complicates an independent quantitative comparison [58].

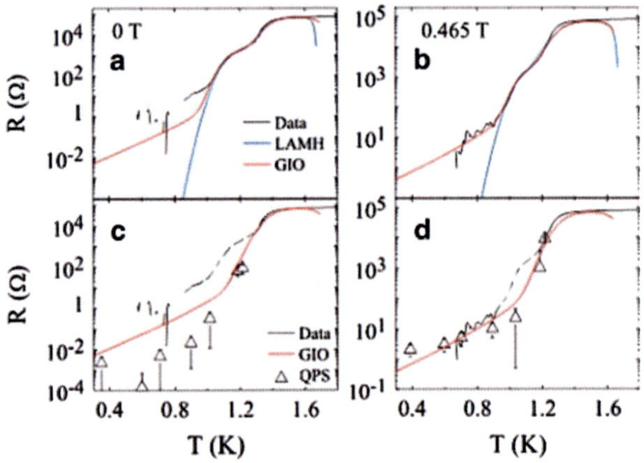

Fig. 2.15 Experimental $R(T)$ dependence for Al nanowires (*symbols and black lines*). *Solid color lines* stand for fits using TAPS (*blue*) and QPS (*red*) contributions [30]

A convenient material to study the phenomena associated with QPS mechanism is aluminum. Its bulk critical temperature $T_C^{\text{bulk}} \sim 1.2$ K is relatively low, hence, the QPS rate should be comparatively high (2.2) enabling pronounced manifestation of the effect. An additional useful feature of aluminum is its peculiar size dependence of T_C. Although the origin of this effect remains unclear, an increase of T_C with reduction of the characteristic dimension of aluminum structures (wire diameter or film thickness) is a well-known experimental fact and can be taken as granted [42]. This peculiarity does not allow to interpret broadening of $R(T)$dependencies at temperatures $T < T_C^{\text{bulk}}$ in terms of sample inhomogeneities, such as constrictions.

The observed broad $R(T)$ dependencies in long Al nanowires (up to 100 μm) were associated with QPS (Fig. 2.15) [30]. Although rather scarce experimental data make quantitative conclusions on the QPS mechanism difficult, the experiments on long Al nanowires [30] have a clear message: (a) the width of the experimental $R(T)$ transitions is much larger than predicted by TAPS model, (b) no correlations between the total normal state wire resistance R_N (compared to the quantum resistance unit $Rq = 6.45$ kΩ) and superconductivity in such wires was found.

Aluminium was chosen for investigations of 1D superconductivity in [33, 34]. It was demonstrated that low energy Ar^+ ion sputtering can progressively and nondestructively reduce dimensions of various nanostructures including nanowires [31, 32]. The penetration depth of Ar^+ ions into Al matrix at acceleration voltages of ~500 eV is about 2 nm and is comparable to the thickness of naturally formed oxide. The accuracy of the effective diameter determination from the normal state resistance by SEM and SPM measurements is about ±2 nm. Only those samples which showed no obvious geometrical imperfections were used for further experiments. To a large extent, the method allows one to study the evolution of the size phenomenon, eliminating artifacts related to uniqueness of samples fabricated in independent processing runs. The ion beam treatment polishes the surface of the

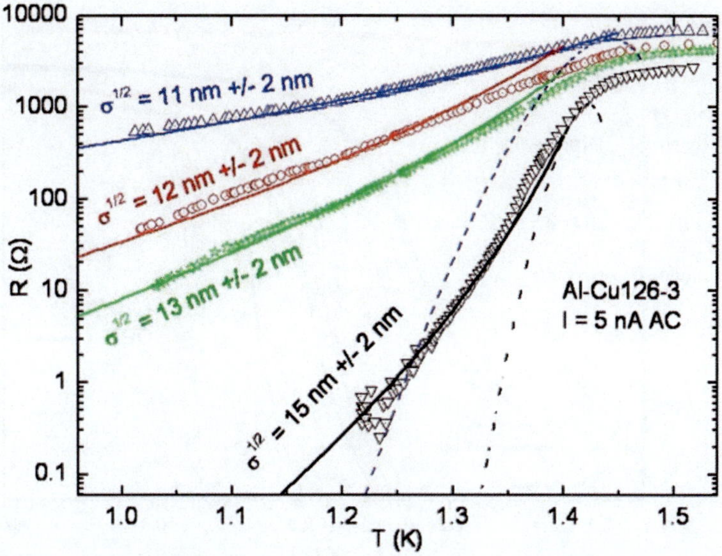

Fig. 2.16 $R(T)$ curves for the thinnest samples obtained by progressive diameter reduction for the same aluminium nanowire with length $L = 10\,\mu\text{m}$. The TAPS model fitting is shown with dashed lines for 11 and 15 nm samples. Fits using QPS mechanism are shown by *solid color lines* [34]

samples removing inevitable roughness just after fabrication (Fig. 2.4). If there were no detectable geometrical imperfections in the original (thick) wires, they could not be introduced in the course of diameter reduction by low energy ion sputtering.

After a sequence of sputterings (alternated with $R(T)$ measurements), the wire diameter was reduced from $\sqrt{\sigma} \sim 100\,\text{nm}$ down to $\sqrt{\sigma} \sim 10\,\text{nm}$. Experiments were performed on several sets of aluminum nanowires with length L equal to 1, 5, and $10\,\mu\text{m}$. For larger diameters $\sqrt{\sigma} \gtrsim 20\,\text{nm}$, the shape of the $R(T)$ dependence is rather "sharp" and can be qualitatively understood by the TAPS mechanism. Note that the abovementioned size-dependent variation of T_C in aluminum nanowires results in broadening of the $R(T)$ transition and significantly reduces applicability of the TAPS model [35] (Fig. 2.5). When the wire diameter is further reduced, deviations from the TAPS behavior become obvious (Fig. 2.16). Fits to the TAPS model fail to provide any reasonable quantitative agreement with experiment for diameter values below $\sqrt{\sigma} \lesssim 20\,\text{nm}$ even if one hypothetically assumes the existence of unrealistically narrow constrictions not observed by SPM. And, as we already discussed, broadening of the $R(T)$ dependencies in aluminum nanowires at $T < T_C^{\text{bulk}}$ can hardly be ascribed to geometrical imperfections, such as constrictions. On the contrary, the broadened $R(T)$ curves of the thines *Al* samples (Fig. 2.16) can be nicely fit with the QPS model (2.2). One can conclude that the most natural interpretation of the results [33, 34] can be associated with quantum fluctuations.

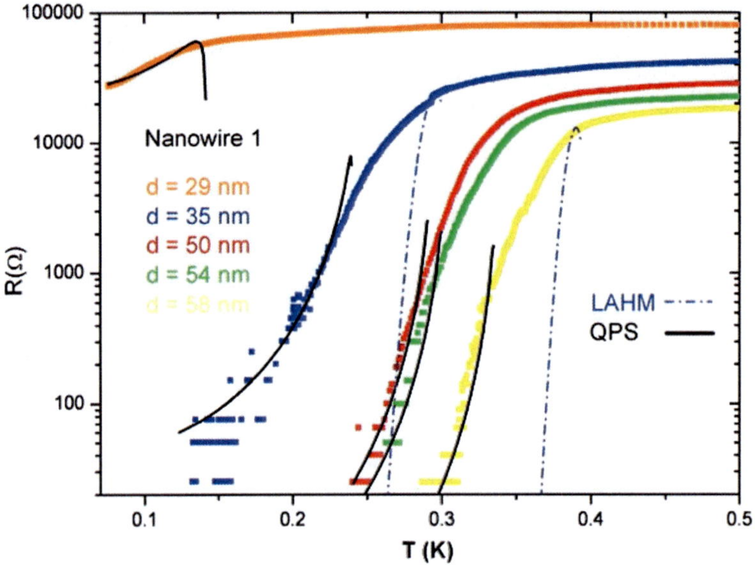

Fig. 2.17 $R(T)$ dependencies for the same Ti nanowire with progressively reduced diameter $d = \sqrt{\sigma}$. *Dot-dashed lines* correspond to fits with TAPS model, and *solid lines* – to QPS model with the realistic fitting parameters: $\ell = 3$nm, $A \simeq 0.4 \pm 0.2$, and normal state resistance and the critical temperature determined from the experiment [59]

In very recent results the same method of progressive cross-section reduction has been applied to titanium nanowires [59], where the impact of quantum fluctuations should be observed on relatively "thick" samples due to the low critical temperature T_C and high normal state resistivity ρ_N (2.2). Indeed, the experiments [59] revealed pronounced broadening of the $R(T)$ dependencies inconsistent with the TAPS model, while the fits with the QPS model provide reasonable agreement (Fig. 2.17).

2.7 Conclusion

It is well established that fluctuations play an important role in structures with reduced dimensionality. In this paper, we addressed fluctuation effects which occur in ultrathin superconducting wires at temperatures below the mean field BCS critical temperature. Superconducting properties of such systems have been intensively studied both theoretically and experimentally. Various techniques were suggested for fabrication of extremely narrow quasi-one-dimensional superconducting structures with the effective diameters down to few nanometers.

In relatively thick samples, the finite width of experimental $R(T)$ transition can be qualitatively understood within the formalism of thermally activated phase slips (TAPS) [9, 10]. Various parasitic contributions impose serious limitations for the quantitative analysis [35]. So far only atomically pure superconducting single crystals (whiskers) can be considered as suitable for establishment of the quantitative agreement between the theory and experiment [18, 19].

In significantly narrower nanowires, the superconducting $R(T)$ transition is much wider than predicted by the TAPS mechanism. On the contrary, the model employing quantum fluctuations (or quantum phase slips – QPS) [11–13] provides reasonable agreement with recent experiments.

The effect of quantum fluctuations should have a universal validity, indicating a breakdown of the zero resistance state in quasi-one-dimensional superconductors. In addition to the significant importance for the basic understanding of the superconductivity phenomenon, quantum fluctuations set fundamental limitations on miniaturization of nanoelectronic components designed to carry a dissipationless supercurrent.

References

1. A. Larkin, A. Varlamov, *Theory of Fluctuations in Superconductors* (Clarendon, Oxford, 2005)
2. K.Yu. Arutyunov, D.G. Golubev, A. Zaikin, Phys. Rep. **464**, 1 (2008). DOI 10.1016/j.physrep.2008.04.009
3. L. Aslamazov, A. Larkin, Fiz. Tverd. Tela **10**, 1140 (1968)
4. K. Maki, Prog. Theor. Phys. **39**, 897 (1968)
5. R. Thompson, Phys. Rev. B **1**, 327 (1970)
6. W. Little, Phys. Rev. **156**, 396 (1967)
7. J. Meyer, G. Minnegerode, Phys. Lett. **38A**, 529 (1972)
8. R. Tidecks, *Current-Induced Nonequilibrium Phenomena in Quasi-One-Dimensional Superconductors* (Springer, New York, 1990)
9. J. Langer, V. Ambegaokar, Phys. Rev. **164**, 498 (1967)
10. D. McCumber, B. Halperin, Phys. Rev. B **1**, 1054 (1970)
11. A.D. Zaikin, D.S. Golubev, A. van Otterlo, G.T. Zimányi, Phys. Rev. Lett. **78**(8), 1552 (1997). DOI 10.1103/PhysRevLett.78.1552
12. A.D. Zaikin, D.S. Golubev, A. van Otterlo, G. Zimanyi, PUsp. Fiz. Nauk **168**, 244 (1998)
13. D. Golubev, A. Zaikin, Phys. Rev. B **64**, 014504 (2001)
14. K.Y. Arutyunov, Recent Patents Nanotechnol. **1**(1), 129 (2007)
15. E.I. Givargizov, *Highly Anisotropic Crystals, Series: Material Science of Minerals and Rocks* (Springer, Germany, 1987)
16. R.M. Fisher, L.S. Darken, K.G. Carrol, Acta Metallurgica **2**, 368 (1954)
17. O.S. Lutes, Phys. Rev. **105**, 1451 (1957)
18. J.E. Lukens, R.J. Warburton, W.W. Webb, Phys. Rev. Lett. **25**(17), 1180 (1970). DOI 10.1103/PhysRevLett.25.1180
19. R.S. Newbower, M.R. Beasley, M. Tinkham, Phys. Rev. B **5**(3), 864 (1972). DOI 10.1103/PhysRevB.5.864
20. Yu. P. Gaidukov, N.P. Danilova, and R. Sh. Georgius-Mankarius, Sov.Phys. JETP, **66**, 605 (1987)
21. K.Y. Arutyunov, T.V. Ryynänen, J.P. Pekola, A.B. Pavolotski, Phys. Rev. B **63**(9), 092506 (2001). DOI 10.1103/PhysRevB.63.092506
22. A. Bezryadin, A. Bollinger, D. Hopkins, M. Murphey, M. Remeika, A. Rogachev, *Dekker Encyclopedia of Nanoscience and Nanotechnology* (Dekker, 2004)
23. D.S. Hopkins, D. Pekker, P.M. Goldbart, A. Bezryadin, Science **308**, 1762 (2005)

24. M. Remeika, A. Bezryadin, Nanotechnology **16**, 1172 (2005)
25. A. Johansson, G. Sambandamurthy, D. Shahar, N. Jacobson, R. Tenne, Phys. Rev. Lett **95**, 116805 (2005)
26. N. Giordano, Phys. Rev. Lett. **61**, 2137 (1988)
27. N. Giordano, E.R. Shuler, Phys. Rev. Lett. **63**, 2417 (1989)
28. S. Sharifi, A.V. Herzog, R.C. Dynes, Phys. Rev. Lett. **71**, 428 (1993)
29. A.V. Herzog, P. Xiong, S. Sharifi, R.C. Dynes, Phys. Rev. Lett. **76**, 668 (1996)
30. F. Altomare, A.M. Chang, M.R. Melloch, Y. Hong, C.W. Tu, Phys. Rev. Lett **97**, 017001 (2006)
31. M. Savolainen, V. Touboltsev, P. Koppinen, K.P. Riikonen, K. Arutyunov, Appl. Phys. Mater. Sci. Process **79**, 1769 (2004). URL http://dx.doi.org/10.1007/s00339-004-2709-8. 10.1007/s00339-004-2709-8
32. M. Zgirski, K.P. Riikonen, V. Tuboltsev, P. Jalkanen, T.T. Hongisto, K.Y. Arutyunov, Nanotechnology **19**(5), 055301 (2008). URL http://stacks.iop.org/0957-4484/19/i=5/a=055301
33. M. Zgirski, K.-P. Riikonen, V. Touboltsev, K. Arutyunov, Nano Lett. **5**(6), 1029 (2005)
34. M. Zgirski, K.-P. Riikonen, V. Touboltsev, K.Yu. Arutyunov, Phys. Rev. B **77**, 054508 (2008). DOI 10.1103/PhysRevB.77.054508
35. M. Zgirski, K.Y. Arutyunov, Phys. Rev. B **75**(17), 172509 (2007). DOI 10.1103/PhysRevB.75.172509. http://link.aps.org/abstract/PRB/v75/e172509
36. M. Tian, N. Kumar, S. Xu, J. Wang, J.S. Kurtz, M.H.W. Chan, Phys. Rev. Lett. **95**(7), 076802 (2005). DOI 10.1103/PhysRevLett.95.076802
37. A. Rogachev, A. Bezryadin, Appl. Phys. Lett. **83**, 512 (2003). DOI 10.1063/1.1592313
38. A. Rogachev, A.T. Bollinger, A. Bezryadin, Phys. Rev. Lett. **94**(1), 017004 (2005). DOI 10.1103/PhysRevLett.94.017004
39. A. Bezryadin, C.N. Lau, M. Tinkham, Nature **404**, 971 (2000). DOI 10.1038/35010060Letter
40. A.T. Bollinger, A. Rogachev, A. Bezryadin, Europhys. Lett. **76**(3), 505 (2006). DOI 10.1209/epl/i2006-10275-5
41. Y. Oreg1, A.M. Finkel'stein, Phys. Rev. Lett. **83**(1), 191 (1999)
42. A.A. Shanenko, M.D. Croitoru, M. Zgirski, F.M. Peeters, K. Arutyunov, Phys. Rev. B **74**(5), 052502 (2006). DOI 10.1103/PhysRevB.74.052502
43. C.N. Lau, N. Markovic, M. Bockrath, A. Bezryadin, M. Tinkham, Phys. Rev. Lett **87**(21), 217003 (2001). DOI 10.1103/PhysRevLett.87.217003
44. Y.O. Dganit Meidan, G. Refael, Phys. Rev. Lett. **98**(18), 187001 (2007). DOI 10.1103/PhysRevLett.98.187001
45. D. Meidan, Y. Oreg, G. Refael, R.A. Smith, Phys. C **468**(5), 341 (2008)
46. J.M. Graybeal1, P.M. Mankiewich, R.C. Dynes, M.R. Beasley, Phys. Rev. Lett **59**(23), 2697 (1987)
47. N. Giordano, Phys. Rev. B **43**(1), 160 (1991)
48. N. Giordano, Phys. Rev. B **41**(10), 6350 (1990)
49. N. Giordano, Physica B **43**(1-3), 460 (1994)
50. A. Caldeira, A. Leggett, Phys. Rev. Lett. **46**(4), 211 (1981). DOI 10.1103/PhysRevLett.46.211
51. P. Xiong, A.V. Herzog, R.C. Dynes, Phys. Rev. Lett. **78**, 927 (1997)
52. Y.V. Nazarov, Phys. Rev. Lett **82**, 1245 (1999)
53. D. Golubev, A. Zaikin, Phys. Rev. Lett. **86**, 4887 (2001)
54. D. Golubev, A. Zaikin, Phys. Rev. B **69**, 075318 (2004)
55. D.A. Bagrets, Y.V. Nazarov, Phys. Rev. Lett. **94**, 056801 (2005)
56. A. Bezryadin, J. Phys.: Cond. Mat. **20**, 1 (2008)
57. Mingliang Tian, Jinguo Wang, James S. Kurtz, Ying Liu, M.H.W. Chan, Theresa S. Maye, Thomas E. Malloukr, Phys. Rev. B **71**, 104521 (2005)
58. S. Michotte, S. Mátéfi-Tempfli, L. Piraux, D.Y. Vodolazov and F.M. Peeters, Phys. Rev. B **69**, 094512 (2004)
59. J.S. Lehtinen, T. Sajavaara, K. Yu. Arutyunov, A. Vasiliev, arXiv:1106.3852.

Chapter 3
Crossed Andreev Reflection and Spin-Resolved Non-local Electron Transport

Mikhail S. Kalenkov and Andrei D. Zaikin

Abstract The phenomenon of crossed Andreev reflection (CAR) is known to play a key role in non-local electron transport across three-terminal normal-superconducting-normal (NSN) devices. Here, we review our general theory of non-local charge transport in three-terminal disordered ferromagnet-superconductor-ferromagnet (FSF) structures. We demonstrate that CAR is highly sensitive to electron spins and yields a rich variety of properties of non-local conductance, which we describe non-perturbatively at arbitrary voltages, temperature, degree of disorder, spin-dependent interface transmissions and their polarizations. We demonstrate that magnetic effects have different implications: While strong exchange field suppresses disorder-induced electron interference in ferromagnetic electrodes, spin-sensitive electron scattering at SF interfaces can drive the total non-local conductance negative at sufficiently low energies. At higher energies, magnetic effects become less important and the non-local resistance behaves similarly to the non-magnetic case. Our results can be applied to multi-terminal hybrid structures with normal, ferromagnetic and half-metallic electrodes and can be directly tested in future experiments.

M.S. Kalenkov
I.E. Tamm Department of Theoretical Physics, P.N. Lebedev Physics Institute, 119991 Moscow, Russia

A.D. Zaikin (✉)
Institut für Nanotechnologie, Karlsruher Institut für Tehnologie (KIT), 76021, Karlsruhe, Germany
and
I.E. Tamm Department of Theoretical Physics, P.N. Lebedev Physics Institute, 119991 Moscow, Russia
e-mail: andrei.zaikin@kit.edu

A. Sidorenko (ed.), *Fundamentals of Superconducting Nanoelectronics*, NanoScience and Technology, DOI 10.1007/978-3-642-20158-5_3,

3.1 Introduction

In hybrid NS structures, quasiparticle current flowing in a normal metal is converted into that of Cooper pairs inside a superconductor. For quasiparticle energies above the superconducting gap $\varepsilon > \Delta$, this conversion is accompanied by electron–hole imbalance, which relaxes deep inside a superconductor. At subgap energies $\varepsilon < \Delta$, the physical picture is entirely different. In this case, quasiparticle-to-Cooper-pair current conversion is provided by the mechanism of Andreev reflection (AR) [1]: A quasiparticle enters the superconductor from the normal metal at a length of order of the superconducting coherence length ξ_S and forms a Cooper pair together with another quasiparticle, while a hole goes back into the normal metal. As a result, the net charge *2e* is transferred through the NS interface, which acquires non-zero subgap conductance down to $T = 0$ [2].

AR remains essentially a local effect provided there exists only one NS interface in the system or, else, if the distance between different NS interfaces greatly exceeds the superconducting coherence length ξ. If, however, the distance L between two adjacent NS interfaces (i.e. the superconductor size) is smaller than (or comparable with) ξ, two additional *non-local* processes come into play (see Fig. 3.1). One such process corresponds to direct electron transfer between two N-metals through a superconductor. Another process is the so-called crossed Andreev reflection [3, 4] (CAR): An electron penetrating into the superconductor from the first N-terminal may form a Cooper pair together with another electron from the second N-terminal. In this case, a hole will go into the second N-metal and AR becomes a non-local effect. This phenomenon of CAR enables direct experimental demonstration of entanglement between electrons in spatially separated N-electrodes and can strongly influence non-local transport of electrons in hybrid NSN systems.

Non-local electron transport in the presence of CAR was recently investigated both experimentally [5–13] and theoretically [14–27] (see also further references therein) demonstrating a rich variety of physical processes involved in the problem. It was shown [14] that in the lowest order in the interface transmission and at $T = 0$ CAR contribution to cross-terminal conductance is exactly cancelled by that from elastic electron cotunneling (EC), i.e. the non-local conductance vanishes in this limit. Taking into account higher order processes in barrier transmissions eliminates this feature and yields non-zero values of cross-conductance [15].

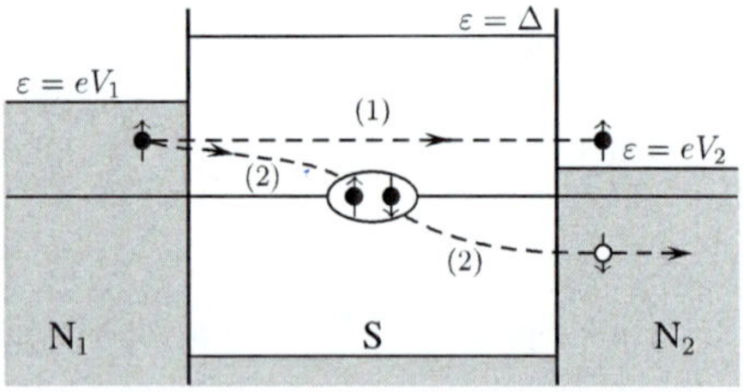

Fig. 3.1 Two elementary processes contributing to non-local conductance of an NSN device: (**1**) direct electron transfer and (**2**) crossed Andreev reflection

Another interesting issue is the effect of disorder. It is well known that disorder enhances interference effects and, hence, can strongly modify local subgap conductance of NS interfaces in the low energy limit [28–30]. Non-local conductance of multi-terminal hybrid NSN structures in the presence of disorder was studied in a number of papers [16–19, 25, 26]. A general quasiclassical theory was constructed by Golubev and the present authors [25]. It was demonstrated that an interplay between CAR, quantum interference of electrons and non-local charge imbalance dominates the behavior of diffusive NSN systems being essential for quantitative interpretation of a number of experimental observations [8, 10, 11]. In particular, strong enhancement of non-local spectral conductance was predicted at low energies due to quantum interference of electrons in disordered N-terminals. At the same time, non-local resistance R_{12} remains smooth at small energies and, furthermore, was found to depend neither on parameters of NS interfaces nor on those of N-terminals. At higher temperatures, R_{12} was shown to exhibit a peak caused by the trade-off between charge imbalance and Andreev reflection.

Yet, another interesting subject is an interplay between CAR and Coulomb interaction. The effect of electron–electron interactions on AR was investigated in a number of papers [30–32]. Interactions should also affect CAR, e.g., by lifting the exact cancellation of EC and CAR contributions [20] already in the lowest order in tunneling. A similar effect can occur in the presence of external ac fields [24]. A general theory of non-local transport in NSN systems with disorder and electron–electron interactions was very recently developed by Golubev and one of the present authors [27] direct relation between Coulomb effects and non-local shot noise. In the tunneling limit, non-local differential conductance is found to have an S-like shape and can turn negative at non-zero bias. At high transmissions, CAR turned out to be responsible both for positive noise cross-correlations and for Coulomb anti-blockade of non-local electron transport.

An important property of both AR and CAR is that these processes should be sensitive to magnetic properties of normal electrodes because these processes essentially depend on spins of scattered electrons. One possible way to demonstrate spin-resolved CAR is to use ferromagnets (F) instead of normal electrodes ferromagnet-superconductor-ferromagnet (FSF) structures. First experiments on such FSF structures [5, 6] illustrated this point by demonstrating the dependence of non-local conductance on the polarization of ferromagnetic terminals. Hence, for better understanding of non-local effects in multi-terminal hybrid proximity structures, it is necessary to construct a theory of *spin-resolved* non-local transport. In the case of ballistic systems in the lowest order in tunneling, this task was accomplished in [14]. Here, we will generalize our quasiclassical approach [15, 25] to explicitly focus on spin effects and construct a theory of non-local electron transport in both ballistic and diffusive NSN and FSF structures with spin-active interfaces beyond lowest order perturbation theory in their transmissions.

The structure of the paper is as follows. In Sect. 3.2, we will describe non-local spin-resolved electron transport in ballistic NSN structures with spin-active interfaces. In Sect. 3.3, we will further extend our formalism and evaluate both local and non-local conductances in SFS structures in the presence of disorder. Our main conclusions are briefly summarized in Sect. 3.4.

Fig. 3.2 Schematics of our NSN device

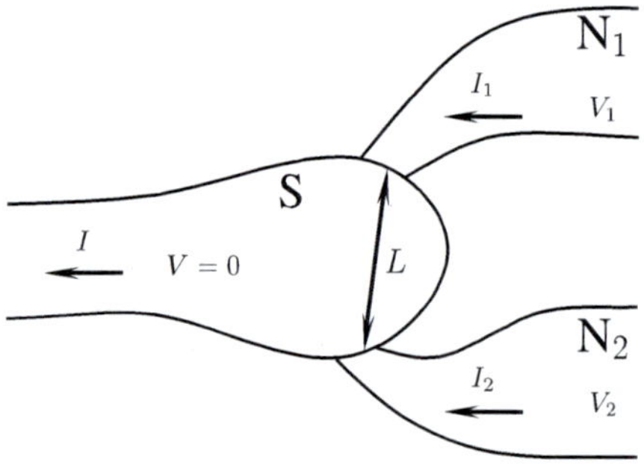

3.2 Spin-Resolved Transport in Ballistic Systems

Let us consider three-terminal NSN structure depicted in Fig. 3.2. We will assume that all three metallic electrodes are non-magnetic and ballistic, i.e. the electron elastic mean free path in each metal is larger than any other relevant size scale. To resolve spin-dependent effects, we will assume that both NS interfaces are spinactive, i.e. we will distinguish "spin-up" and "spin-down" transmissions of the first ($D_{1\uparrow}$ and $D_{1\downarrow}$) and the second ($D_{2\uparrow}$ and $D_{2\downarrow}$) SN interface. All these four transmissions may take any value from zero to one. We also introduce the angle φ between polarizations of two interfaces, which can take any value between 0 and 2π.

In what follows effective cross-sections of the two interfaces will be denoted, respectively, as $\mathcal{A}_1$ and $\mathcal{A}_2$. The distance between these interfaces L and other geometric parameters is assumed to be much larger than $\sqrt{\mathcal{A}_{1,2}}$, i.e. effectively both contacts are metallic constrictions. In this case, the voltage drops only across SN interfaces and not inside large metallic electrodes.

For convenience, we will set the electric potential of the S-electrode equal to zero, $V = 0$. In the presence of bias voltages, V_1 and V_2 applied to two normal electrodes (see Fig. 3.2) the currents I_1 and I_2 will flow through SN_1 and SN_2 interfaces. These currents can be evaluated with the aid of the quasiclassical formalism of nonequilibrium Green–Eilenberger–Keldysh functions [33] $\hat{g}^{R,A,K}$, which we briefly specify below.

3.2.1 *Quasiclassical Equations*

In the ballistic limit, the corresponding Eilenberger equations take the form

$$\begin{aligned}&\left[\varepsilon\hat{\tau}_3 + eV(\boldsymbol{r},t) - \hat{\Delta}(\boldsymbol{r},t), \hat{g}^{R,A,K}(\boldsymbol{p}_F,\varepsilon,\boldsymbol{r},t)\right]\\&\quad + \mathrm{i}\boldsymbol{v}_F\nabla\hat{g}^{R,A,K}(\boldsymbol{p}_F,\varepsilon,\boldsymbol{r},t) = 0,\end{aligned} \tag{3.1}$$

where $[\hat{a}, \hat{b}] = \hat{a}\hat{b} - \hat{b}\hat{a}$, ε is the quasiparticle energy, $\boldsymbol{p}_F = m\boldsymbol{v}_F$ is the electron Fermi momentum vector and $\hat{\tau}_3$ is the Pauli matrix in Nambu space. The functions $\hat{g}^{R,A,K}$ also obey the normalization conditions $(\hat{g}^R)^2 = (\hat{g}^A)^2 = 1$ and $\hat{g}^R\hat{g}^K + \hat{g}^K\hat{g}^A = 0$. Here and below the product of matrices is defined as time convolution.

Green functions $\hat{g}^{R,A,K}$ and $\hat{\Delta}$ are 4×4 matrices in Nambu and spin spaces. In Nambu space, they can be parameterized as

$$\hat{g}^{R,A,K} = \begin{pmatrix} g^{R,A,K} & f^{R,A,K} \\ \tilde{f}^{R,A,K} & \tilde{g}^{R,A,K} \end{pmatrix}, \quad \hat{\Delta} = \begin{pmatrix} 0 & \Delta \mathrm{i}\sigma_2 \\ \Delta^* \mathrm{i}\sigma_2 & 0 \end{pmatrix}, \tag{3.2}$$

where $g^{R,A,K}$, $f^{R,A,K}$, $\tilde{f}^{R,A,K}$, $\tilde{g}^{R,A,K}$ are 2×2 matrices in the spin space, Δ is the BCS order parameter and σ_i are Pauli matrices. For simplicity, we will only consider the case of spin-singlet isotropic pairing in the superconducting electrode. The current density is related to the Keldysh function $\hat{g}^K$ according to the standard relation

$$\boldsymbol{j}(\boldsymbol{r}, t) = \frac{eN_0}{8} \int \mathrm{d}\varepsilon \left\langle \boldsymbol{v}_F \mathrm{Sp}[\hat{\tau}_3 \hat{g}^K(\boldsymbol{p}_F, \varepsilon, \boldsymbol{r}, t)] \right\rangle, \tag{3.3}$$

where $N_0 = mp_F/2\pi^2$ is the density of state at the Fermi level and angular brackets $\langle ... \rangle$ denote averaging over the Fermi momentum.

3.2.2 *Riccati Parameterization*

The above matrix Green–Keldysh functions can be conveniently parameterized by four Riccati amplitudes $\gamma^{R,A}$, $\tilde{\gamma}^{R,A}$ and two "distribution functions" x^K, $\tilde{x}^K$ (here and below we chose to follow the notations [34]):

$$\hat{g}^K = 2\hat{N}^R \begin{pmatrix} x^K - \gamma^R \tilde{x}^K \tilde{\gamma}^A & -\gamma^R \tilde{x}^K + x^K \gamma^A \\ -\tilde{\gamma}^R x^K + \tilde{x}^K \tilde{\gamma}^A & \tilde{x}^K - \tilde{\gamma}^R x^K \gamma^A \end{pmatrix} \hat{N}^A, \tag{3.4}$$

where functions $\gamma^{R,A}$ and $\tilde{\gamma}^{R,A}$ are Riccati amplitudes

$$\hat{g}^{R,A} = \pm \hat{N}^{R,A} \begin{pmatrix} 1 + \gamma^{R,A}\tilde{\gamma}^{R,A} & 2\gamma^{R,A} \\ -2\tilde{\gamma}^{R,A} & -1 - \tilde{\gamma}^{R,A}\gamma^{R,A} \end{pmatrix} \tag{3.5}$$

and $\hat{N}^{R,A}$ are the following matrices

$$\hat{N}^{R,A} = \begin{pmatrix} (1 - \gamma^{R,A}\tilde{\gamma}^{R,A})^{-1} & 0 \\ 0 & (1 - \tilde{\gamma}^{R,A}\gamma^{R,A})^{-1} \end{pmatrix}. \tag{3.6}$$

With the aid of the above parameterization, one can identically transform the quasiclassical equations (3.1) into the following set of effectively decoupled equations for Riccati amplitudes and distribution functions [34]

$$\begin{aligned} &\mathrm{i}\boldsymbol{v}_F\nabla\gamma^{R,A} + [\varepsilon + eV(\boldsymbol{r},t)]\gamma^{R,A} + \gamma^{R,A}[\varepsilon - eV(\boldsymbol{r},t)] \\ &\qquad = \gamma^{R,A}\Delta^*\mathrm{i}\sigma_2\gamma^{R,A} - \Delta\mathrm{i}\sigma_2, \end{aligned} \tag{3.7}$$

$$\begin{aligned} &\mathrm{i}\boldsymbol{v}_F\nabla\tilde{\gamma}^{R,A} - [\varepsilon - eV(\boldsymbol{r},t)]\tilde{\gamma}^{R,A} - \tilde{\gamma}^{R,A}[\varepsilon + eV(\boldsymbol{r},t)] \\ &\qquad = \tilde{\gamma}^{R,A}\Delta\mathrm{i}\sigma_2\tilde{\gamma}^{R,A} - \Delta^*\mathrm{i}\sigma_2, \end{aligned} \tag{3.8}$$

$$\begin{aligned} &\mathrm{i}\boldsymbol{v}_F\nabla x^K + [\varepsilon + eV(\boldsymbol{r},t)]x^K - x^K[\varepsilon + eV(\boldsymbol{r},t)] \\ &\qquad -\gamma^R\Delta^*\mathrm{i}\sigma_2 x^K - x^K\Delta\mathrm{i}\sigma_2\tilde{\gamma}^A = 0, \end{aligned} \tag{3.9}$$

$$\begin{aligned} &\mathrm{i}\boldsymbol{v}_F\nabla\tilde{x}^K - [\varepsilon - eV(\boldsymbol{r},t)]\tilde{x}^K + \tilde{x}^K[\varepsilon - eV(\boldsymbol{r},t)] \\ &\qquad -\tilde{\gamma}^R\Delta\mathrm{i}\sigma_2\tilde{x}^K - \mathrm{i}\sigma_2\gamma^A = 0. \end{aligned} \tag{3.10}$$

Depending on the particular trajectory, it is also convenient to introduce a "replica" of both Riccati amplitudes and distribution functions which – again following the notations [34,35] – will be denoted by capital letters Γ and X. These "capital" Riccati amplitudes and distribution functions obey the same (3.7)–(3.10) with the replacement $\gamma \to \Gamma$ and $x \to X$. The distinction between different Riccati amplitudes and distribution functions will be made explicit below.

3.2.3 Boundary Conditions

Quasiclassical equations should be supplemented by appropriate boundary conditions at metallic interfaces. In the case of specularly reflecting spin-degenerate interfaces, these conditions were derived by Zaitsev [36] and later generalized to spin-active interfaces [37], see also [38] for recent review on this subject.

Before specifying these conditions, it is important to emphasize that the applicability of the Eilenberger quasiclassical formalism with appropriate boundary conditions to hybrid structures with two or more barriers is, in general, a non-trivial issue [39, 40]. Electrons scattered at different barriers interfere and form bound states (resonances), which cannot be correctly described within the quasiclassical formalism employing Zaitsev boundary conditions or their direct generalization. Here, we avoid this problem by choosing the appropriate geometry of our NSN device, see Fig. 3.2. In our system any relevant trajectory reaches each NS interface only once whereas the probability of multiple reflections at both interfaces is small in the parameter $\mathcal{A}_1\mathcal{A}_2/L^4 \ll 1$. Hence, resonances formed by multiply reflected electron waves can be neglected, and our formalism remains adequate for the problem in question.

It will be convenient for us to formulate the boundary conditions directly in terms of Riccati amplitudes and the distribution functions. Let us consider the

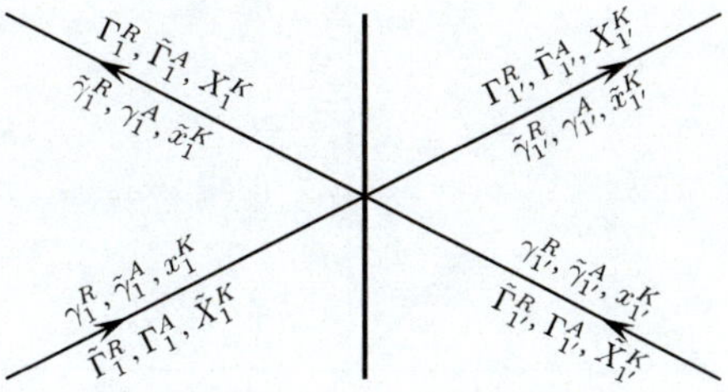

Fig. 3.3 Riccati amplitudes for incoming and outgoing trajectories from the both sides of the interface

first NS interface and explicitly specify the relations between Riccati amplitudes and distribution functions for incoming and outgoing trajectories, see Fig. 3.3. The boundary conditions for Γ_1^R, $\tilde{\Gamma}_1^A$ and X_1^K can be written in the form [35]

$$\Gamma_1^R = r_{1l}^R \gamma_1^R \underline{S}_{11}^+ + t_{1l}^R \gamma_{1'}^R \underline{S}_{11'}^+, \tag{3.11}$$

$$\tilde{\Gamma}_1^A = \underline{S}_{11} \tilde{\gamma}_1^A \tilde{r}_{1r}^A + \underline{S}_{11'} \tilde{\gamma}_{1'}^A \tilde{t}_{1r}^A, \tag{3.12}$$

$$X_1^K = r_{1l}^R x_1^K \tilde{r}_{1r}^A + t_{1l}^R x_{1'}^K \tilde{t}_{1r}^A - a_{1l}^R \tilde{x}_1^K \tilde{a}_{1r}^A. \tag{3.13}$$

Here, we defined the transmission (t), reflection (r), and branch-conversion (a) amplitudes as:

$$r_{1l}^R = +[(\beta_{1'1}^R)^{-1} S_{11}^+ - (\beta_{1'1'}^R)^{-1} S_{11'}^+]^{-1} (\beta_{1'1}^R)^{-1}, \tag{3.14}$$

$$t_{1l}^R = -[(\beta_{1'1}^R)^{-1} S_{11}^+ - (\beta_{1'1'}^R)^{-1} S_{11'}^+]^{-1} (\beta_{1'1'}^R)^{-1}, \tag{3.15}$$

$$\tilde{r}_{1r}^A = +(\beta_{1'1}^A)^{-1} [S_{11} (\beta_{1'1}^A)^{-1} - S_{11'} (\beta_{1'1'}^A)^{-1}]^{-1}, \tag{3.16}$$

$$\tilde{t}_{1r}^A = -(\beta_{1'1'}^A)^{-1} [S_{11} (\beta_{1'1}^A)^{-1} - S_{11'} (\beta_{1'1'}^A)^{-1}]^{-1}, \tag{3.17}$$

$$a_{1l}^R = (\Gamma_1^R \underline{S}_{11} - S_{11} \gamma_1^R)(\tilde{\beta}_{11'}^R)^{-1}, \tag{3.18}$$

$$\tilde{a}_{1r}^A = (\tilde{\beta}_{11'}^A)^{-1} (\underline{S}_{11}^+ \tilde{\Gamma}_1^A - \tilde{\gamma}_1^A S_{11}^+), \tag{3.19}$$

where

$$\beta_{\mathrm{ij}}^R = S_{\mathrm{ij}}^+ - \gamma_j^R \underline{S}_{\mathrm{ij}}^+ \tilde{\gamma}_i^R, \; \tilde{\beta}_{\mathrm{ij}}^R = \underline{S}_{\mathrm{ji}} - \tilde{\gamma}_j^R S_{\mathrm{ji}} \gamma_i^R, \tag{3.20}$$

$$\beta_{\mathrm{ij}}^A = S_{\mathrm{ij}} - \gamma_i^A \underline{S}_{\mathrm{ij}} \tilde{\gamma}_j^A, \; \tilde{\beta}_{\mathrm{ij}}^A = \underline{S}_{\mathrm{ji}}^+ - \tilde{\gamma}_i^A S_{\mathrm{ji}}^+ \gamma_j^A. \tag{3.21}$$

Similarly, the boundary conditions for $\tilde{\Gamma}_1^R$, Γ_1^A, and $\tilde{X}_1^K$ take the form:

$$\tilde{\Gamma}_1^R = \tilde{r}_{1l}^R \tilde{\gamma}_1^R S_{11} + \tilde{t}_{1l}^R \tilde{\gamma}_{1'}^R S_{1'1}, \tag{3.22}$$

$$\Gamma_1^A = S_{11}^+ \gamma_1^A r_{1r}^A + S_{1'1}^+ \gamma_{1'}^A t_{1r}^A, \tag{3.23}$$

$$\tilde{X}_1^K = \tilde{r}_{1l}^R \tilde{x}_1^K r_{1r}^A + \tilde{t}_{1l}^R \tilde{x}_{1'}^K t_{1r}^A - \tilde{a}_{1l}^R x_{1'}^K a_{1r}^A, \tag{3.24}$$

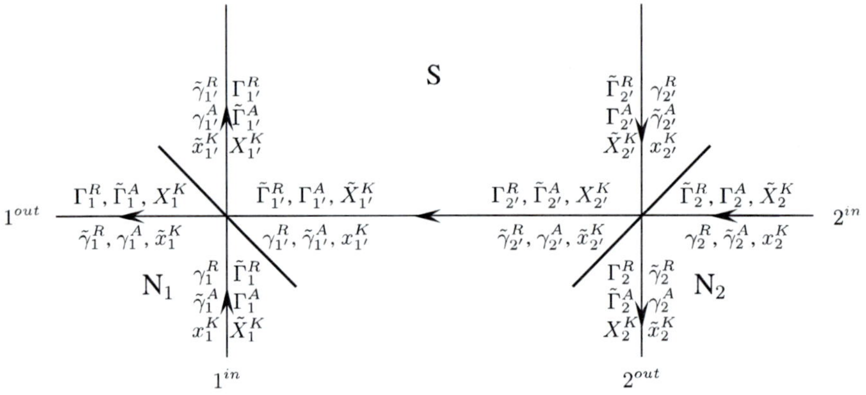

Fig. 3.4 Riccati amplitudes for incoming and outgoing trajectories for an NSN structure with two barriers. The *arrows* define quasiparticle momentum directions. We also indicate relevant Riccati amplitudes and distribution functions parameterizing the Green–Keldysh function for the corresponding trajectory

where

$$\tilde r^R_{1l} = +[(\tilde\beta^R_{1'1})^{-1}\underline{S}_{11} - (\tilde\beta^R_{1'1'})^{-1}\underline{S}_{1'1}]^{-1}(\tilde\beta^R_{1'1})^{-1}, \tag{3.25}$$

$$t^R_{1l} = -[(\tilde\beta^R_{1'1})^{-1}\underline{S}_{11} - (\tilde\beta^R_{1'1'})^{-1}\underline{S}_{1'1}]^{-1}(\tilde\beta^R_{1'1'})^{-1}, \tag{3.26}$$

$$r^A_{1r} = +(\tilde\beta^A_{1'1})^{-1}[\underline{S}^+_{11}(\tilde\beta^A_{1'1})^{-1} - \underline{S}^+_{1'1}(\tilde\beta^A_{1'1'})^{-1}]^{-1}, \tag{3.27}$$

$$\tilde t^A_{1r} = -(\tilde\beta^A_{1'1'})^{-1}[\underline{S}^+_{11}(\tilde\beta^A_{1'1})^{-1} - \underline{S}^+_{1'1}(\tilde\beta^A_{1'1'})^{-1}]^{-1}, \tag{3.28}$$

$$\tilde a^R_{1l} = (\tilde\Gamma^R_1 S^+_{11} - \underline{S}^+_{11}\tilde\gamma^R_1)(\beta^R_{11'})^{-1}, \tag{3.29}$$

$$a^A_{1r} = (\beta^A_{11'})^{-1}(S_{11}\Gamma^A_1 - \gamma^A_1\underline{S}_{11}). \tag{3.30}$$

Boundary conditions for $\Gamma^{R,A}_{1'}$, $\tilde\Gamma^{R,A}_{1'}$, $X^K_{1'}$ and $\tilde X^K_{1'}$ can be obtained from the above equations simply by replacing $1 \leftrightarrow 1'$.

The matrices S_{11}, $S_{11'}$, $S_{1'1}$ and $S_{1'1'}$ constitute the components of the $\mathcal{S}$-matrix describing electron scattering at the first interface:

$$\mathcal{S} = \begin{pmatrix} S_{11} & S_{11'} \\ S_{1'1} & S_{1'1'} \end{pmatrix}, \quad \mathcal{S}\mathcal{S}^+ = 1 \tag{3.31}$$

In our three terminal geometry non-local conductance arises only from trajectories that cross both interfaces, as illustrated in Fig. 3.4. Accordingly, the above boundary conditions should be employed at both NS interfaces.

Finally, one needs to specify the asymptotic boundary conditions far from NS interfaces. Deep in metallic electrodes we have

$$\gamma_1^R = \tilde{\gamma}_1^R = \gamma_1^A = \tilde{\gamma}_1^A = 0, \tag{3.32}$$

$$x_1^K = h_0(\varepsilon + eV_1), \quad \tilde{x}_1^K = -h_0(\varepsilon - eV_1), \tag{3.33}$$

$$\gamma_2^R = \tilde{\gamma}_2^R = \gamma_2^A = \tilde{\gamma}_2^A = 0, \tag{3.34}$$

$$x_2^K = h_0(\varepsilon + eV_2), \quad \tilde{x}_2^K = -h_0(\varepsilon - eV_2), \tag{3.35}$$

where $h_0(\varepsilon) = \tanh(\varepsilon/2T)$ – equilibrium distribution function. In the bulk of superconducting electrode, we have

$$\tilde{\gamma}_{1'}^R = -a(\varepsilon)\mathrm{i}\sigma_2, \quad \gamma_{1'}^A = a^*(\varepsilon)\mathrm{i}\sigma_2, \tag{3.36}$$

$$\tilde{x}_{1'}^K = -[1 - |a(\varepsilon)|^2]h_0(\varepsilon), \tag{3.37}$$

$$\gamma_{2'}^R = a(\varepsilon)\mathrm{i}\sigma_2, \quad \tilde{\gamma}_{2'}^A = -a^*(\varepsilon)\mathrm{i}\sigma_2, \tag{3.38}$$

$$x_{2'}^K = [1 - |a(\varepsilon)|^2]h_0(\varepsilon), \tag{3.39}$$

where we denoted $a(\varepsilon) = -(\varepsilon - \sqrt{\varepsilon^2 - \Delta^2})/\Delta$.

3.2.4 Green Functions

With the aid of the above equations and boundary conditions, it is straightforward to evaluate the quasiclassical Green–Keldysh functions for our three-terminal device along any trajectory of interest. For instance, from the boundary conditions at the second interface, we find

$$\Gamma_{2'}^R = \mathrm{i}a(\varepsilon)A_2\sigma_2, \tag{3.40}$$

where $A_2 = S_{2'2'}\sigma_2\underline{S}_{2'2'}^+\sigma_2$. Integrating (3.7) along the trajectory connecting both interfaces and using (3.40) as the initial condition we immediately evaluate the Riccati amplitude at the first interface:

$$\gamma_{1'}^R = i\frac{aA_2 + (aA_2\varepsilon + \Delta)Q}{1 - (aA_2\Delta + \varepsilon)Q}\sigma_2, \tag{3.41}$$

$$Q = \frac{\tanh\left[\mathrm{i}\Omega L/v_F\right]}{\Omega}, \quad \Omega = \sqrt{\varepsilon^2 - \Delta^2}. \tag{3.42}$$

Employing the boundary conditions again we obtain

$$\Gamma_1^R = \mathrm{i}S_{11'}K_{21}^{-1}\left[aA_2 + (aA_2\varepsilon + \Delta)Q\right]\sigma_2\underline{S}_{11'}^+, \tag{3.43}$$

$$\tilde{\Gamma}_1^R = -\mathrm{i}a\underline{S}_{1'1}^+\sigma_2 S_{1'1'}K_{21}^{-1}\left[1 - (aA_2\Delta + \varepsilon)Q\right]S_{1'1'}^{-1}S_{1'1}, \tag{3.44}$$

where

$$K_{ij} = (1 - a^2 A_i A_j) - \left[\varepsilon(1 + a^2 A_i A_j) + \Delta a(A_i + A_j)\right] Q, \tag{3.45}$$

$$A_1 = \sigma_2 \underline{S}^+_{1'1'} \sigma_2 S_{1'1'}. \tag{3.46}$$

We also note that the relation $(\Gamma^{R,A})^+ = \tilde{\Gamma}^{A,R}$ and $(\gamma^{R,A})^+ = \tilde{\gamma}^{A,R}$ makes it unnecessary (while redundant) to separately calculate the advanced Riccati amplitudes.

Let us now evaluate the distribution functions at both interfaces. With the aid of the boundary conditions at the second interface, we obtain

$$\begin{aligned} X^K_{2'} &= S_{2'2'} S^+_{2'2'} \left(1 - |a|^2\right) h_0(\varepsilon) + S_{2'2} S^+_{2'2} x^K_2 \\ &\quad - |a|^2 S_{2'2'} \sigma_2 \underline{S}^+_{22'} \underline{S}_{22'} \sigma_2 S^+_{2'2'} \tilde{x}^K_2 . \end{aligned} \tag{3.47}$$

Integrating (3.9) along the trajectory connecting both interfaces with initial condition for $X^K_{2'}$, we arrive at the expression for $x^K_{1'}$

$$\begin{aligned} x^K_{1'} &= [1 - (aA_2\Delta + \varepsilon)Q]^{-1} X^K_{2'} \\ &\quad \times (1 - \tanh^2 iL\Omega/v_F) \, [1 - (aA_2\Delta + \varepsilon)Q]^{+^{-1}} . \end{aligned} \tag{3.48}$$

Then we can find distribution functions at the first interface. On the normal metal side of the interface, we find

$$X^K_1 = r^R_{1l} x^K_1 r^{R+}_{1l} + t^R_{1l} x^K_{1'} t^{R+}_{1l} + a^R_{1l} a^{R+}_{1l} \left(1 - |a|^2\right) h_0(\varepsilon) \tag{3.49}$$

where

$$\begin{aligned} r^R_{1l} &= S_{11'} K^{-1}_{21} \Big[(1 - (aA_2\Delta + \varepsilon)Q) S^+_{1'1'} S^{+^{-1}}_{1'1} \\ &\quad - a(aA_2 + (aA_2\varepsilon + \Delta)Q) \sigma_2 \underline{S}^+_{1'1'} \sigma_2 S^{+^{-1}}_{1'1} \Big], \end{aligned} \tag{3.50}$$

$$t^R_{1l} = S_{11'} K^{-1}_{21} (1 - (aA_2\Delta + \varepsilon)Q), \tag{3.51}$$

$$a^R_{1l} = i S_{11'} K^{-1}_{21} (aA_2 + (aA_2\varepsilon + \Delta)Q) \sigma_2 \underline{S}^+_{1'1'}. \tag{3.52}$$

The corresponding expression for $\tilde{X}^K_1$ is obtained analogously. We get

$$\tilde{X}^K_1 = \tilde{r}^R_{1l} \tilde{x}^K_1 \tilde{r}^{R+}_{1l} - \tilde{t}^R_{1l} \tilde{t}^{R+}_{1l} \left(1 - |a|^2\right) h_0(\varepsilon) - \tilde{a}^R_{1l} x^K_{1'} \tilde{a}^{R+}_{1l} , \tag{3.53}$$

where

$$\begin{aligned} \tilde{r}^R_{1l} &= -\Big[\underline{S}^{-1}_{1'1} \underline{S}_{1'1'} \sigma_2 (1 - (aA_2\Delta + \varepsilon)Q) \\ &\quad - \underline{S}^{-1}_{1'1} \sigma_2 S_{1'1'} a(aA_2 + (aA_2\varepsilon + \Delta)Q) \Big] K^{-1}_{12} \sigma_2 \underline{S}^+_{11'}, \end{aligned} \tag{3.54}$$

$$\tilde{t}_{1l}^{R} = \underline{S}_{1'1}^{+}\underline{S}_{1'1'}^{+-1}\sigma_2(1-(aA_2\Delta+\varepsilon)Q)K_{12}^{-1}\sigma_2\underline{S}_{1'1'}^{+}, \tag{3.55}$$

$$\tilde{a}_{1l}^{R} = ia\underline{S}_{1'1}^{+}\sigma_2 S_{1'1'}K_{21}^{-1}(1-(aA_2\Delta+\varepsilon)Q). \tag{3.56}$$

Combining the above results for the Riccati amplitudes and the distribution functions, we can easily evaluate the Keldysh–Green function at the first interface. For instance, for the trajectory 1^{out} (see Fig. 3.4), we obtain

$$g_{1^{\text{out}}}^{K} = 2(X_1^K - \Gamma_1^R\tilde{x}_1^K\Gamma_1^{R+}), \quad \tilde{g}_{1^{\text{out}}}^{K} = 2\tilde{x}_1^K. \tag{3.57}$$

The Keldysh–Green function for the trajectory 1^{in} is evaluated analogously, and we get

$$g_{1^{\text{in}}}^{K} = 2x_1^K, \quad \tilde{g}_{1^{\text{in}}}^{K} = 2(\tilde{X}_1^K - \tilde{\Gamma}_1^R x_1^K\tilde{\Gamma}_1^{R+}). \tag{3.58}$$

3.2.5 *Non-local Conductance: General Results*

Now we are ready to evaluate the current I_1 across the first interface. This current takes the form:

$$I_1 = I_1^{\text{BTK}}(V_1) - \frac{G_0}{8e}\int d\varepsilon\, \text{Sp}(\hat{\tau}_3\hat{g}_{1^{\text{out}}}^{K} - \hat{\tau}_3\hat{g}_{1^{\text{in}}}^{K}), \tag{3.59}$$

where

$$G_0 = \frac{8\gamma_1\gamma_2\mathcal{N}_1\mathcal{N}_2}{R_q p_F^2 L^2} \tag{3.60}$$

is the normal state conductance of our device at fully transparent interfaces, $p_F\gamma_{1(2)}$ is normal to the first (second) interface component of the Fermi momentum for electrons propagating straight between the interfaces, $\mathcal{N}_{1,2} = p_F^2\mathcal{A}_{1,2}/4\pi$ define the number of conducting channels of the corresponding interface, $R_q = 2\pi/\text{e}^2$ is the quantum resistance unit.

Here, $I_1^{\text{BTK}}(V_1)$ stands for the contribution to the current through the first interface coming from trajectories that never cross the second interface. This is just the standard BTK contribution [2, 35]. The non-trivial contribution is represented by the last term in (3.59), which accounts for the presence of the second NS interface. We observe that this non-local contribution to the current is small as $\propto 1p_F^2L^2$. This term will be analyzed in detail below.

The functions $\hat{g}_{1^{\text{in}}}^{K}$ and $\hat{g}_{1^{\text{out}}}^{K}$ are the Keldysh Green functions evaluated on the trajectories 1^{in} and 1^{out}, respectively. Using the above expression for the Riccati amplitudes and the distribution functions, we find

$$
\begin{aligned}
\mathrm{Sp}(\hat{\tau}_3\hat{g}^K_{1\mathrm{out}} - \hat{\tau}_3\hat{g}^K_{1\mathrm{in}}) &= 2\,\mathrm{Sp}[r^R_{1l} r^{R+}_{1l} - \tilde{\Gamma}^R_1 \tilde{\Gamma}^{R+}_1 - 1](h_0(\varepsilon + eV_1) - h_0(\varepsilon)) \\
&\quad - 2\,\mathrm{Sp}[\tilde{r}^R_{1l} \tilde{r}^{R+}_{1l} - \Gamma^R_1 \Gamma^{R+}_1 - 1](h_0(\varepsilon - eV_1) - h_0(\varepsilon)) \\
&\quad 2(1 - \tanh^2 \mathrm{i}L\Omega/v_F)\,\mathrm{Sp}[K^{-1}_{21}\{S_{2'2}S^+_{2'2}(h_0(\varepsilon + eV_2) - h_0(\varepsilon)) \\
&\quad + |a|^2 S_{2'2'}\sigma_2 \underline{S}^+_{22'} \underline{S}_{22'}\sigma_2 S^+_{2'2'}(h_0(\varepsilon - eV_2) - h_0(\varepsilon))\}K^{+-1}_{21} \\
&\quad \times (S^+_{11'}S_{11'} - |a^2| S^+_{1'1'}\sigma_2 \underline{S}_{1'1} \underline{S}^+_{1'1}\sigma_2 S_{1'1'})],
\end{aligned}
\tag{3.61}
$$

where we explicitly used the fact that in equilibrium $\mathrm{Sp}(\hat{\tau}_3\hat{g}^K_{1\mathrm{out}} - \hat{\tau}_3\hat{g}^K_{1\mathrm{in}}) \equiv 0$. Substituting (3.61) into (3.59), we finally obtain

$$
I_1 = I^{\mathrm{BTK}}_1(V_1) + I_{11}(V_1) + I_{12}(V_2). \tag{3.62}
$$

The correction to the local BTK current (arising from trajectories crossing also the second NS interface) has the following form

$$
\begin{aligned}
I_{11}(V_1) = -\frac{G_0}{4e}\int \mathrm{d}\varepsilon \{ &\mathrm{Sp}[r^R_{1l} r^{R+}_{1l} - \tilde{\Gamma}^R_1 \tilde{\Gamma}^{R+}_1 - 1](h_0(\varepsilon + eV_1) - h_0(\varepsilon)) \\
- &\mathrm{Sp}[\tilde{r}^R_{1l} \tilde{r}^{R+}_{1l} - \Gamma^R_1 \Gamma^{R+}_1 - 1](h_0(\varepsilon - eV_1) - h_0(\varepsilon))\},
\end{aligned}
\tag{3.63}
$$

while for the cross-current we obtain

$$
\begin{aligned}
I_{12}(V_2) = -\frac{G_0}{4e}\int &\mathrm{d}\varepsilon (1 - \tanh^2 \mathrm{i}L\Omega/v_F) \\
\times\ &\mathrm{Sp}[K^{-1}_{21}\{S_{2'2}S^+_{2'2}(h_0(\varepsilon + eV_2) - h_0(\varepsilon)) \\
+\ &|a|^2 S_{2'2'}\sigma_2 \underline{S}^+_{22'} \underline{S}_{22'}\sigma_2 S^+_{2'2'}(h_0(\varepsilon - eV_2) - h_0(\varepsilon))\}K^{+-1}_{21} \\
\times\ &(S^+_{11'}S_{11'} - |a^2| S^+_{1'1'}\sigma_2 \underline{S}_{1'1} \underline{S}^+_{1'1}\sigma_2 S_{1'1'})].
\end{aligned}
\tag{3.64}
$$

Equations (3.62)–(3.64) fully determine the current across the first interface at arbitrary voltages, temperature and spin-dependent interface transmissions.

In right-hand side of (3.64), we can distinguish four contributions with different products of S-matrices. Each of these terms corresponds to a certain sequence of elementary events, such as transmission, reflection, Andreev reflection and propagation between interfaces. Diagrammatic representation of these four terms is offered in Fig. 3.5. The amplitude of each of the processes is given by the product of the amplitudes of the corresponding elementary events. For instance, the amplitude of the process in Fig. 3.5c is $f = -\mathrm{i}S_{11'}t_{21}S_{2'2'}a\sigma_2\underline{S}^+_{22'}$. In (3.64), this process is identified by the term $\mathrm{Sp}(ff^+)$ with the hole distribution function as a prefactor. It is straightforward to observe that the processes of Fig. 3.5a, b and d correspond to the other three terms in (3.64). We also note that the processes of

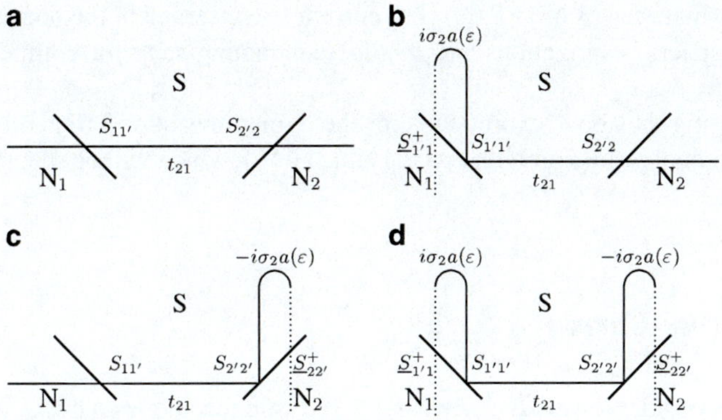

Fig. 3.5 Diagrams representing four different contributions to the cross-current I_{12} (3.64). Solid (*dotted*) lines correspond to propagating electron-like (hole-like) excitations and $t_{21} = K_{21}^{-1}/\cosh(\mathrm{i}L\Omega/v_F)$

Fig. 3.5a, d describe direct electron (hole) transport, while the processes of Fig. 3.5b, c correspond to the contribution of CAR.

Assuming that both interfaces possess inversion symmetry and reflection symmetry in the plane normal to the corresponding interface, we can choose S-matrices in the following form

$$S_{11} = S_{1'1'} = \underline{S}_{11}^{T} = \underline{S}_{1'1'}^{T} = U(\varphi)\begin{pmatrix} \sqrt{R_{1\uparrow}}\mathrm{e}^{\mathrm{i}\theta_1/2} & 0 \\ 0 & \sqrt{R_{1\downarrow}}\mathrm{e}^{-\mathrm{i}\theta_1/2} \end{pmatrix} U^{+}(\varphi), \quad (3.65)$$

$$S_{11'} = S_{1'1} = \underline{S}_{11'}^{T} = \underline{S}_{1'1}^{T} = U(\varphi)\mathrm{i}\begin{pmatrix} \sqrt{D_{1\uparrow}}\mathrm{e}^{\mathrm{i}\theta_1/2} & 0 \\ 0 & \sqrt{D_{1\downarrow}}\mathrm{e}^{-\mathrm{i}\theta_1/2} \end{pmatrix} U^{+}(\varphi), \quad (3.66)$$

$$S_{22} = S_{2'2'} = \underline{S}_{22} = \underline{S}_{2'2'} = \begin{pmatrix} \sqrt{R_{2\uparrow}}\mathrm{e}^{\mathrm{i}\theta_2/2} & 0 \\ 0 & \sqrt{R_{2\downarrow}}\mathrm{e}^{-\mathrm{i}\theta_2/2} \end{pmatrix}, \quad (3.67)$$

$$S_{22'} = S_{2'2} = \underline{S}_{22'} = \underline{S}_{2'2} = \mathrm{i}\begin{pmatrix} \sqrt{D_{2\uparrow}}\mathrm{e}^{\mathrm{i}\theta_2/2} & 0 \\ 0 & \sqrt{D_{2\downarrow}}\mathrm{e}^{-\mathrm{i}\theta_2/2} \end{pmatrix}. \quad (3.68)$$

Here, $R_{1(2)\uparrow(\downarrow)} = 1 - D_{1(2)\uparrow(\downarrow)}$ are the spin-dependent reflection coefficients of both NS interfaces, $\theta_{1,2}$ are spin-mixing angles and $U(\varphi)$ is the rotation matrix in the spin space, which depends on the angle φ between polarizations of the two interfaces,

$$U(\varphi) = \exp(-\mathrm{i}\varphi\sigma_1/2) = \begin{pmatrix} \cos(\varphi/2) & -\mathrm{i}\sin(\varphi/2) \\ -\mathrm{i}\sin(\varphi/2) & \cos(\varphi/2) \end{pmatrix}. \quad (3.69)$$

In general, spin current is not conserved in heterostructures with spin-active interfaces. However, single barrier with S-matrix (3.67)–(3.68) does not violate spin current conservation [41]. It is easy to show that in our two barrier structure with interface S-matrices (3.65)–(3.68) spin current is conserved in the normal state for arbitrary barriers polarizations and in superconducting state for collinear barriers polarizations.

Substituting the above expressions for the S-matrices into (3.63) and (3.64), we arrive at the final results for both $I_{11}(V_1)$ and $I_{12}(V_2)$, which will be specified further below.

3.2.6 Cross-Current

First let us consider the cross-current $I_{12}(V_2)$. From the above analysis, we obtain

$$
\begin{aligned}
I_{12}(V_2) = -\frac{G_0}{4e}\int d\varepsilon \left[\tanh\frac{\varepsilon+eV_2}{2T} - \tanh\frac{\varepsilon}{2T}\right]\frac{1-\tanh^2 iL\Omega/v_F}{W(z_1,z_2,\varepsilon,\varphi)} \\
\times \Big\{\big[D_{1\downarrow}D_{2\downarrow} - |a|^2 D_{1\uparrow}D_{2\downarrow}(R_{1\downarrow}+R_{2\uparrow}) + |a|^4 D_{1\downarrow}R_{1\uparrow}D_{2\downarrow}R_{2\uparrow}\big]|K(z_1,z_2,\varepsilon)|^2\tilde{c} \\
+ \big[D_{1\uparrow}D_{2\uparrow} - |a|^2 D_{1\downarrow}D_{2\uparrow}(R_{1\uparrow}+R_{2\downarrow}) + |a|^4 D_{1\uparrow}R_{1\downarrow}D_{2\uparrow}R_{2\downarrow}\big]|K(z_1^*,z_2^*,\varepsilon)|^2\tilde{c} \\
+ \big[D_{1\uparrow}D_{2\downarrow} - |a|^2 D_{1\downarrow}D_{2\downarrow}(R_{1\uparrow}+R_{2\uparrow}) + |a|^4 D_{1\uparrow}R_{1\downarrow}D_{2\downarrow}R_{2\uparrow}\big]|K(z_1^*,z_2,\varepsilon)|^2\tilde{s} \\
+ \big[D_{1\downarrow}D_{2\uparrow} - |a|^2 D_{1\uparrow}D_{2\uparrow}(R_{1\downarrow}+R_{2\downarrow}) + |a|^4 D_{1\downarrow}R_{1\uparrow}D_{2\uparrow}R_{2\downarrow}\big]|K(z_1,z_2^*,\varepsilon)|^2\tilde{s}\Big\},
\end{aligned}
\tag{3.70}
$$

where we define $\tilde{c} = \cos^2(\varphi/2)$, $\tilde{s} = \sin^2(\varphi/2)$,

$$
K(z_1,z_2,\varepsilon) = (1-a^2 z_1 z_2) - \left[\varepsilon(1+a^2 z_1 z_2) + \Delta a(z_1+z_2)\right] Q, \tag{3.71}
$$

$$
\begin{aligned}
W(z_1,z_2,\varepsilon,\varphi) = |K(z_1,z_2,\varepsilon)K(z_1^*,z_2^*,\varepsilon)\cos^2(\varphi/2) \\
+ K(z_1^*,z_2,\varepsilon)K(z_1,z_2^*,\varepsilon)\sin^2(\varphi/2)|^2
\end{aligned}
\tag{3.72}
$$

and $z_i = \sqrt{R_{i\uparrow}R_{i\downarrow}}\exp(i\theta_i)$ ($i = 1, 2$).

Equation (3.70) represents our central result. It fully determines the non-local spin-dependent current in our three-terminal ballistic NSN structure at arbitrary voltages, temperature, interface transmissions and polarizations.

Let us introduce the non-local differential conductance

$$
G_{12}(V_2) = -\frac{\partial I_1}{\partial V_2} = -\frac{\partial I_{12}(V_2)}{\partial V_2}. \tag{3.73}
$$

Before specifying this quantity further it is important to observe that in general the conductance $G_{12}(V_2)$ is not an even function of the applied voltage V_2. This asymmetry arises due to formation of Andreev bound states in the vicinity of a spin-active interface [42, 43]. It disappears provided the spin-mixing angles θ_1 and θ_2 remain equal to 0 or π.

In the normal state, we have $I_{12}(V_2) = -G_{N_{12}} V_2$, where

$$G_{N_{12}} = \frac{G_0}{2}\big[(D_{1\downarrow} D_{2\downarrow} + D_{1\uparrow} D_{2\uparrow}) \cos^2(\varphi/2) + (D_{1\uparrow} D_{2\downarrow} + D_{1\downarrow} D_{2\uparrow}) \sin^2(\varphi/2)\big]. \tag{3.74}$$

Turning to the superconducting state, let us consider the limit of low temperatures and voltage $T, V_2 \ll \Delta$. In this limit, only subgap quasiparticles contribute to the cross-current and the differential conductance becomes voltage-independent, i.e. $I_{12} = -G_{12} V_2$, where

$$G_{12} = G_0(1 - \tanh^2 L\Delta/v_F)\left\{ \frac{D_{1\uparrow} D_{1\downarrow} D_{2\uparrow} D_{2\downarrow}}{|K(z_1, z_2, 0)|^2 \tilde{c} + |K(z_1, z_2^*, 0)|^2 \tilde{s}} + (D_{1\uparrow} - D_{1\downarrow})(D_{2\uparrow} - D_{2\downarrow}) \frac{|K(z_1, z_2, 0)|^2 \tilde{c} - |K(z_1, z_2^*, 0)|^2 \tilde{s}}{\left(|K(z_1, z_2, 0)|^2 \tilde{c} + |K(z_1, z_2^*, 0)|^2 \tilde{s}\right)^2} \right\}. \tag{3.75}$$

where, as before, $\tilde{c} = \cos^2(\varphi/2)$ and $\tilde{s} = \sin^2(\varphi/2)$. In the case of spin-isotropic interfaces, (3.75) and (3.70) reduce to the results [15].

Provided at least one of the interfaces is spin-isotropic, the conductance (3.75) is proportional to the product of all four transmissions $D_{1\uparrow} D_{1\downarrow} D_{2\uparrow} D_{2\downarrow}$, i.e. it differs from zero only due to processes involving scattering with both spin projections at both NS interfaces. As in the case of spin-isotropic interfaces [15], the value G_{12} (3.75) gets strongly suppressed with increasing L, and at sufficiently high interface transmissions this dependence is in general non-exponential in L. In the spin-degenerate case for a given L the non-local conductance reaches its maximum for reflectionless barriers $D_{1,2} = 1$. In this case, we arrive at a simple formula

$$G_{12} = G_0(1 - \tanh^2 L\Delta/v_F). \tag{3.76}$$

We observe that for small $L \ll v_F/\Delta$ the conductance G_{12} identically coincides with its normal state value $G_{N_{12}} \equiv G_0$ at any temperature and voltage [15]. This result implies that CAR *vanishes for fully open barriers*. Actually, this conclusion is general and applies not only to small but also to any value of L, i.e. the result (3.76) is determined solely by the process of direct electron transfer between N-terminals for all L.

At the first sight, this result might appear counterintuitive since the behaviour of ordinary (local) AR is just the opposite: It reaches its maximum at full barrier transmissions. The physics behind vanishing of CAR for perfectly transparent NS interfaces is simple. One observes (cf. Fig. 3.1) that CAR inevitably implies the flow of Cooper pairs out of the contact area into the superconducting terminal. This flow is described by electron trajectories which end deep in the superconductor. On the contrary, it is obvious that CAR requires "mixing" of these trajectories with those going straight between two normal terminals. Provided there exists no normal electron reflection at both NS interfaces such mixing does not occur, CAR vanishes and the only remaining contribution to the non-local conductance is one from direct electron transfer between N-terminals.

This situation is illustrated by the diagrams in Fig. 3.5. It is obvious that in the case of non-reflecting NS interfaces, only the process of Fig. 3.5a survives, whereas all other processes (Fig. 3.5b, c and d) vanish for reflectionless barriers with $R_{1(2)\uparrow(\downarrow)} = 0$. The situation changes provided at least one of the transmissions is smaller than one. In this case, scattering at SN interfaces mixes up trajectories connecting N_1 and N_2 terminals with ones going deep into and coming from the superconductor. As a result, all four processes depicted in Fig. 3.5 contribute to the cross-current and CAR contribution to G_{12} does not vanish.

In the limit $|eV_2|, T \ll \Delta$ and at zero spin-mixing angles $\theta_{1,2} = 0$ from (3.75), we obtain

$$G_{12} = G_0 \frac{1 - \tanh^2 L\Delta/v_F}{|K(z_1, z_2, 0)|^2} \{D_{1\uparrow} D_{1\downarrow} D_{2\uparrow} D_{2\downarrow} + (D_{1\uparrow} - D_{1\downarrow})(D_{2\uparrow} - D_{2\downarrow}) \cos\varphi\}. \tag{3.77}$$

In the lowest (first order) order in the transmissions of both interfaces and for collinear interface polarizations (3.77) reduces to the result by Falci et al. [14] provided we identify the tunneling density of states $N_0 D_{1\uparrow}$, $N_0 D_{1\downarrow}$, $N_0 D_{2\uparrow}$, and $N_0 D_{2\downarrow}$ with the corresponding spin-resolved densities of states in the ferromagnetic electrodes. For zero spin-mixing angles and low voltages, the L-dependence of the non-local conductance G_{12} reduces to the exponential form $G_{12} \propto \exp(-2L\Delta/v_F)$ either in the limit of small transmissions or large $L \gg v_F/\Delta$.

At arbitrary voltages and temperatures, the cross-current has a simple φ dependence in the limit of zero spin mixing angles ($\theta_{1,2} = 0$)

$$I_{12}(\varphi, V_2) = I_{12}(\varphi = 0, V_2) \cos^2(\varphi/2) + I_{12}(\varphi = \pi, V_2) \sin^2(\varphi/2), \tag{3.78}$$

i.e. in this limit at any φ the non-local current is equal to a proper superposition of the two contributions corresponding to parallel ($\varphi = 0$) and antiparallel ($\varphi = \pi$) interface polarizations. Some typical curves for the differential non-local conductance are presented in Fig. 3.6 at sufficiently high interface transmissions and zero spin mixing angles $\theta_{1,2} = 0$.

Let us now turn to the limit of highly polarized interfaces, which is accounted for by taking the limit of vanishing spin-up (or spin-down) transmission of each interface. In this limit, our model describes an HSH structure, where H stands for fully spin-polarized half-metallic electrodes. In this case, we obtain ($D_{1\uparrow} = D_1$, $D_{1\downarrow} = 0$, $D_{2\uparrow} = D_2$, and $D_{2\downarrow} = 0$)

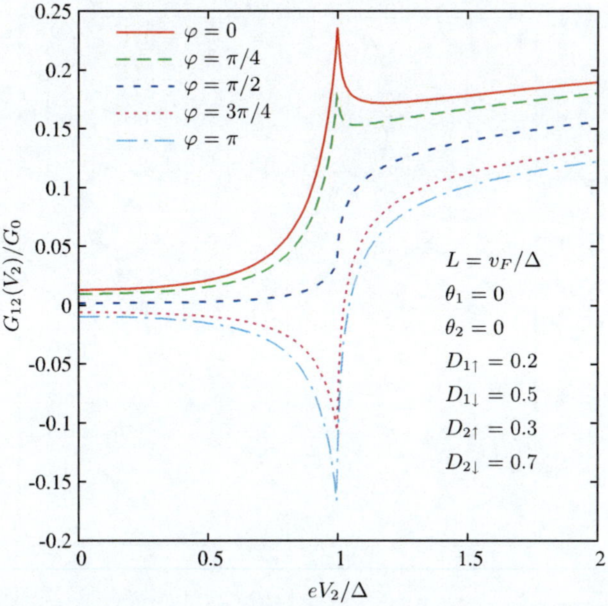

Fig. 3.6 Zero temperature differential non-local conductance as a function of voltage at zero spin-mixing angles $\theta_{1,2} = 0$.

$$I_{12}(V_2) = -\frac{G_0}{4e}\int d\varepsilon\,[h_0(\varepsilon + eV_2) - h_0(\varepsilon)]\,\frac{1-\tanh^2 iL\Omega/v_F}{W(z_1, z_2, \varepsilon, \varphi)}D_1D_2$$
$$\times\Big\{\big[1+|a|^4\big]\,|K(z_1^*, z_2^*, \varepsilon)|^2\tilde{c} - 2|a|^2|K(z_1, z_2^*, \varepsilon)|^2\tilde{s}\Big\}. \tag{3.79}$$

We observe that the non-local conductance has *opposite signs* for parallel ($\varphi = 0$) and antiparallel ($\varphi = \pi$) interface polarizations. We also emphasize that, as it is also clear from (3.77), the cross-conductance G_{12} of HSH structures – in contrast to that for NSN structures – does not vanish already in the lowest order in barrier transmissions $D_{1\uparrow}D_{2\uparrow}$.

In general, the non-local conductance is very sensitive to particular values of the spin-mixing angles θ_1 and θ_2, as illustrated, e.g., in Fig. 3.7. Comparing the voltage dependencies of the non-local conductance evaluated for the same transmissions and presented in Figs. 3.6 and 3.7, we observe that they can differ drastically at zero and non-zero values of $\theta_{1,2}$.

At low voltages and temperatures and at zero-spin mixing angles, the non-local conductance of HSH structures is determined by (3.77) with $D_{1\downarrow} = D_{2\downarrow} = 0$. For fully open barriers (for "spin-up" electrons) $D_{1\uparrow} = D_{2\uparrow} = 1$, we obtain

$$G_{12} = G_0(1 - \tanh^2 L\Delta/v_F)\cos\varphi. \tag{3.80}$$

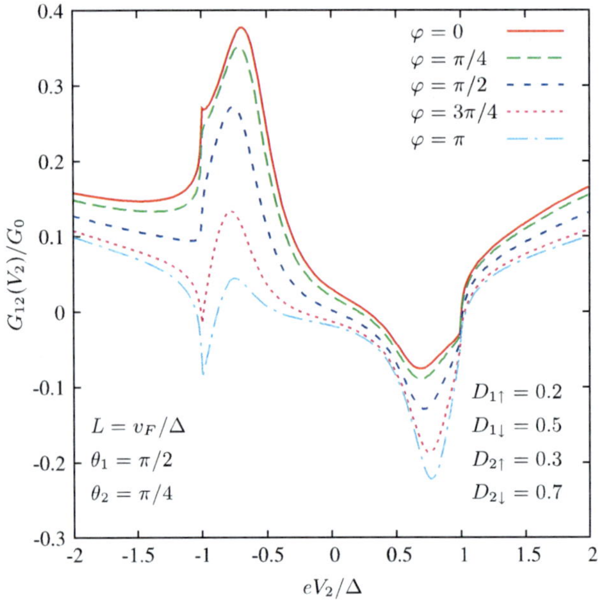

Fig. 3.7 The same as in Fig. 3.6 for $\theta_1 = \pi/2$, $\theta_2 = \pi/4$

Interestingly, for $\varphi = 0$ this expression exactly coincides with that for fully open NSN structures, (3.76). At the same time for small L, the result (3.80) turns out to be 2 times bigger than the analogous expression in the normal case, i.e. for fully open HNH structures, cf. (3.74). This result can easily be interpreted in terms of diagrams in Fig. 3.5. We observe that – exactly as for the spin degenerate case – CAR diagrams of Fig. 3.5b,c vanish for reflectionless barriers, whereas diagrams of Fig. 3.5a,d describing direct electron transfer survive and both contribute to G_{12}. Thus, *CAR vanishes identically also for fully open HSH structures*. The factor of 2 difference with the normal case is due to the fact that the diagram of Fig. 3.5d vanishes in the normal limit.

3.2.7 Correction to BTK

Using the above formalism, one can easily generalize the BTK result to the case of spin-polarized interfaces [35]. For the first interface, we have

$$I_1^{\mathrm{BTK}}(V_1) = \frac{\mathcal{N}_1}{R_q e} \int d\varepsilon [h_0(\varepsilon + eV_1) - h_0(\varepsilon)](1 + |a|^2)$$
$$\times \left\langle \frac{|v_{x_1}|}{v_F} \left(D_{1\uparrow} \frac{1 - R_{1\downarrow}|a|^2}{|1 - z_1 a^2|^2} + D_{1\downarrow} \frac{1 - R_{1\uparrow}|a|^2}{|1 - z_1^* a^2|^2} \right) \right\rangle. \quad (3.81)$$

Here, transmission and reflection coefficients as well as the spin mixing angle depend on the direction of the Fermi momentum. In the spin-degenerate case, the above expression reduces to the standard BTK result [2].

Evaluating the non-local correction to the BTK current due to the presence of the second interface, we arrive at a somewhat lengthy general expression

$$
\begin{aligned}
I_{11}(V_1) = {} & \frac{G_0}{2e}\int d\varepsilon (h_0(\varepsilon + eV_1) - h_0(\varepsilon))\frac{1}{W(z_1, z_2, \varepsilon, \varphi)}\Big\{2W(z_1, z_2, \varepsilon, \varphi) \\
& - R_{1\uparrow}\left|\tilde{c}K(z_1/R_{1\uparrow}, z_2, \varepsilon)K(z_1^*, z_2^*, \varepsilon) + \tilde{s}K(z_1/R_{1\uparrow}, z_2^*, \varepsilon)K(z_1^*, z_2, \varepsilon)\right|^2 \\
& - R_{1\downarrow}\left|\tilde{c}K(z_1^*/R_{1\downarrow}, z_2^*, \varepsilon)K(z_1, z_2, \varepsilon) + \tilde{s}K(z_1^*/R_{1\downarrow}, z_2, \varepsilon)K(z_1, z_2^*, \varepsilon)\right|^2\Big\} \\
& + \frac{G_0}{4e}\int d\varepsilon (h_0(\varepsilon + eV_1) - h_0(\varepsilon))\frac{D_{1\uparrow}D_{1\downarrow}}{W(z_1, z_2, \varepsilon, \varphi)} \\
& \times \Big\{|a|^2\left|\tilde{c}K(0, z_2, \varepsilon)K(z_1^*, z_2^*, \varepsilon) + \tilde{s}K(0, z_2^*, \varepsilon)K(z_1^*, z_2, \varepsilon)\right|^2 \\
& + |a|^2\left|\tilde{c}K(0, z_2^*, \varepsilon)K(z_1, z_2, \varepsilon) + \tilde{s}K(0, z_2, \varepsilon)K(z_1, z_2^*, \varepsilon)\right|^2 \\
& + \frac{1}{|a|^2}\left|\tilde{c}K'(z_2^*, \varepsilon)K(z_1, z_2, \varepsilon) + \tilde{s}K'(z_2, \varepsilon)K(z_1, z_2^*, \varepsilon)\right|^2 \\
& + \frac{1}{|a|^2}\left|\tilde{c}K'(z_2, \varepsilon)K(z_1^*, z_2^*, \varepsilon) + \tilde{s}K'(z_2^*, \varepsilon)K(z_1^*, z_2, \varepsilon)\right|^2\Big\} \\
& + \frac{G_0}{e}R_{2\uparrow}R_{2\downarrow}\sin^2(\theta_2/2)\tilde{s}\tilde{c} \\
& \times \int d\varepsilon (h_0(\varepsilon + eV_1) - h_0(\varepsilon))\frac{(1 - \tanh^2 iL\Omega/v_F)^2}{W(z_1, z_2, \varepsilon, \varphi)} \\
& \times \Big[|a|^2(D_{1\uparrow}^2 + D_{1\downarrow}^2) - 2|a|^4 D_{1\uparrow}D_{1\downarrow}(R_{1\uparrow} + R_{1\downarrow}) \\
& + |a|^6(D_{1\uparrow}^2 R_{1\downarrow}^2 + D_{1\downarrow}^2 R_{1\uparrow}^2)\Big],
\end{aligned}
\tag{3.82}
$$

where $K'(z_2, \varepsilon) = \partial K(z_1, z_2, \varepsilon)/\partial z_1$. This expression gets significantly simplified in the limit of zero spin-mixing angles $\theta_{1,2} = 0$ in which case we obtain

$$
\begin{aligned}
I_{11}(V_1) = {} & \frac{G_0}{2e}\int d\varepsilon (h_0(\varepsilon + eV_1) - h_0(\varepsilon)) \\
& \times \Big\{2 - R_{1\uparrow}\frac{|K(z_1/R_{1\uparrow}, z_2, \varepsilon)|^2}{|K(z_1, z_2, \varepsilon)|^2} - R_{1\downarrow}\frac{|K(z_1/R_{1\downarrow}, z_2, \varepsilon)|^2}{|K(z_1, z_2, \varepsilon)|^2} \\
& + D_{1\uparrow}D_{1\downarrow}\frac{|a(\varepsilon)|^2|K(0, z_2, \varepsilon)|^2 + |K'(z_2, \varepsilon)|^2/|a(\varepsilon)|^2}{|K(z_1, z_2, \varepsilon)|^2}\Big\}.
\end{aligned}
\tag{3.83}
$$

In contrast to the expression for the cross-current I_{12} (cf. (3.78)), in the limit of zero spin-mixing angles the correction I_{11} to the BTK current does not depend on the angle φ between the interface polarizations. In particular, at $|eV_1|, T \ll \Delta$ we have $I_{11} = G_{11}V_1$ where

$$G_{11} = G_0(D_{1\uparrow} + D_{1\downarrow})\frac{(1-z_2^2)(1-\tanh^2 L\Delta/v_F)}{[1+z_1z_2+(z_1+z_2)\tanh L\Delta/v_F]^2}$$

$$+ G_0 D_{1\uparrow}D_{1\downarrow}\frac{(1+z_2\tanh L\Delta/v_F)^2 + 3(z_2+\tanh L\Delta/v_F)^2}{[1+z_1z_2+(z_1+z_2)\tanh L\Delta/v_F]^2}. \qquad (3.84)$$

In the tunneling limit $D_{1\uparrow}, D_{1\downarrow}, D_{2\uparrow}, D_{2\downarrow} \ll 1$ we reproduce the result of Ref. [14]

$$G_{11} = \frac{G_0}{4}(D_{1\uparrow} + D_{1\downarrow})(D_{2\uparrow} + D_{2\downarrow})\exp(-2L\Delta/v_F), \qquad (3.85)$$

which turns out to hold at any value φ.

As compared to the BTK conductance, the CAR correction (3.82) contains an extra small factor $\mathcal{A}_2/L^2$ and, hence, in many cases remains small and can be neglected. On the contrary, since CAR involves tunneling of *one* electron through each interface, for strongly asymmetric structures with $D_{1\uparrow}, D_{1\downarrow} \ll 1$ and $D_{2\uparrow}, D_{2\downarrow} \sim 1$, it can actually *strongly exceed* the BTK conductance. Indeed, for $D_{1\uparrow\downarrow} \ll 1$, $R_{2\uparrow}R_{2\downarrow} \ll 1$ and provided the spin mixing angle θ_1 is not very close to π from (3.82) we get

$$G_{11} = \frac{G_0(D_{1\uparrow} + D_{1\downarrow})}{\cosh(2L\Delta/v_F) + \cos\theta_1 \sinh(2L\Delta/v_F)}, \qquad (3.86)$$

i.e. for

$$\frac{D_{1\uparrow}D_{1\downarrow}}{(D_{1\uparrow} + D_{1\downarrow})} < \frac{\mathcal{A}_2}{L^2}\exp(-2L\Delta/v_F)$$

the contribution (3.86) may well exceed the BTK term $G_1^{\mathrm{BTK}} \propto D_{1\uparrow}D_{1\downarrow}$. The existence of such a non-trivial regime further emphasizes the importance of the mechanism of non-local Andreev reflection in multi-terminal hybrid NSN structures.

3.3 Diffusive FSF Structures

Let us now turn to the effect of disorder. In what follows we will consider a three-terminal diffusive FSF structure schematically shown in Fig. 3.8. Two ferromagnetic terminals F_1 and F_2 with resistances r_{N_1} and r_{N_2} and electric potentials V_1 and V_2 are connected to a superconducting electrode of length L with normal state (Drude) resistance r_L and electric potential $V = 0$ via tunnel barriers. The magnitude

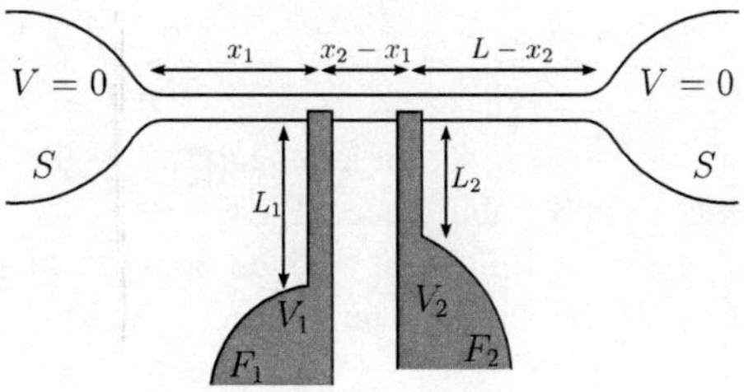

Fig. 3.8 FSF structure under consideration

of the exchange field $h_{1,2} = |\boldsymbol{h}_{1,2}|$ in both ferromagnets F_1 and F_2 is assumed to be much bigger than the superconducting order parameter Δ of the S-terminal and, on the contrary, much smaller than the Fermi energy, i.e. $\Delta \ll h_{1,2} \ll \epsilon_F$. The latter condition allows to perform the analysis of our FSF system within the quasiclassical formalism of Usadel equations for the Green–Keldysh matrix functions G formulated below.

3.3.1 Quasiclassical Equations

In each of our metallic terminals, the Usadel equations can be written in the form [33]

$$iD\nabla(\check{G}\nabla\check{G}) = [\check{\Omega} + eV, \check{G}], \quad \check{G}^2 = 1, \tag{3.87}$$

where D is the diffusion constant, V is the electric potential, $\check{G}$ and $\check{\Omega}$ are 8×8 matrices in Keldysh-Nambu-spin space (denoted by check symbol)

$$\check{G} = \begin{pmatrix} \check{G}^R & \check{G}^K \\ 0 & \check{G}^A \end{pmatrix}, \quad \check{\Omega} = \begin{pmatrix} \check{\Omega}^R & 0 \\ 0 & \check{\Omega}^A \end{pmatrix}, \tag{3.88}$$

$$\check{\Omega}^R = \check{\Omega}^A = \begin{pmatrix} \varepsilon - \hat{\boldsymbol{\sigma}}\boldsymbol{h} & \Delta \\ -\Delta^* & -\varepsilon + \hat{\boldsymbol{\sigma}}\boldsymbol{h} \end{pmatrix}, \tag{3.89}$$

ε is the quasiparticle energy, $\Delta(T)$ is the superconducting order parameter, which will be considered real in a superconductor and zero in both ferromagnets, $\boldsymbol{h} \equiv \boldsymbol{h}_{1(2)}$ in the first (second) ferromagnetic terminal, $\boldsymbol{h} \equiv 0$ outside these terminals and $\hat{\boldsymbol{\sigma}} = (\hat{\sigma}_1, \hat{\sigma}_2, \hat{\sigma}_3)$ are Pauli matrices in spin space.

Retarded and advanced Green functions $\check{G}^R$ and $\check{G}^A$ have the following matrix structure

$$\check{G}^{R,A} = \begin{pmatrix} \hat{G}^{R,A} & \hat{F}^{R,A} \\ -\hat{F}^{R,A} & -\hat{G}^{R,A} \end{pmatrix}. \tag{3.90}$$

Here and below 2×2 matrices in spin space are denoted by hat symbol.

Having obtained the expressions for the Green–Keldysh functions $\check{G}$, one can easily evaluate the current density $\boldsymbol{j}$ in our system with the aid of the standard relation

$$j = -\frac{\sigma}{16e} \int \mathrm{Sp}[\tau_3 (\check{G} \nabla \check{G})^K] \mathrm{d}\varepsilon, \tag{3.91}$$

where σ is the Drude conductivity of the corresponding metal and τ_3 is the Pauli matrix in Nambu space.

In what follows it will be convenient for us to employ the so-called Larkin–Ovchinnikov parameterization of the Keldysh Green function

$$\check{G}^K = \check{G}^R \check{f} - \check{f} \check{G}^A, \quad \check{f} = \hat{f}_L + \tau_3 \hat{f}_T, \tag{3.92}$$

where the distribution functions $\hat{f}_L$ and $\hat{f}_T$ are 2×2 matrices in the spin space.

For the sake of simplicity, we will assume that magnetizations of both ferromagnets and the interfaces (see below) are collinear. Within this approximation, the Green functions and the matrix $\check{\Omega}$ are diagonal in the spin space and the diffusion-like equations for the distribution function matrices $\hat{f}_L$ and $\hat{f}_T$ take the form

$$-D\nabla \left(\hat{D}^T(\boldsymbol{r}, \varepsilon) \nabla \hat{f}_T(\boldsymbol{r}, \varepsilon) \right) + 2\hat{\Sigma}(\boldsymbol{r}, \varepsilon) \hat{f}_T(\boldsymbol{r}, \varepsilon) = 0, \tag{3.93}$$

$$-D\nabla \left(\hat{D}^L(\boldsymbol{r}, \varepsilon) \nabla \hat{f}_L(\boldsymbol{r}, \varepsilon) \right) = 0, \tag{3.94}$$

where

$$\hat{\Sigma}(\boldsymbol{r}, \varepsilon) = -\mathrm{i}\Delta \,\mathrm{Im}\, \hat{F}^R, \tag{3.95}$$

$$\hat{D}^T = \left(\mathrm{Re}\, \hat{G}^R \right)^2 + \left(\mathrm{Im}\, \hat{F}^R \right)^2, \tag{3.96}$$

$$\hat{D}^L = \left(\mathrm{Re}\, \hat{G}^R \right)^2 - \left(\mathrm{Re}\, \hat{F}^R \right)^2. \tag{3.97}$$

The function $\hat{\Sigma}(\boldsymbol{r}, \varepsilon)$ differs from zero only inside the superconductor. It accounts both for energy relaxation of quasiparticles and for their conversion to Cooper pairs due to Andreev reflection. The functions $\hat{D}^T$ and $\hat{D}^L$ acquire space and energy dependencies due to the presence of the superconducting wire and renormalize the diffusion coefficient D.

The solution of (3.93)–(3.94) can be expressed in terms of the diffusion-like functions $\hat{\mathcal{D}}^T$ and $\hat{\mathcal{D}}^L$, which obey the following equations

$$-D\nabla \left[\hat{D}^T(\boldsymbol{r}, \varepsilon) \nabla \hat{\mathcal{D}}^T(\boldsymbol{r}, \boldsymbol{r}', \varepsilon) \right] + 2\hat{\Sigma}(\boldsymbol{r}, \varepsilon) \hat{\mathcal{D}}^T(\boldsymbol{r}, \boldsymbol{r}', \varepsilon) = \delta(\boldsymbol{r} - \boldsymbol{r}'), \tag{3.98}$$

$$-D\nabla \left[\hat{D}^L(\boldsymbol{r}, \varepsilon) \nabla \hat{\mathcal{D}}^L(\boldsymbol{r}, \boldsymbol{r}', \varepsilon) \right] = \delta(\boldsymbol{r} - \boldsymbol{r}'). \tag{3.99}$$

3.3.2 Boundary Conditions

The solutions of Usadel equation (3.87) in each of the metals should be matched at FS-interfaces by means of appropriate boundary conditions which account for electron tunneling between these terminals. The form of these boundary conditions essentially depends on the adopted model describing electron scattering at FS-interfaces. As before, we stick to the model of the so-called spin-active interfaces, which takes into account possibly different barrier transmissions for spin-up and spin-down electrons. Here, we employ this model in the case of diffusive electrodes and also restrict our analysis to the case of tunnel barriers with channel transmissions much smaller than one. In this case, the corresponding boundary conditions read [44, 45]

$$\mathcal{A}\sigma_+ \check{G}_+ \partial_x \check{G}_+ = \frac{G_T}{2}[\check{G}_-, \check{G}_+] + \frac{G_m}{4}[\{\hat{\boldsymbol{\sigma}}\boldsymbol{m}\tau_3, \check{G}_-\}, \check{G}_+] + \mathrm{i}\frac{G_\varphi}{2}[\hat{\boldsymbol{\sigma}}\boldsymbol{m}\tau_3, \check{G}_+], \tag{3.100}$$

$$-\mathcal{A}\sigma_- \check{G}_- \partial_x \check{G}_- = \frac{G_T}{2}[\check{G}_+, \check{G}_-] + \frac{G_m}{4}[\{\hat{\boldsymbol{\sigma}}\boldsymbol{m}\tau_3, \check{G}_+\}, \check{G}_-] + \mathrm{i}\frac{G_\varphi}{2}[\hat{\boldsymbol{\sigma}}\boldsymbol{m}\tau_3, \check{G}_-], \tag{3.101}$$

where $\check{G}_-$ and $\check{G}_+$ are the Green–Keldysh functions from the left ($x < 0$) and from the right ($x > 0$) side of the interface, $\mathcal{A}$ is the effective contact area, $\boldsymbol{m}$ is the unit vector in the direction of the interface magnetization, $\sigma_\pm$ are Drude conductivities of the left and right terminals and G_T is the spin-independent part of the interface conductance. Along with G_T there also exists the spin-sensitive contribution to the interface conductance, which is accounted for by the G_m-term. The value G_m equals to the difference between interface conductances for spin-up and spin-down conduction bands in the normal state. The G_φ-term arises due to different phase shifts acquired by scattered quasiparticles with opposite spin directions.

Employing the above boundary conditions we can establish the following linear relations between the distribution functions at both sides of the interface

$$\mathcal{A}\sigma_+ \hat{D}^T_+ \partial_x \hat{f}_{+T} = \mathcal{A}\sigma_- \hat{D}^T_- \partial_x \hat{f}_{-T} = \hat{g}_T(\hat{f}_{+T} - \hat{f}_{-T}) + \hat{g}_m(\hat{f}_{+L} - \hat{f}_{-L}), \tag{3.102}$$

$$\mathcal{A}\sigma_+ \hat{D}^L_+ \partial_x \hat{f}_{+L} = \mathcal{A}\sigma_- \hat{D}^L_- \partial_x \hat{f}_{-L} = \hat{g}_L(\hat{f}_{+L} - \hat{f}_{-L}) + \hat{g}_m(\hat{f}_{+T} - \hat{f}_{-T}), \tag{3.103}$$

where $\hat{g}_T$, $\hat{g}_L$, and $\hat{g}_m$ are matrix interface conductances, which depend on the retarded and advanced Green functions at the interface

$$\hat{g}_T = G_T\left[\left(\mathrm{Re}\,\hat{G}^R_+\right)\left(\mathrm{Re}\,\hat{G}^R_-\right) + \left(\mathrm{Im}\,\hat{F}^R_+\right)\left(\mathrm{Im}\,\hat{F}^R_-\right)\right], \tag{3.104}$$

$$\hat{g}_L = G_T \left[\left(\mathrm{Re}\,\hat{G}_+^R\right)\left(\mathrm{Re}\,\hat{G}_-^R\right) - \left(\mathrm{Re}\,\hat{F}_+^R\right)\left(\mathrm{Re}\,\hat{F}_-^R\right)\right], \tag{3.105}$$

$$\hat{g}_m = G_m \hat{\boldsymbol{\sigma}}\boldsymbol{m} \left(\mathrm{Re}\,\hat{G}_+^R\right)\left(\mathrm{Re}\,\hat{G}_-^R\right). \tag{3.106}$$

Note that the above boundary conditions for the distribution functions do not contain the G_φ-term explicitly since this term in (3.100)–(3.101) does not mix Green functions from both sides of the interface.

The current density (3.91) can then be expressed in terms of the distribution function $\hat{f}_T$ as

$$\boldsymbol{j} = -\frac{\sigma}{4e}\int \mathrm{Sp}[\hat{D}^T \nabla \hat{f}_T]\mathrm{d}\varepsilon. \tag{3.107}$$

3.3.3 Spectral Conductances

Let us now employ the above formalism to evaluate electric currents in our FSF device. The current across the first (SF_1) interface can be written as

$$\begin{aligned} I_1 = &\frac{1}{e}\int g_{11}(\varepsilon)\,[f_0(\varepsilon + eV_1) - f_0(\varepsilon)]\,\mathrm{d}\varepsilon \\ &-\frac{1}{e}\int g_{12}(\varepsilon)\,[f_0(\varepsilon + eV_2) - f_0(\varepsilon)]\,\mathrm{d}\varepsilon, \end{aligned} \tag{3.108}$$

where $f_0(\varepsilon) = \tanh(\varepsilon/2T)$, g_{11} and g_{12} are local and non-local spectral electric conductances. Expression for the current across the second interface can be obtained from the above equation by interchanging the indices $1 \leftrightarrow 2$. Solving (3.93)–(3.94) with boundary conditions (3.102)–(3.103) we express both local and non-local conductances $\hat{g}_{\mathrm{ij}}(\varepsilon)$ in terms of the interface conductances and the function $\hat{\mathcal{D}}$. The corresponding results read

$$\hat{g}_{11}(\varepsilon) = \left(\hat{R}_2^T\hat{\mathcal{M}}^L + \hat{R}_2^T\hat{R}_2^L\hat{R}_{1m} - \hat{R}_1^L\hat{R}_{2m}^2 + \hat{R}_{12}^T\hat{R}_{12}^L\hat{R}_{2m} - \hat{R}_{1m}\hat{R}_{2m}^2\right)\hat{\mathcal{K}}, \tag{3.109}$$

$$\hat{g}_{12}(\varepsilon) = \hat{g}_{21}(\varepsilon) = \left(\hat{R}_{12}^T\hat{\mathcal{M}}^L + \hat{R}_2^T\hat{R}_{12}^L\hat{R}_{1m} + \hat{R}_{12}^L\hat{R}_{1m}\hat{R}_{2m} + \hat{R}_{12}^T\hat{R}_1^L\hat{R}_{2m}\right)\hat{\mathcal{K}}, \tag{3.110}$$

where we defined

$$\hat{\mathcal{M}}^{T,L} = \hat{R}_1^{T,L}\hat{R}_2^{T,L} - (\hat{R}_{12}^{T,L})^2, \tag{3.111}$$

$$\hat{\mathcal{K}}^{-1} = \hat{\mathcal{M}}^T\hat{\mathcal{M}}^L + \hat{R}_{1m}^2\hat{R}_{2m}^2 - \hat{R}_2^T\hat{R}_2^L\hat{R}_{1m}^2 - 2\hat{R}_{12}^T\hat{R}_{12}^L\hat{R}_{1m}\hat{R}_{2m} - \hat{R}_1^T\hat{R}_1^L\hat{R}_{2m}^2 \tag{3.112}$$

and introduced the auxiliary resistance matrix

$$\hat{R}_1^T = \hat{g}_{1T}(\varepsilon)[\hat{g}_{1T}(\varepsilon)\hat{g}_{1L}(\varepsilon) - \hat{g}_{1m}^2(\varepsilon)]^{-1} + \frac{D_1\hat{\mathcal{D}}_1^T(\boldsymbol{r}_1, \boldsymbol{r}_1, \varepsilon)}{\sigma_1} + \frac{D_S\hat{\mathcal{D}}_S^T(\boldsymbol{r}_1, \boldsymbol{r}_1, \varepsilon)}{\sigma_S}, \quad (3.113)$$

The resistance matrices $\hat{R}_2^T$, $\hat{R}_1^L$ and $\hat{R}_2^L$ can be obtained by interchanging the indices $1 \leftrightarrow 2$ and $T \leftrightarrow L$ in (3.113). The remaining resistance matrices $\hat{R}_{12}^{T,L}$ and $\hat{R}_{\rm jm}$ are defined as

$$\hat{R}_{12}^{T,L} = \hat{R}_{21}^{T,L} = \frac{D_S\hat{\mathcal{D}}_S^{T,L}(\boldsymbol{r}_1, \boldsymbol{r}_2, \varepsilon)}{\sigma_S}, \quad (3.114)$$

$$\hat{R}_{\rm jm} = \hat{g}_{\rm jm}(\varepsilon)[\hat{g}_{\rm jT}(\varepsilon)\hat{g}_{\rm jL}(\varepsilon) - \hat{g}_{\rm jm}^2(\varepsilon)]^{-1}, \quad (3.115)$$

where $j = 1, 2$. The spectral conductance $g_{\rm ij}$ can be recovered from the matrix $\hat{g}_{\rm ij}$ simply by summing up over the spin states

$$g_{\rm ij}(\varepsilon) = \frac{1}{2}\,\mathrm{Sp}\left[\hat{g}_{\rm ij}(\varepsilon)\right]. \quad (3.116)$$

It is worth pointing out that (3.109), (3.110) defining, respectively, local and non-local spectral conductances are presented with excess accuracy. This is because the boundary conditions (3.100)–(3.101) employed here remain applicable only in the tunneling limit and for weak spin-dependent scattering $|G_m|, |G_\varphi| \ll G_T$. Hence, strictly speaking only the lowest order terms in $G_{m_{1,2}}$ and $G_{\varphi_{1,2}}$ need to be kept in our final results.

To proceed it is necessary to evaluate the interface conductances as well as the matrix functions $\hat{\mathcal{D}}_{1,2,S}^{T,L}$. Restricting ourselves to the second order in the interface transmissions we obtain

$$\hat{g}_{1T}(\varepsilon) = G_{T_1}\hat{\nu}_S(\boldsymbol{r}_1, \varepsilon) + G_{T_1}^2\frac{\Delta^2\theta(\Delta^2 - \varepsilon^2)}{\Delta^2 - \varepsilon^2}\hat{U}_1(\varepsilon), \quad (3.117)$$

$$\hat{g}_{1L}(\varepsilon) = G_{T_1}\hat{\nu}_S(\boldsymbol{r}_1, \varepsilon) - G_{T_1}^2\frac{\Delta^2\theta(\varepsilon^2 - \Delta^2)}{\varepsilon^2 - \Delta^2}\hat{U}_1(\varepsilon), \quad (3.118)$$

$$\hat{g}_{1m}(\varepsilon) = G_{m_1}\hat{\nu}_S(\boldsymbol{r}_1, \varepsilon)\hat{\boldsymbol{\sigma}}\boldsymbol{m}_1, \quad (3.119)$$

and analogous expressions for the interface conductances of the second interface. The matrix function

$$\hat{U}_1(\varepsilon) = \frac{D_1}{2\sigma_1}\Big\{\mathrm{Re}\left[\mathcal{C}_1(\boldsymbol{r}_1, \boldsymbol{r}_1, 2h_1^+) + \mathcal{C}_1(\boldsymbol{r}_1, \boldsymbol{r}_1, 2h_1^-)\right] - \hat{\boldsymbol{\sigma}}\boldsymbol{m}_1\,\mathrm{Re}\left[\mathcal{C}_1(\boldsymbol{r}_1, \boldsymbol{r}_1, 2h_1^+) - \mathcal{C}_1(\boldsymbol{r}_1, \boldsymbol{r}_1, 2h_1^-)\right]\Big\} \quad (3.120)$$

with $h_1^{\pm} = h_1 \pm \varepsilon$ defines the correction due to the proximity effect in the normal metal.

Taking into account the first-order corrections in the interface transmissions, one can derive the density of states inside the superconductor in the following form

$$\hat{\nu}_S(\boldsymbol{r}, \varepsilon) = \frac{|\varepsilon|\theta(\varepsilon^2 - \Delta^2)}{\sqrt{|\varepsilon^2 - \Delta^2|}} + \frac{D_S}{\sigma_S}\frac{\Delta^2}{\Delta^2 - \varepsilon^2}\sum_{i=1,2}\Big[G_{T_i}\,\mathrm{Re}\,\mathcal{C}_S(\boldsymbol{r}, \boldsymbol{r}_i, 2\omega^R) - \hat{\boldsymbol{\sigma}}\boldsymbol{m}_i G_{\varphi_i}\,\mathrm{Im}\,\mathcal{C}_S(\boldsymbol{r}, \boldsymbol{r}_i, 2\omega^R)\Big], \tag{3.121}$$

where

$$\omega^R = \begin{cases} \sqrt{\varepsilon^2 - \Delta^2}, & \varepsilon > \Delta, \\ \mathrm{i}\sqrt{\Delta^2 - \varepsilon^2}, & |\varepsilon| < \Delta, \\ -\sqrt{\varepsilon^2 - \Delta^2}, & \varepsilon < \Delta, \end{cases} \tag{3.122}$$

and the Cooperon $\mathcal{C}_j(\boldsymbol{r}, \boldsymbol{r}', \varepsilon)$ represents the solution of the equation

$$\left(-D\nabla^2 - \mathrm{i}\varepsilon\right)\mathcal{C}(\boldsymbol{r}, \boldsymbol{r}', \varepsilon) = \delta(\boldsymbol{r} - \boldsymbol{r}') \tag{3.123}$$

in the normal metal leads ($j = 1, 2$) and the superconductor ($j = S$). In the quasi-one-dimensional geometry, the corresponding solutions take the form

$$\mathcal{C}_j(x_j, x_j, \varepsilon) = \frac{\tanh\left(k_j L_j\right)}{S_j D_j k_j}, \quad j = 1, 2, \tag{3.124}$$

$$\mathcal{C}_S(x, x', \varepsilon) = \frac{\sinh[k_S(L - x')]\sinh k_S x}{k_S S_S D_S \sinh(k_S L)}, \quad x' > x, \tag{3.125}$$

where $S_{S,1,2}$ are the wire cross sections and $k_{1,2,S} = \sqrt{-\mathrm{i}\varepsilon/D_{1,2,S}}$.

Substituting (3.121) into (3.117) and (3.118) and comparing the terms $\propto G_{T_1}^2$, we observe that the tunneling correction to the density of states dominates over the terms proportional to $\hat{U}_1$ which contain an extra small factor $\sqrt{\Delta/h} \ll 1$. Hence, the latter terms in (3.117) and (3.118) can be safely neglected. In addition, in (3.121), we also neglect small tunneling corrections to the superconducting density of states at energies exceeding the superconducting gap Δ. Within this approximation, the density of states inside the superconducting wire becomes spin-independent $\hat{\nu}_S(\boldsymbol{r}, \varepsilon) = \hat{\sigma}_0 \nu_S(\boldsymbol{r}, \varepsilon)$. It can then be written as

$$\nu_S(\boldsymbol{r}, \varepsilon) = \frac{|\varepsilon|}{\sqrt{|\varepsilon^2 - \Delta^2|}}\theta(\varepsilon^2 - \Delta^2) + \frac{D_S}{\sigma_S}\frac{\Delta^2\theta(\Delta^2 - \varepsilon^2)}{\Delta^2 - \varepsilon^2}\sum_{i=1,2} G_{T_i}\,\mathrm{Re}\,\mathcal{C}_S(\boldsymbol{r}, \boldsymbol{r}_i, 2\omega^R). \tag{3.126}$$

Accordingly, the interface conductances take the form

$$\hat{g}_{1T}(\varepsilon) = \hat{g}_{1L}(\varepsilon) = G_{T_1}\nu_S(\boldsymbol{r}_1,\varepsilon), \tag{3.127}$$

$$\hat{g}_{1m}(\varepsilon) = G_{m_1}\nu_S(\boldsymbol{r}_1,\varepsilon)\hat{\boldsymbol{\sigma}}\boldsymbol{m}_1. \tag{3.128}$$

Let us emphasize again that within our approximation the G_φ-term does not enter into expressions for the interface conductances (3.127)–(3.128) and, hence, does not appear in the final expressions for the conductances $g_{\mathrm{ij}}(\varepsilon)$.

In the limit of strong exchange fields $h_{1,2} \gg \Delta$ and small interface transmissions considered here the proximity effect in the ferromagnets remains weak and can be neglected. Hence, the functions $\hat{\mathcal{D}}_1^{T,L}(\boldsymbol{r}_1,\boldsymbol{r}_1,\varepsilon)$ and $\hat{\mathcal{D}}_2^{T,L}(\boldsymbol{r}_2,\boldsymbol{r}_2,\varepsilon)$ can be approximated by their normal state values

$$\hat{\mathcal{D}}_1^{T,L}(\boldsymbol{r}_1,\boldsymbol{r}_1,\varepsilon) = \sigma_1 r_{N_1}\hat{1}/D_1, \tag{3.129}$$

$$\hat{\mathcal{D}}_2^{T,L}(\boldsymbol{r}_2,\boldsymbol{r}_2,\varepsilon) = \sigma_2 r_{N_2}\hat{1}/D_2, \tag{3.130}$$

$$r_{N_j} = L_j/(\sigma_j S_j), \quad j = 1,2, \tag{3.131}$$

where r_{N_1} and r_{N_2} are the normal state resistances of ferromagnetic terminals. In the superconducting region, an effective expansion parameter is $G_{T_{1,2}} r_{\xi_S}(\varepsilon)$, where $r_{\xi_S}(\varepsilon) = \xi_S(\varepsilon)/(\sigma_S S_S)$ is the Drude resistance of the superconducting wire segment of length $\xi_S(\varepsilon) = \sqrt{D_S/2|\omega^R|}$ and ω^R is the function of ε according to (3.122). In the limit

$$G_{T_{1,2}} r_{\xi_S}(\varepsilon) \ll 1, \tag{3.132}$$

which is typically well satisfied for realistic system parameters, it suffices to evaluate the function $\hat{\mathcal{D}}_S^T(x,x',\varepsilon)$ for impenetrable interfaces. In this case, we find

$$\hat{\mathcal{D}}_S^T(x,x',\varepsilon) = \begin{cases} \dfrac{\Delta^2-\varepsilon^2}{\Delta^2}\mathcal{C}_S(x,x',2\omega^R), & |\varepsilon| < \Delta, \\ \dfrac{\varepsilon^2-\Delta^2}{\varepsilon^2}\mathcal{C}_S(x,x',0), & |\varepsilon| > \Delta. \end{cases} \tag{3.133}$$

We note that special care should be taken while calculating $\mathcal{D}_S^L(x,x',\varepsilon)$ at subgap energies, since the coefficient D^L in (3.94) tends to zero deep inside the superconductor. Accordingly, the function $\mathcal{D}_S^L(x,x',\varepsilon)$ becomes singular in this case. Nevertheless, the combinations $\hat{R}_j^L(\mathcal{M}^L)^{-1}$ and $\hat{R}_{12}^L(\mathcal{M}^L)^{-1}$ remain finite also in this limit. At subgap energies, we obtain

$$\hat{R}_1^L(\hat{\mathcal{M}}^L)^{-1} = \hat{R}_2^L(\hat{\mathcal{M}}^L)^{-1} = \hat{R}_{12}^L(\hat{\mathcal{M}}^L)^{-1} = \frac{1}{r_{N_1} + r_{N_2} + \dfrac{2\kappa e^{d/\xi_S(\varepsilon)}}{r_{\xi_S}(\varepsilon) G_{T_1} G_{T_2}}}, \tag{3.134}$$

where $\kappa = 1 - \varepsilon^2/\Delta^2$ and $d = |x_2 - x_1|$ is the distance between two FS contacts. Substituting the above relations into (3.110), we arrive at the final result for the non-local spectral conductance of our device at subgap energies ($|\varepsilon| < \Delta$)

$$g_{12}(\varepsilon) = g_{21}(\varepsilon)$$
$$= \frac{\kappa r_{\xi_S}(\varepsilon) \exp[-d/\xi_S(\varepsilon)]}{2[r_{N_1} + 1/g_{T1}(\varepsilon)][r_{N_2} + 1/g_{T2}(\varepsilon)]}$$
$$\times \left[1 + \frac{\boldsymbol{m}_1 \boldsymbol{m}_2}{\kappa} \frac{G_{m1}}{g_{T1}(\varepsilon)} \frac{G_{m2}}{g_{T2}(\varepsilon)} \frac{1}{\kappa + \frac{r_{N_1} + r_{N_2}}{2} r_{\xi_S}(\varepsilon) G_{T_1} G_{T_2} \mathrm{e}^{-d/\xi_S(\varepsilon)}} \right]. \tag{3.135}$$

Equation (3.135) represents the central result of this section. It consists of two different contributions. The first of them is independent of the interface polarizations $\boldsymbol{m}_{1,2}$. This term represents direct generalization of the result [25] in two different aspects. First, the analysis [25] was carried out under the assumption $r_{N_{1,2}} g_{T1,2}(\varepsilon) \ll 1$, which is abandoned here. Second (and more importantly), sufficiently large exchange fields $h_{1,2} \gg \Delta$ of ferromagnetic electrodes suppress disorder-induced electron interference in these electrodes and, hence, eliminate the corresponding zero-bias anomaly both in local [28–30] and in non-local [25] spectral conductances. In this case, with sufficient accuracy one can set $g_{\mathrm{Ti}}(\varepsilon) = G_{\mathrm{Ti}} \nu_S(x_i, \varepsilon)$ implying that at subgap energies $g_{\mathrm{Ti}}(\varepsilon)$ is entirely determined by the second term in (3.126), which yields in the case of quasi-one-dimensional electrodes

$$g_{T1}(\varepsilon) = \frac{\Delta^2 G_{T_1} r_{\xi_S}(\varepsilon)}{2(\Delta^2 - \varepsilon^2)} \left[G_{T_1} + G_{T_2} \mathrm{e}^{-d/\xi_S(\varepsilon)} \right], \tag{3.136}$$

$$g_{T2}(\varepsilon) = \frac{\Delta^2 G_{T_2} r_{\xi_S}(\varepsilon)}{2(\Delta^2 - \varepsilon^2)} \left[G_{T_2} + G_{T_1} \mathrm{e}^{-d/\xi_S(\varepsilon)} \right]. \tag{3.137}$$

Note, that if the exchange field $h_{1,2}$ in both normal electrodes is reduced well below Δ and eventually is set equal to zero, the term containing $\hat{U}_1(\varepsilon)$ in (3.117), (3.118) becomes important and should be taken into account. In this case, we again recover the zero-bias anomaly [28–30] $g_{\mathrm{Ti}}(\varepsilon) \propto 1/\sqrt{\varepsilon}$ and from the first term in (3.135) we reproduce the results [25] derived in the limit $h_{1,2} \to 0$.

The second term in (3.135) is proportional to the product $\boldsymbol{m}_1 \boldsymbol{m}_2 G_{m1} G_{m2}$ and describes non-local magnetoconductance effect in our system emerging due to spin-sensitive electron scattering at FS interfaces. It is important that – despite the strong inequality $|G_{\mathrm{mi}}| \ll G_{\mathrm{Ti}}$ – both terms in (3.135) can be of the same order, i.e. the second (magnetic) contribution can significantly modify the non-local conductance of our device.

In the limit of large interface resistances $r_{N_{1,2}} g_{T1,2}(\varepsilon) \ll 1$, the formula (3.135) reduces to a much simpler one

$$g_{12}(\varepsilon) = g_{21}(\varepsilon) = \frac{r_{\xi_S}(\varepsilon)}{2} \exp[-d/\xi_S(\varepsilon)]$$
$$\times \left[\frac{\Delta^2 - \varepsilon^2}{\Delta^2} g_{T1}(\varepsilon) g_{T2}(\varepsilon) + \boldsymbol{m}_1 \boldsymbol{m}_2 G_{m1} G_{m2} \frac{\Delta^2}{\Delta^2 - \varepsilon^2} \right]. \quad (3.138)$$

Interestingly, (3.138) remains applicable for arbitrary values of the angle between interface polarizations $\boldsymbol{m}_1$ and $\boldsymbol{m}_2$ and strongly resembles the analogous result for the non-local conductance in ballistic FSF systems (cf., e.g., (3.77) in the previous section). The first term in the square brackets in (3.138) describes the fourth order contribution in the interface transmissions, which remains non-zero also in the limit of the non-ferromagnetic leads [25]. In contrast, the second term is proportional to the product of transmissions of both interfaces, i.e. only to the second order in barrier transmissions. This term vanishes identically provided at least one of the interfaces is spin-isotropic.

Contrary to the non-local conductance at subgap energies, both local conductance (at all energies) and non-local spectral conductance at energies above the superconducting gap are only weakly affected by magnetic effects. Neglecting small corrections due to G_m term in the boundary conditions, we obtain

$$\hat{g}_{11}(\varepsilon) = \hat{R}_1^T (\hat{\mathcal{M}}^T)^{-1}, \quad \hat{g}_{22}(\varepsilon) = \hat{R}_2^T (\hat{\mathcal{M}}^T)^{-1}, \quad (3.139)$$

$$\hat{g}_{12}(\varepsilon) = g_{21}(\varepsilon) = \hat{R}_{12}^T (\hat{\mathcal{M}}^T)^{-1}, \quad |\varepsilon| > \Delta. \quad (3.140)$$

Equations (3.139) and (3.140) together with the above expressions for the non-local subgap conductance enable one to recover both local and non-local spectral conductances of our system at all energies. Typical energy dependencies for both $g_{11}(\varepsilon)$ and $g_{12}(\varepsilon)$ are displayed in Fig. 3.9. For instance, we observe that at subgap energies the non-local conductance g_{12} changes its sign being positive for parallel and negative for antiparallel interface polarizations.

3.3.4 I–V *Curves*

Having established the spectral conductance matrix $g_{ij}(\varepsilon)$, one can easily recover the complete $I - V$ curves for our hybrid FSF structure. In the limit of low bias voltages, these $I - V$ characteristics become linear, i.e.

$$I_1 = G_{11}(T)V_1 - G_{12}(T)V_2, \quad (3.141)$$

$$I_2 = -G_{21}(T)V_1 + G_{22}(T)V_2, \quad (3.142)$$

where $G_{ij}(T)$ represent the linear conductance matrix defined as

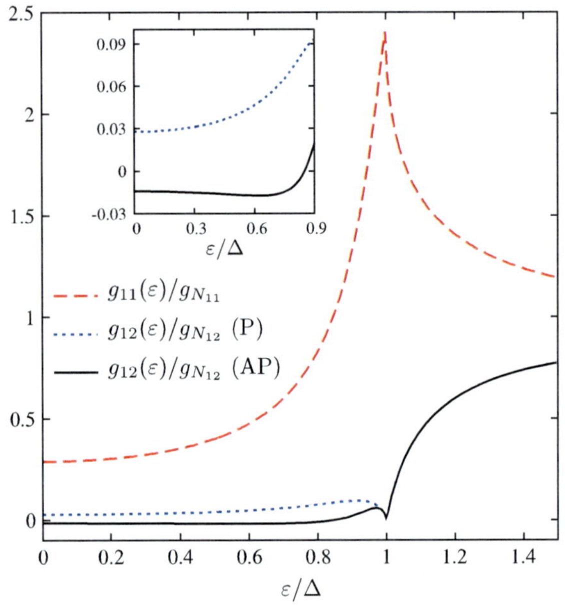

Fig. 3.9 Local (*long-dashed line*) and non-local (*short-dashed and solid lines*) spectral conductances normalized to its normal state values. Here, we choose $r_{N_1} = r_{N_2} = 5r_{\xi_S}(0)$, $x_1 = L - x_2 = 5\xi_S(0)$, $x_2 - x_1 = \xi_S(0)$, $G_{T_1} = G_{T_2} = 4G_{m_1} = 4G_{m_2} = 0.2/r_{\xi_S}(0)$. Energy dependence of non-local conductance is displayed for parallel (P) $\boldsymbol{m}_1\boldsymbol{m}_2 = 1$ and antiparallel (AP) $\boldsymbol{m}_1\boldsymbol{m}_2 = -1$ interface magnetizations. Inset: The same in the limit of low energies

$$G_{ij}(T) = \frac{1}{4T}\int g_{ij}(\varepsilon)\frac{d\varepsilon}{\cosh^2\dfrac{\varepsilon}{2T}}. \tag{3.143}$$

It may also be convenient to invert the relations (3.141)–(3.142), thus expressing induced voltages $V_{1,2}$ in terms of injected currents $I_{1,2}$:

$$V_1 = R_{11}(T)I_1 + R_{12}(T)I_2, \tag{3.144}$$

$$V_2 = R_{21}(T)I_1 + R_{22}(T)I_2, \tag{3.145}$$

where the coefficients $R_{ij}(T)$ define local ($i = j$) and non-local ($i \neq j$) resistances

$$R_{11}(T) = \frac{G_{22}(T)}{G_{11}(T)G_{22}(T) - G_{12}^2(T)}, \tag{3.146}$$

$$R_{12}(T) = R_{21}(T) = \frac{G_{12}(T)}{G_{11}(T)G_{22}(T) - G_{12}^2(T)} \tag{3.147}$$

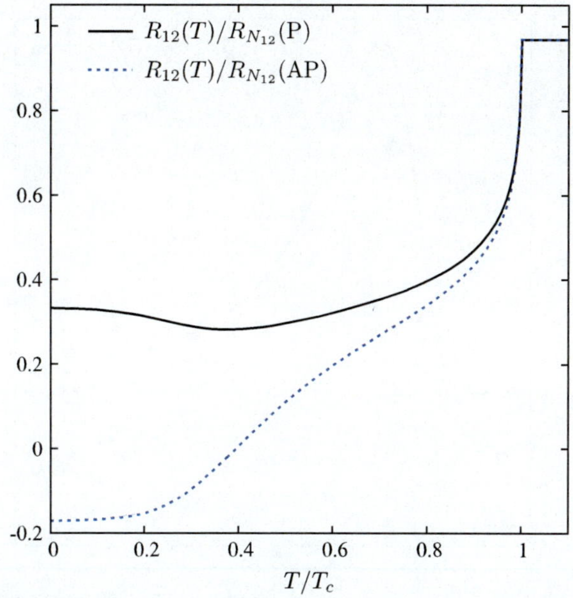

Fig. 3.10 Non-local resistance (normalized to its normal state value) versus temperature (normalized to the superconducting critical temperature T_C) for parallel (P) and antiparallel (AP) interface magnetizations. The parameters are the same as in Fig. 3.9

and similarly for $R_{22}(T)$. In non-ferromagnetic NSN structures, the low temperature non-local resistance $R_{12}(T \to 0)$ turns out to be independent of both the interface conductances and the parameters of the normal leads [25]. However, this universality of R_{12} does not hold anymore provided non-magnetic normal metal leads are substituted by ferromagnets. Non-local linear resistance R_{12} of our FSF structure is displayed in Figs. 3.10 and 3.11 as a function of temperature for parallel ($\boldsymbol{m}_1\boldsymbol{m}_2 = 1$) and antiparallel ($\boldsymbol{m}_1\boldsymbol{m}_2 = -1$) interface magnetizations. In Fig. 3.10, we show typical temperature behaviour of the non-local resistance for sufficiently transparent interfaces. For both mutual interface magnetizations R_{12} first decreases with temperature below T_C similarly to the non-magnetic case. However, at lower T important differences occur: While in the case of parallel magnetizations R_{12} always remains positive and even shows a noticeable upturn at sufficiently low T, the non-local resistance for antiparallel magnetizations keeps monotonously decreasing with T and may become negative in the low temperature limit. In the limit of very low interface transmissions, the temperature dependence of the non-local resistance exhibits a well-pronounced charge imbalance peak (see Fig. 3.11) which physics is similar to that analyzed in the case of non-ferromagnetic NSN structures [19,23,25]. Let us point out that the above behaviour of the non-local resistance is qualitatively consistent with available experimental observations [5,6]. More experiments would be desirable to quantitatively verify our theoretical predictions.

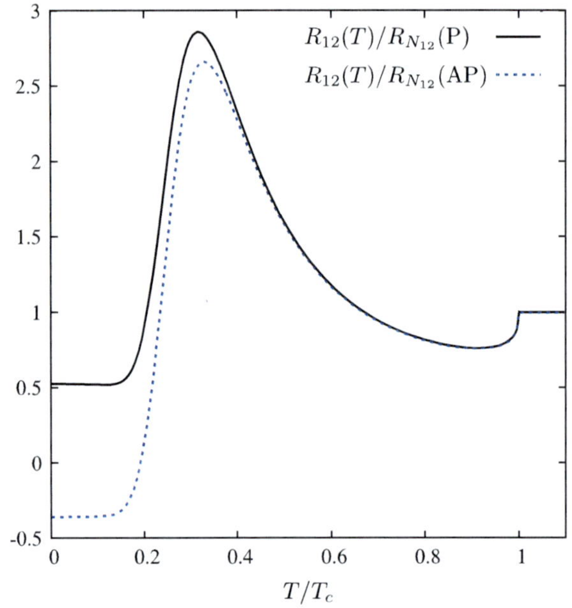

Fig. 3.11 The same as in Fig. 3.10 for the following parameter values: $r_{N_1} = r_{N_2} = 5r_{\xi_S}(0)$, $x_1 = L - x_2 = 5\xi_S(0)$, $x_2 - x_1 = \xi_S(0)$, $G_{T_1} = G_{T_2} = 25G_{m_1} = 25G_{m_2} = 0.025/r_{\xi_S}(0)$

3.4 Concluding Remarks

In this paper, we developed a non-perturbative theory of non-local electron transport in both ballistic and diffusive NSN and FSF three-terminal structures with spin-active interfaces. Our theory is based on the quasiclassical formalism of energy-integrated Green-Eilenberger functions supplemented by appropriate boundary conditions describing spin-dependent scattering at NS and FS interfaces. Our approach applies at arbitrary interface transmissions and allows to fully describe non-trivial interplay between spin-sensitive normal scattering, local and non-local Andreev reflection at NS and FS interfaces.

In the case of ballistic structures, our main results are the general expressions for the non-local cross-current I_{12}, (3.70), and for the non-local correction I_{11} to the BTK current, (3.82). These expressions provide complete description of the conductance matrix of our three-terminal NSN device at arbitrary voltages, temperature, spin-dependent transmissions of NS interfaces and their polarizations. One of our important observations is that in the case of ballistic electrodes no crossed Andreev reflection can occur in both NSN and HSH structures with fully open interfaces. Beyond the tunneling limit, the dependence of the non-local conductance on the size of the S-electrode L is in general non-exponential and reduces to $G_{12} \propto \exp(-2L\Delta/v_F)$ only in the limit of large L. For hybrid structures, half-metal-superconductor-half-metal we predict that the low energy non-local conductance does not vanish already in the lowest order in barrier transmissions $G_{12} \propto D_{1\uparrow} D_{2\uparrow}$.

In the second part of our paper, we addressed spin-resolved non-local electron transport in FSF structures in the presence of disorder in the electrodes. Within our model transfer of electrons across FS interfaces is described in the tunneling limit and magnetic properties of the system are accounted for by introducing (a) exchange fields $\boldsymbol{h}_{1,2}$ in both normal metal electrodes and (b) magnetizations $\boldsymbol{m}_{1,2}$ of both FS interfaces. The two ingredients (a) and (b) of our model are in general independent from each other and have different physical implications. While the role of (comparatively large) exchange fields $h_{1,2} \gg \Delta$ is merely to suppress disorder-induced interference of electrons [28–30] penetrating from a superconductor into ferromagnetic electrodes, spin-sensitive electron scattering at FS interfaces yields an extra contribution to the non-local conductance, which essentially depends on relative orientations of the interface magnetizations. For anti-parallel magnetizations, the total non-local conductance g_{12} and resistance R_{12} can turn negative at sufficiently low energies/temperatures. At higher temperatures, the difference between the values of R_{12} evaluated for parallel and anti-parallel magnetizations becomes less important. At such temperatures, the non-local resistance behaves similarly to the non-magnetic case demonstrating, e.g., a well-pronounced charge imbalance peak [25] in the limit of low-interface transmissions.

Our predictions can be directly used for quantitative analysis of experiments on non-local electron transport in hybrid FSF structures.

Acknowledgements This work was supported, in part, by DFG and by RFBR grant 09-02-00886. M.S.K. also acknowledges support from the Council for grants of the Russian President (Grant No. 89.2009.2) and from the Dynasty Foundation.

References

1. A.F. Andreev, Zh. Eksp. Teor. Fiz. **46**, 1823 (1964) [Sov. Phys. JETP **19**, 1228 (1964)]
2. G.E. Blonder, M. Tinkham, T.M. Klapwijk, Phys. Rev. B **25**, 4515 (1982)
3. J.M. Byers, M.E. Flatte, Phys. Rev. Lett. **74**, 306 (1995)
4. G. Deutscher, D. Feinberg, Appl. Phys. Lett. **76**, 487 (2000)
5. D. Beckmann, H.B. Weber, H.v. Löhneysen, Phys. Rev. Lett. **93**, 197003 (2004)
6. D. Beckmann, H.v. Löhneysen, Appl. Phys. A **89**, 603 (2007)
7. S. Russo, M. Kroug, T.M. Klapwijk, A.F. Morpurgo, Phys. Rev. Lett. **95**, 027002 (2005)
8. P. Cadden-Zimansky, V. Chandrasekhar, Phys. Rev. Lett. **97**, 237003 (2006)
9. P. Cadden-Zimansky, Z. Jiang, V. Chandrasekhar, New J. Phys. **9**, 116 (2007)
10. A. Kleine , A. Baumgartner, J. Trbovic, C. Schönenberger, Europhys. Lett. **87**, 27011 (2009)
11. B. Almog, S. Hacohen-Gourgy, A. Tsukernik, G. Deutscher, Phys. Rev. B **80**, 220512(R) (2009)
12. A. Kleine, A. Baumgartner, J. Trbovic, D.S. Golubev, A.D. Zaikin, C. Schönenberger, Nanotechnology **21**, 274002 (2010)
13. J. Brauer, F. Hübler, M. Smetanin, D. Beckmann, H.v. Löhneysen, Phys. Rev. B **81**, 024515 (2010)

14. G. Falci, D. Feinberg, F.W.J. Hekking, Europhys. Lett. **54**, 255 (2001)
15. M.S. Kalenkov, A.D. Zaikin, Phys. Rev. B **75**, 172503 (2007)
16. A. Brinkman, A.A. Golubov, Phys. Rev. B **74**, 214512 (2006)
17. J.P. Morten, A. Brataas, W. Belzig, Phys. Rev. B **74**, 214510 (2006)
18. S. Duhot, R. Melin, Phys. Rev. B **75**, 184531 (2007)
19. D.S. Golubev, A.D. Zaikin, Phys. Rev. B **76**, 184510 (2007)
20. A. Levy Yeyati, F.S. Bergeret, A. Martin-Rodero, T.M. Klapwijk, Nat. Phys. **3**, 455 (2007)
21. M.S. Kalenkov, A.D. Zaikin, Phys. Rev. B **76**, 224506 (2007)
22. M.S. Kalenkov, A.D. Zaikin, Physica E **40**, 147 (2007)
23. M.S. Kalenkov, A.D. Zaikin, JETP Lett. **87**, 140 (2008)
24. D.S. Golubev, A.D. Zaikin, Europhys. Lett. **86**, 37009 (2009)
25. D.S. Golubev, M.S. Kalenkov, A.D. Zaikin, Phys. Rev. Lett. **103**, 067006 (2009)
26. F.S. Bergeret, A. Levy Yeyati, Phys. Rev. B **80**, 174508 (2009)
27. D.S. Golubev, A.D. Zaikin, Phys. Rev. B **82**, (2010)
28. A.F. Volkov, A.V. Zaitsev, T.M. Klapwijk, Physica C **210**, 21 (1993)
29. F.W.J. Hekking, Yu.V. Nazarov, Phys. Rev. Lett. **71**, 1625 (1993)
30. A.D. Zaikin, Physica B **203**, 255 (1994)
31. A. Huck, F.W.J. Hekking, B. Kramer, Europhys. Lett. **41**, 201 (1998)
32. A.V. Galaktionov, A.D. Zaikin, Phys. Rev. B **73**, 184522 (2006)
33. W. Belzig, F.K. Wilhelm, C. Bruder, G. Schön, A.D. Zaikin, Superlatt. Microstruct. **25**, 1251 (1999)
34. M. Eschrig, Phys. Rev. B **61**, 9061 (2000)
35. E. Zhao, T. Löfwander, J.A. Sauls, Phys. Rev. B **70** 134510 (2004)
36. A.V. Zaitsev, Sov. Phys. JETP **59**, 1015 (1984)
37. A. Millis, D. Rainer, J.A. Sauls, Phys. Rev. B **38** 4504 (1988)
38. M. Eschrig, Phys. Rev. B **80**, 134511 (2009)
39. A.V. Galaktionov, A.D. Zaikin, Phys. Rev. B **65**, 184507 (2002)
40. M. Ozana, A. Shelankov, Phys. Rev. B **65**, 014510 (2002)
41. E. Zhao, J.A. Sauls, Phys. Rev. Lett. **98**, 206601 (2007)
42. M. Fogelström, Phys. Rev. B **62**, 11 812 (2000)
43. Yu.S. Barash, I.V. Bobkova, Phys. Rev. B **65**, 144502 (2002)
44. D. Huertas-Hernando, Yu.V. Nazarov, W. Belzig, Phys. Rev. Lett. **88**, 047003 (2002)
45. A. Cottet, D. Huertas-Hernando, W. Belzig, Yu.V. Nazarov, Phys. Rev. B **80**, 184511 (2009)

Chapter 4
Non-local Transport in Superconductor–Ferromagnet Hybrid Structures

D. Beckmann

Abstract We review recent experimental results on non-local transport in superconductor–ferromagnet hybrid structures. We will focus mainly on crossed Andreev reflection and its relation to competing non-local transport phenomena such as elastic cotunneling, charge imbalance and spin diffusion.

4.1 Introduction

Non-local transport experiments have proven to be a powerful tool to investigate non-equilibrium conditions in nanostructures, such as charge imbalance (CI) in superconductors [1–3], or spin accumulation in normal metals [4]. In this chapter, we review recent experimental investigations of coherent non-local transport in multi-terminal superconductor hybrid structures [5–16]. The impetus for these experiments comes mainly from the prediction of crossed Andreev reflection (CAR) [17, 18], a non-local transport process, where an electron entering a superconductor from one normal-metal contact is reflected as a hole into a second nearby contact, creating a Cooper pair in the superconductor. This process is possible at energies below the energy gap Δ of the superconductor, if the distance d between the two contacts is not much larger than the coherence length ξ_S, as shown schematically in Fig. 4.1a. One of the consequences of CAR is the creation of spatially separated, entangled electron pairs in the two normal-metal wires. Building efficient superconducting solid-state entangler devices requires a comprehensive understanding not only of CAR, but also of competing non-local processes, such as the transfer of an electron from one electrode to the other at subgap energies via virtual

D. Beckmann (✉)
Institut für Nanotechnologie, Karlsruher Institut für Technologie, P.O. Box 3640, 76021 Karlsruhe, Germany
e-mail: detlef.beckmann@kit.edu

A. Sidorenko (ed.), *Fundamentals of Superconducting Nanoelectronics*, NanoScience and Technology, DOI 10.1007/978-3-642-20158-5_4,

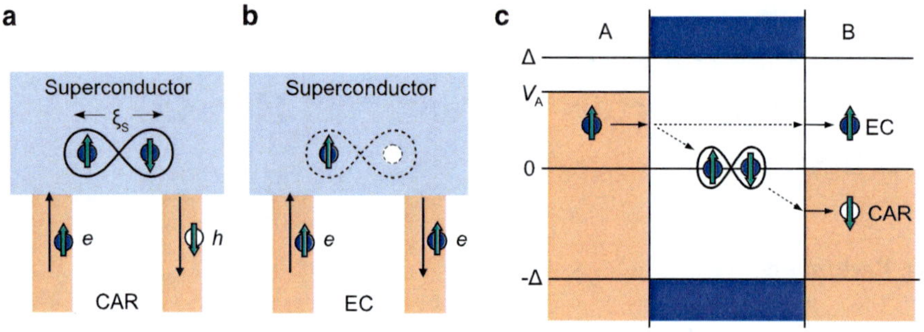

Fig. 4.1 Schematic view of crossed Andreev reflection (**a**) and elastic cotunneling (**b**). *Filled symbols* represent electrons and *open symbols* represent holes. The *arrows* inside the symbols indicate spin bands. (**c**) Energy scheme of CAR and EC

states in the superconductor. To lowest order in transmission probabilities, i.e., in tunnel junctions, this process is similar to cotunneling in a quantum dot, and is therefore usually referred to as elastic cotunneling (EC). Upon increasing contact transparency, higher order processes start to contribute, and the term cotunneling becomes slightly misleading. However, since it has been widely used in the literature, we will stick to it here. A schematic view of EC is shown in Fig. 4.1b. Figure 4.1c shows a representation of CAR and EC in the semiconductor model of superconductivity. If a bias V_A is applied to one contact (A), CAR and EC lead to the emission of a holes or electrons, respectively, into the second contact B. Consequently, both contribute with opposite sign to the non-local conductance.

In Fig. 4.2a, a generic multi-terminal setup for non-local transport experiments is shown. Two terminals A and B are in contact with a grounded central node C. Bias voltages V_A and V_B are applied to the two terminals, and the currents I_A and I_B flowing into the central node are measured. In this geometry, transport can be described by the conductance matrix

$$\begin{pmatrix} I_A \\ I_B \end{pmatrix} = \begin{pmatrix} G_{AA} & G_{AB} \\ G_{BA} & G_{BB} \end{pmatrix} \begin{pmatrix} V_A \\ V_B \end{pmatrix}, \tag{4.1}$$

where the off-diagonal elements G_{AB} and G_{BA} describe non-local transport, while local conductances are represented by the diagonal elements. The generalization to differential conductances $g_{ij} = dI_i/dV_j$ is straightforward. The output of theories of coherent transport processes are typically the elements of the conductance matrix (4.1). In actual experiments, however, often non-local resistance is measured: current bias I_A is applied to one contact, and the non-local voltage V_B is measured under the condition $I_B = 0$, as shown in Fig. 4.2b. Therefore, it is often necessary to convert between non-local conductance and non-local resistance by inverting the conductance matrix. The non-local resistance R_{AB} is given by

$$R_{AB} = \frac{V_B}{I_A} = -\frac{G_{AB}}{G_{AA}G_{BB} - G_{AB}G_{BA}} \approx -\frac{G_{AB}}{G_{AA}G_{BB}}. \tag{4.2}$$

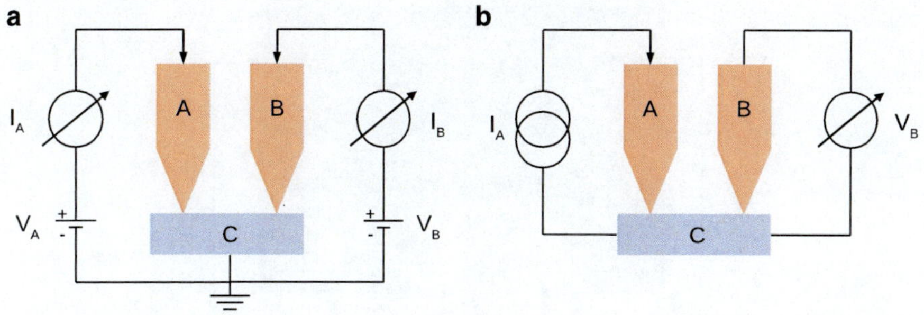

Fig. 4.2 (**a**) Idealized view of a non-local conductance setup. Two contacts A and B are attached to a grounded central normal-metal or superconductor node (S/N). The contacts are voltage biased, and the resulting currents are measured. *Arrows* indicate current polarities. (**b**) Schematic view of many actual experiments. Contact A is currentbiased, and the resulting non-local voltage is measured at contact B

The approximate expression holds if the non-local conductances are negligible compared to the local conductances, which is usually the case.

4.2 Experiments

In this section, we discuss in detail our recent experimental investigations of non-local transport in superconductor/ferromagnet and superconductor/normal-metal hybrid structures. We focus mainly on the discussion of coherent subgap transport, and possible implications for superconducting entangler devices.

4.2.1 F/S Point Contacts

Local Andreev reflection dominates transport in high-transparency contacts, and therefore multiterminal structures with transparent point contacts are a natural starting point for the investigation of crossed Andreev reflection. Figure 4.3 shows a multi-terminal superconductor–ferromagnet hybrid structure with three ferromagnetic wires forming small contacts to an aluminum bar [5]. The structure has been fabricated by electron-beam lithography using a two-layer shadow mask. Metal films were deposited through the mask under different angles relative to the substrate normal in a UHV evaporation chamber. First, a 20-nm thick iron film is evaporated to form the ferromagnetic tips. Then, an 80-nm thick aluminum film is evaporated. The angle is chosen such that the aluminum film slightly touches the iron tips. Thus, small contacts of about $20 \times 50\ \text{nm}^2$ are formed. We will show here results of non-local resistance measurements on two samples, T2 (shown in Fig. 4.3) and S5, which has six rather than three contacts at varying distance.

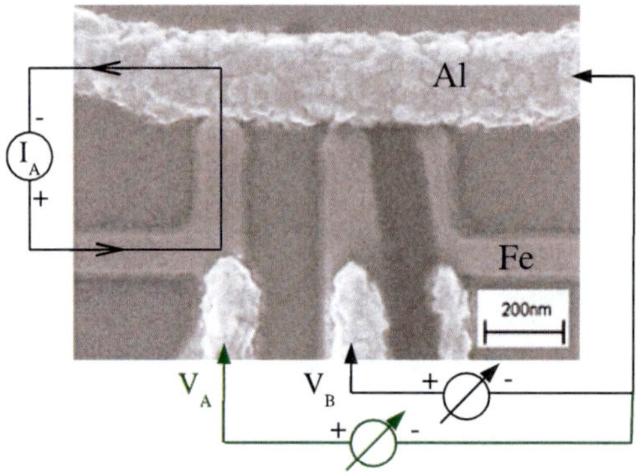

Fig. 4.3 SEM image of a ferromagnet–superconductor multiterminal structure (sample T2). Three iron wires (*vertical*) of slightly different width form small contacts to the aluminum strip (*top*). The two outer iron wires have additional lateral wires attached for use as voltage probes. The measurement scheme for local and non-local resistance is also shown for a pair of contacts. A current I_A is injected into one contact, and the local (V_A) or non-local (V_B) voltage is measured

4.2.2 *Spin Accumulation*

We first characterize the spin-valve behaviour in the normal state, and establish control over the magnetization configuration. Figure 4.4a shows the non-local resistance for a pair of contacts of sample T2 at $T = 4.2$ K, i.e. with the aluminum strip in the normal state, as a function of magnetic field B applied along the ferromagnetic wires. The traces exhibit sharp, well-defined jumps at two distinct coercive fields, $B_{c1} = 15$ mT and $B_{c2} = 50$ mT. These jumps correspond to magnetization reversal of the two ferromagnetic wires. The magnetization reversal is inferred from the resistance measurements only indirectly. Therefore, micromagnetics simulations have been performed to study the magnetization process of a two-dimensional iron film with the actual shape of sample T2 taken from the SEM image (Fig. 4.3), using the micromagnetics simulation package OOMMF.[1] The results indicate that the wires leading to the contacts indeed have single-domain configuration with coherent magnetization reversal, compatible with the observed step-like resistance change. From the hysteresis loops, the non-local resistance difference $\Delta R_N = R_N^{\text{parallel}} - R_N^{\text{antiparallel}}$ at zero applied magnetic field, i.e. the spin-valve signal, can be obtained.

Spin injection in metallic spin valves is governed by the characteristic resistances $\mathscr{R}_{F,N} = \rho_{F,N}\lambda_{F,N}/\mathscr{A}_{F,N}$, where ρ, λ and $\mathscr{A}$ are the resistivities, spin-diffusion lengths, and cross-sections of the ferromagnet (F) and normal metal (N), respectively. Since λ_F is usually only a few nanometers [19, 20], $\mathscr{A}_F$ is actually given by the contact cross-section. Despite the large resistivity of ferromagnets in typical

[1] http://math.nist.gov/oommf.

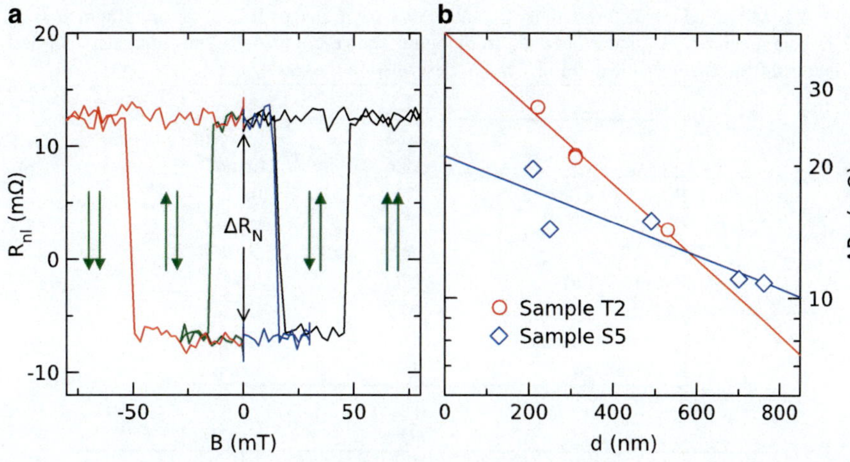

Fig. 4.4 (**a**) Non-local resistance R_{nl} for a contact pair of sample T2 at $T = 4.2$ K as a function of magnetic field B applied along the direction of the ferromagnetic wires. Both major (*black*, *red*) and minor (*blue*, *green*) hysteresis loops are shown. The sample shows well-defined switching between parallel and antiparallel magnetization states. Magnetization states are indicated by *green arrows*. (**b**) Non-local resistance difference ΔR_{N} between parallel and antiparallel magnetization state for different contact pairs of two samples (T2 and S5) at $T = 4.2$ K as a function of contact distance d. The *solid lines* are fits to (4.3)

spin valves with large-area overlap junctions, one consequently finds $\mathscr{R}_{\mathrm{F}} \ll \mathscr{R}_{\mathrm{N}}$, the so-called resistance mismatch [21]. In our samples, we estimate $\mathscr{R}_{\mathrm{F}} \approx 3\Omega$ and $\mathscr{R}_{\mathrm{N}} \approx 1\Omega$. Due to the small contact area, we are in the opposite limit, $\mathscr{R}_{\mathrm{F}} > \mathscr{R}_{\mathrm{N}}$. In this case,

$$\Delta R_{\mathrm{N}} = P_\sigma^2 \mathscr{R}_{\mathrm{N}} \exp\left(-\frac{\mathrm{d}}{\lambda_{\mathrm{N}}}\right) \tag{4.3}$$

can be obtained from the general expression given in [22]. The formula has the same structure as for the tunnel limit, but contains the spin polarization P_σ of the bulk conductivity rather than the polarization P_G of the contact conductance. In contrast to the case of metallic overlap junctions, spin-injection by point contacts does not depend explicitly on the spin-diffusion length λ_{F} of the ferromagnet. Therefore, it can be used to obtain P_σ directly.

The spin-valve signal ΔR_{N} is shown in Fig. 4.4b for two different samples T2 and S5 as a function of contact distance d at $T = 4.2$ K, together with fits to (4.3). From these fits, and with the known resistivities and sample geometry, P_σ and λ_{N} can be obtained (see Table 4.1).

In paramagnetic metals, the dominant spin-flip scattering process is spin-orbit scattering via the Elliott-Yafet mechanism [23, 24]. Spin-orbit scattering introduces a small admixture of spin-down electrons to spin-up bands, and vice-versa. Consequently, any momentum scattering will lead to some amount of spin scattering as well. With this assumption, the spin-flip time can be expressed as $\tau_{\mathrm{sf}} = \alpha\tau$,

Table 4.1 Measured residual resistivity ρ_N and elastic mean free path l_{el} of the aluminum at $T = 4.2$ K, parameters P_σ, λ_N and α obtained from the fits shown in Fig. 4.4b, and parameters P_G and ξ_S obtained from the fits shown in Fig. 4.6b, as described in the text

Sample	T2	S5
ρ_N	2.85	1.41 μΩcm
l_{el}	15.6	31.6 nm
P_σ	22	16 %
λ_N	505	1,150 nm
α	3.2	2.5×10^{-4}
P_G, fit to (4.6)	51	47 %
ξ_S, fit to (4.6)	275	345 nm
ξ_S, fit to exponential	160	190 nm
ξ_S, calculated	160	230 nm

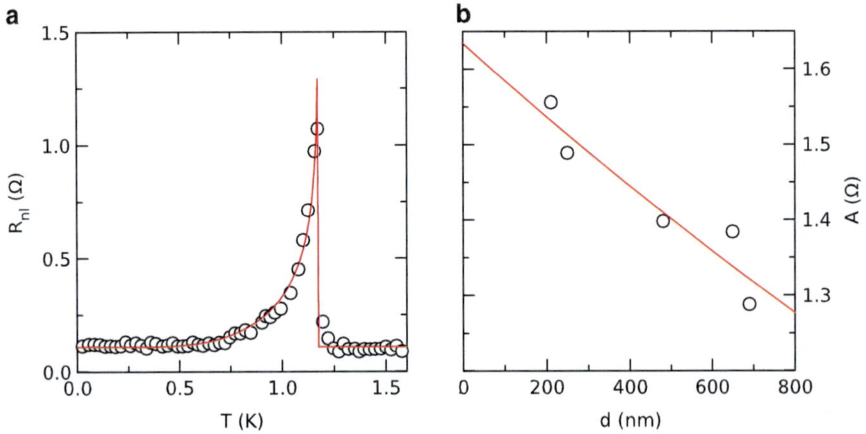

Fig. 4.5 (**a**) Non-local resistance R_{nl} for a contact pair of sample S5 as a function of temperature T with parallel magnetization aligment. The *red line* is a fit to (4.4). (**b**) Prefactor A obtained from the fit as a function of contact distance d

where τ is the momentum scattering time, and α is a proportionality constant that depends on material and momentum scattering mechanism. With $\lambda_{sf} = \sqrt{D\tau_{sf}}$, and $D = v_F\tau/3$, we obtain $\lambda_{sf} = l\sqrt{1/3\alpha}$, where l is the transport mean free path, i.e. $\lambda_{sf} \propto \rho^{-1}$. The value of $\alpha \approx 3 \times 10^{-4}$ for elastic impurity scattering in aluminum obtained from our samples agrees well with the results reported in the literature [20].

4.2.3 Charge Imbalance

Figure 4.5a shows the temperature dependence of the non-local resistance for a contact pair of sample S5 in parallel magnetization alignment as the sample is cooled through the superconducting transition at zero-applied magnetic field. At

$T_c \approx 1.2$ K, a large peak due to charge imbalance is observed [1, 2]. As T is lowered further and the gap opens, the fraction of current that is injected into the superconductor via quasiparticle transmission decreases, and consequently charge imbalance disappears. At temperatures below about 0.5 K, current injection is dominated by Andreev reflection, and the non-local signal becomes independent of temperature.

To describe our experiment, we decompose the charge-imbalance contribution to the non-local resistance into a normalized temperature dependence $Z(T)$, and a distance-dependent amplitude $A(d)$

$$R_{ci}(T, d) = A(d)Z(T). \tag{4.4}$$

For the amplitude, we assume that CI spreads over the length scale λ_{Q^*} from the injector, which yields [25–28]

$$A(d) = \frac{\rho_N \lambda_{Q^*}}{2\mathscr{A}_N} \exp\left(-\frac{d}{\lambda_{Q^*}}\right). \tag{4.5}$$

For the temperature dependence, we must consider that a fraction of the injected current is converted to Cooper pairs by Andreev reflection, and therefore does not contribute to charge imbalance. We use the fraction of quasiparticle injection $F^*(T)$ as given by the BTK model in the limit of small injector bias [29]. For high transparency contacts, F^* approaches unity near the critical temperature, and drops rapidly to zero as temperature is lowered due to the opening of the superconducting gap. It is practically zero below $T_c/2$, as observed in the experiment. Further, the relaxation time τ_{Q^*} is strongly temperature dependent, and diverges near the critical temperature T_c. However, using an analytical approximation for τ_{Q^*} for low-bias injection close to T_c [30] did not improve fitting the temperature dependence of our signals significantly over simply setting $Z(T) = F^*(T)$. Therefore, this approximation was employed for the fit shown as solid line in Fig. 4.5a.

The dependence of the signal amplitude A on contact distance is shown in Fig. 4.5b. From the fit of the data to (4.5), we obtain $\lambda_{Q^*} = 3.2$ µm, and with the known sample parameters, we can calculate $\rho_N \lambda_{Q^*}/2\mathscr{A}_N = 1.2\ \Omega$, which is compatible with the extrapolation $A(d = 0) \approx 1.6\ \Omega$ obtained from the fit. Our value of λ_{Q^*}, which is effectively an average over the temperature range $0.5T_c < T < T_c$, yields a relaxation time $\tau_{Q^*} = \lambda_{Q^*}^2/D = 740$ ps. This is considerably shorter than typical values of $\tau_{Q^*} \approx 5 - 10$ ns found for aluminum in the literature [28, 31, 32]. A few recent experiments [8] (A. Kleine, A. Baumgartner, J. Trbovic, D.S. Golubev, A.D. Zaikin, C. Schönenberger, unpublished) have reported similar discrepancies, and also a non-exponential decay as a function of contact distance [8]. Since our contact distances were all much smaller than λ_{Q^*}, we cannot say whether the decay is actually exponential, as assumed by (4.5). A possible explanation for the small τ_{Q^*} at least in our experiment is relaxation via the contacts. A tunnel contact leads to charge relaxation at a rate $\tau_{tun}^{-1} = 1/2N_0\Omega e^2 R_N$ [33], where $\Omega = 2\lambda_{Q^*}/\mathscr{A}_N$ is the effective non-equilibrium volume. Assuming a similar

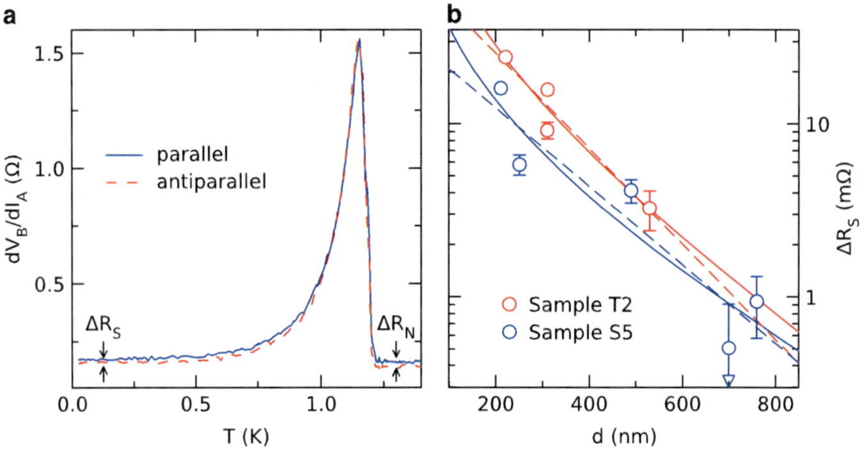

Fig. 4.6 (**a**) Non-local resistance dV_B/dI_A of a contact pair of sample T2 as a function of temperature T in both parallel (*blue*) and antiparallel (*red*) magnetization states. (**b**) Non-local resistance difference ΔR_S between parallel and antiparallel magnetization state for different contact pairs of two samples (T2 and S5) in the superconducting state at lowest temperature as a function of contact distance d. The *solid lines* are fits to (4.6), the *dashed lines* are fits to an exponential decay

relaxation rate for our metallic contacts with $R_N \approx 5\ \Omega$ (see below), we estimate $\tau_{tun} \approx 3$ ns. Since all six contacts are within the non-equilibrium volume, this leads to a total relaxation time $\tau_{tun}/6 \approx 500$ ps, close to the estimated τ_{Q^*}. We therefore conclude that relaxation in our samples is dominated by the contacts rather than intrinsic mechanisms, which might also explain the findings of [8].

4.2.4 Coherent Subgap Transport

Figure 4.6a shows the non-local resistance of a contact pair of sample T2 as a function of temperature for parallel and antiparallel magnetization alignment. The resistance difference ΔR_N already seen in Fig. 4.4a is clearly visible for $T > T_c$. The charge-imbalance signal near $T > T_c$ is nearly independent of magnetization alignment. At temperatures below 0.5 K, where charge imbalance does not play a role any more, a non-local resistance difference ΔR_S is observed, which has the same sign as ΔR_N and is only slightly smaller.

Figure 4.6b shows the resistance difference ΔR_S averaged over temperatures below 250 mK as a function of contact distance d for the two samples T2 and S5. The signal is slightly smaller than the normal-state spin-valve signal at the smallest distance, and falls off rapidly as d increases. As discussed above, the probability of non-local processes depends on the diffusion of virtual quasiparticles over the length scale of ξ_S. In our case, the dimensionality of the sample is not well defined, since $\xi_S \approx 150-200$ nm is of the same order as the width (200 nm) and thickness (80 nm) of the aluminum wire. Therefore, we have attempted two different fit functions. First, the prediction for CAR and EC for point contacts attached to a three-dimensional superconductor, which in lowest order in the tunnel coupling is given by [34]

$$\Delta R_S \approx P_G^2 \frac{\pi}{2} \rho_N \frac{\exp(-d/\xi_S)}{d} \tag{4.6}$$

and, second, a simple exponential decay. Fits of these functions to the data are shown as solid lines and dashed lines in Fig. 4.6b. Parameters are given in Table 4.1. Both fits lie within the scatter of the data. The polarisation $P_G \approx 50\%$ obtained from (4.6) is compatible with the expectation for an iron-aluminum contact [35], whereas ξ_S is a little larger than the value calculated from the residual resistivity using $\xi_S = \sqrt{\hbar D/\Delta}$. ξ_S as obtained from the simple exponential fit agrees very well with the calculated value. We conclude that ξ_S is the relevant length scale for the decay of the signal in the superconducting state, as opposed to the spin-diffusion length for the normal state signal. It can also be noted that the dependence of the decay length on the mean free path is much weaker than for the normal-state signal, which is expected since $\xi_S \propto \sqrt{l_{el}}$, while $\lambda_N \propto l_{el}$.

Before we continue the discussion of the non-local signals, we characterize the (local) contact resistance, and the differential resistances. The temperature dependence of the resistance of a single contact (the left-most contact of sample T2, see Fig. 4.3) is shown in Fig. 4.7a. The resistance is measured in a four-probe

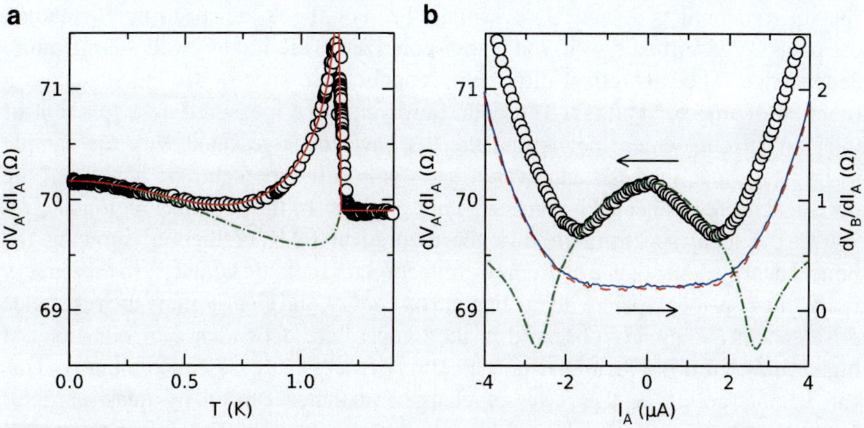

Fig. 4.7 (**a**) Differential local resistance dV_A/dI_A (*symbols*) of one contact of sample T2 as a function of temperature T. The *solid lines* is a fit to (4.7), the *dash-dotted line* is the contribution $R_0 + R_{BTK}(T)$, excluding the charge-imbalance signal (see text). (**b**) Left-hand scale: differential local resistance dV_A/dI_A (symbols) of the same contact. The *dash-dotted line* is the same model as in (**a**). Right-hand scale: differential non-local resistance dV_B/dI_A for parallel (*solid line*) and antiparallel (*dashed line*) magnetization alignment. Data are taken at $T = 15$ mK and zero magnetic field as a function of injector bias current I_A

configuration, with two probes (I- and V-) attached to the aluminum strip, and the other two (I+ and V+) attached to the splitting of the ferromagnetic lead. Thus, the series resistance of the ferromagnetic strip leading up to the contact (R_0), and the contact itself, is measured. We model the data by

$$R(T) = R_0 + R_{\mathrm{BTK}}(T) + R_{\mathrm{ci}}(T), \tag{4.7}$$

where $R_{\mathrm{BTK}}(T)$ is the point-contact resistance according to a spin-polarized BTK model [36], and $R_{\mathrm{ci}}(T)$ is the charge-imbalance contribution described above. The fit to (4.7) is shown as a solid line in Fig. 4.7a. Parameters for the BTK model are $R_{\mathrm{pc}} = 6.2\ \Omega$, $P_G = 0.45$, $Z = 0.22$ and $\Delta = 177\ \mu eV$ for the normal state point-contact resistance, the conductance spin polarization, the interface transparency, and the gap, respectively. Since most of the BTK contribution is obscured by the charge-imbalance peak, Δ and P_G had to be assumed rather than fitted. Δ was calculated from T_c, and P_G was taken from fits to similar samples, which did not show a large charge-imbalance peak. The remaining parameters (R_{pc}, Z and $R_0 = 63.7\ \Omega$) were fitted. From R_0 and the geometry, we obtain $\rho_{\mathrm{F}} = 51\ \mu\Omega\mathrm{cm}$, and a mean free path of about 1 nm for the iron film. Consequently, our contacts with a cross-section area $\mathscr{A} \approx 50 \times 20\ \mathrm{nm}^2$ are in the Maxwell limit [37], and we can estimate the Maxwell point contact resistance $R_{\mathrm{pc}} \approx \rho_{\mathrm{F}}/\sqrt{\mathscr{A}} = 16\ \Omega$. Since only half of the contact region is made of iron, and the other half of aluminum with a much lower resistivity, the actual value should be half of the estimate, in good agreement with the fitted value $R_{\mathrm{pc}} = 6.2\ \Omega$.

Figure 4.7b is a combined plot of the local and non-local differential resistance spectra for contacts A and B of sample T2. For the local spectrum (symbols), the data are described by the same spin-polarized BTK model as the temperature dependence. The theoretical differential conductance spectra are calculated as a function of injector voltage. The data, however, were measured as a function of injector current, which means that the fits have to be rescaled with the sample resistance. Since the resistance variation is only a few percent, we have used the normal state resistance as a constant scaling factor. In the low-bias region $|I_{\mathrm{A}}| \leq 2\ \mu\mathrm{A}$, the local spectrum follows the theoretical BTK prediction, showing the typical energy dependence of Andreev reflection for metallic contacts. In this energy range, the non-local spectra do not depend on energy, and clearly show the resistance difference ΔR_{S} already observed in the temperature dependence in Fig. 4.6a. At higher bias, both the local and non-local resistances increase concomitantly. This can be interpreted as the onset of charge imbalance caused by quasi-particles injected at energies beyond the gap. The onset of charge-imbalance injection appears at a markedly lower bias than predicted by the BTK model, i.e. the minima of the dash-dotted line. This may be attributed to self-heating in the highly resistive ferromagnetic strip due to the large injector current.

4.2.5 F/S Tunnel Contacts

The non-local resistance signals observed in point contacts were too small to reliably determine the sign of the resistances. Only resistance differences could be resolved. Also, for high-transparency contacts, EC is predicted to dominate over CAR. We have therefore investigated F/S/F structures with tunnel contacts, where CAR and EC should have equal probability without spin selection. The study of coherent non-local transport in structures with tunnel contacts may seem surprising, since non-local transport is always at least of second order in transmission probability, and thereby strongly suppressed in the tunnel limit. However, since the product of the conductance of injector and detector contact appears in the denominator of (4.2), the non-local resistance can actually become quite large in low-transparency contacts, despite the small non-local conductance. A large detector voltage is generated if the non-locally transmitted electrons (or holes) have a low probability to return into the superconductor via local Andreev reflection.

For positive injector bias, crossed Andreev reflection leads to a negative voltage at the detector contact. This has the surprising consequence that the voltage of the detector contact may lie outside the voltage window that is spanned by the (grounded) superconductor and the injector bias [38]. This situation is different from the negative voltage, which may occur in a non-local spin-valve structure, where the voltage probes are attached outside the current path. In this case, the chemical potentials at both voltage probes lie inside the source-drain voltage window, but their difference may be either positive or negative.

Figure 4.8 shows the scheme of an experiment designed to measure the negative four-probe voltage due to crossed Andreev reflection [6]. Voltage in this scheme is detected inside the current path, rather than outside as shown in Fig. 4.2b. For the interpretation of the experiment, it is crucial that the voltage of the detector contact B is measured relative to the equilibrium chemical potential of the superconductor. Since current injection into contact A might generate charge imbalance, the drain voltage probe (V_B^-) is attached to the superconductor at a distance of $10 - 20\ \mu$m from the injector contact A. This distance exceeds by far both λ_{Q^*} and λ_N, and therefore the voltage measured there is a good reference for the equilibrium chemical potential of the superconductor. The detector voltage V_B is measured relative to the drain probe.

The samples for these experiments were made of an aluminum wire of 25 nm thickness, which was exposed to an atmosphere of 60 Pa of pure oxygen for 1 min to form tunnel barriers. Subsequently, ferromagnetic wires of 15 nm thick iron were evaporated to form overlap tunnel junctions. The experiments were performed using a battery-powered DC current source and a nanovoltmeter.

Figure 4.9a shows the non-local resistance $R_{AB} = V_B/I_A$ at fixed injector bias current $I_A = 11$ nA as a function of temperature. Above the critical temperature, the signal is always positive. In the normal state, the voltage drop mainly comes from the resistance of the aluminum wire between contact B and the drain voltage probe, with a small spin-accumulation signal superimposed. At $T_c \approx 1.3$ K, the signal

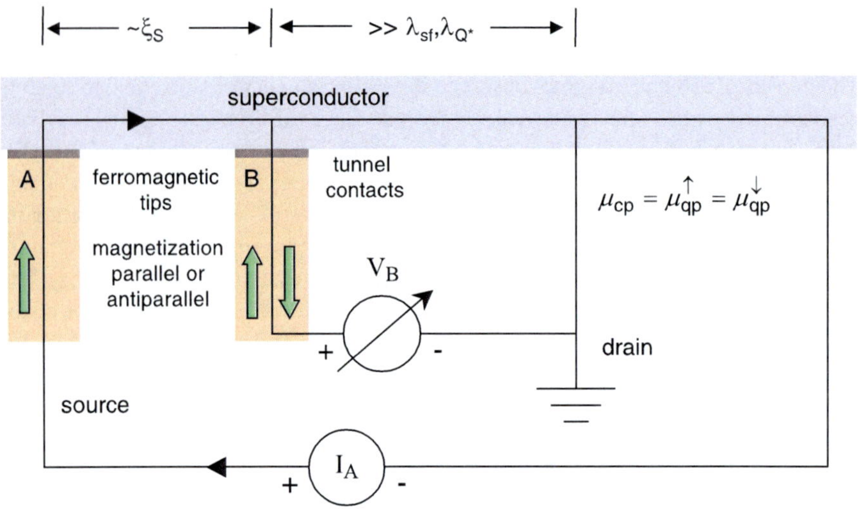

Fig. 4.8 Experimental scheme for measuring the negative four-probe voltage due to crossed Andreev reflection

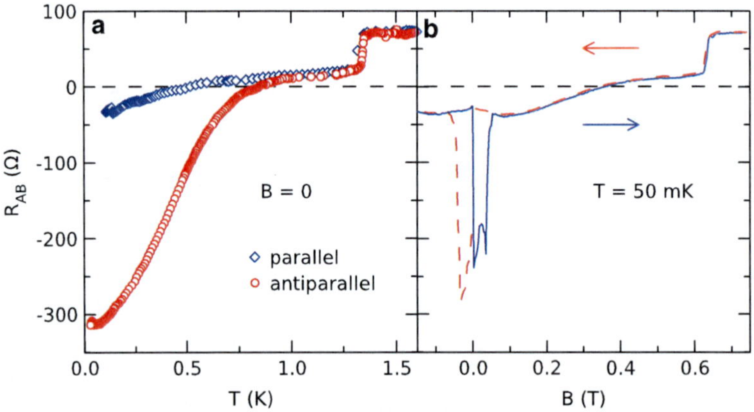

Fig. 4.9 (**a**) Non-local resistance R_{AB} at fixed injector bias current $I_A = 11$ nA as a function of temperature for both parallel (*diamonds*) and antiparallel (*circles*) magnetization alignment at $B = 0$. (**b**) Non-local resistance R_{AB} at fixed injector bias current $I_A = 15$ nA as a function of magnetic field at $T = 50$ mK. *Arrows* indicate sweep directions

decreases abruptly mainly because the aluminum wire becomes superconducting, and the Ohmic voltage drop is eliminated. At low temperature, the detector voltage becomes negative. There is a positive difference ΔR_S between parallel and antiparallel alignment, as in the case of high-transparency contacts. Figure 4.9b shows the non-local resistance at lowest temperature as a function of magnetic field B applied along the ferromagnetic wires for two sweeps in different directions.

As in the normal state, two well-defined jumps associated with the magnetization reversals of the two ferromagnetic electrodes are observed at $|B| < 0.1$ T. Above $|B| \approx 0.1$ T, both electrodes are magnetized parallel, and the traces coincide. The signal becomes positive around 0.3 T, and a jump at $B = 0.63$ T indicates the critical field B_c. The critical field is largely enhanced over the bulk value due to the thin-film geometry.

As expected, in the normal state the resistance is always positive for our four-probe geometry with voltage detection inside the current path. In the superconducting state, however, the voltage becomes negative, which is an unambigous sign of crossed Andreev reflection. The negative signal is larger in magnitude in the anti-parallel alignment, as demonstrated both in the temperature and in the magnetic-field dependent measurements, and expected for singlet Cooper pairs. We can therefore conclude that subgap transport at low temperature and magnetic field is dominated by crossed Andreev reflection rather than elastic cotunneling. While this is expected for antiparallel magnetization, it is surprising that the signal remains negative even for parallel alignment. In the tunneling limit, the "bare" probabilities for CAR and EC are expected to be equal, and spin selection should lead to a positive (parallel) and negative (antiparallel) signal of equal magnitude. We will discuss the possible origin of this discrepancy below.

Figure 4.10a shows the non-local voltage V_B as a function of injector bias current I_A at low temperature for the antiparallel magnetization alignment. In the low-bias region $I_A \leq 30$ nA, a steep negative slope is observed, corresponding to the negative signal seen in Fig. 4.9. With increasing bias, the slope becomes positive. The steep positive slope occurs at a bias which corresponds to the superconducting energy

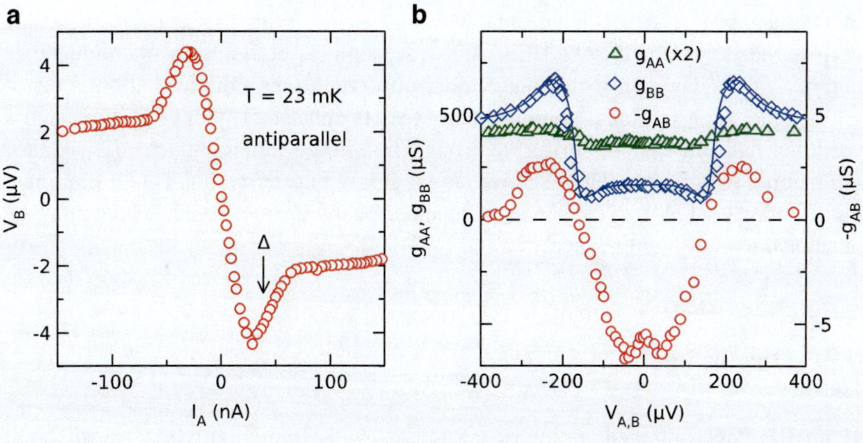

Fig. 4.10 (**a**) Non-local voltage V_B as a function of injector bias current I_A at $T = 23$ mK in the antiparallel magnetization alignment. The *arrow* indicates the bias current that corresponds to the superconducting gap. (**b**) Local differential conductances g_{AA} and g_{BB} (*left scale*), and non-local differential conductance $-g_{AB}$ (*right scale*) as a function of injector voltage. g_{AA} is multiplied by 2 for clarity. The *line* is a fit as described in the text

gap. The change from negative to positive slope near the gap can be interpreted as a crossover from supgap transport dominated by crossed Andreev reflection to charge imbalance above the gap.

In Fig. 4.10b, a combined view of local and non-local differential conductance spectra is shown. The local conductances g_{AA} and g_{BB} are the numerical derivatives of the raw data. The local conductance of the detector contact B shows coherence peaks at the gap, and a reduced subgap conductance, in qualitative agreement with the expectation for tunnel contacts. The subgap conductance is, however, much larger than expected, and shows a downward curvature towards the gap. The spectrum of the injector contact is almost Ohmic. The enhanced subgap conductance cannot be due to thermal smearing or pair-breaking, since both would only broaden the coherence peaks, leading to an upward curvature below the gap. An enhanced subgap conductance due to reflectionless tunneling [39] would explain the downward curvature, but is not expected in ferromagnetic junctions due to the lack of time-reversal symmetry. A possible explanation are pin-holes in conjunction with finite spin-mixing angles, which would lead to the formation of Andreev bound states at subgap energies. The solid line in Fig. 4.10b is a fit of g_{BB} to a model with two contributions from a large-area tunnel contact and high-transparency pin holes, the latter including a finite spin-mixing angle [40]. The model is purely ballistic, while our junctions are diffusive, but similar spectra were predicted for diffusive systems [41]. Also, the number of parameters is too large for a quantitative analysis, and we can only conclude that our model gives a qualitative explanation of the observed spectra.

The the non-local conductance spectrum in Fig. 4.10b is calculated from the non-local voltage shown in Fig. 4.10a using (4.2). As discussed, the non-local conductance has the opposite sign of non-local resistance. To facilitate the comparison of conductance and resistance data, we plot $-g_{AB}$. With this convention, CAR corresponds to a negative and EC to a positive sign for both non-local conductance and resistance. The non-local conductance observed in the sample is about a factor 100 smaller than the local conductances. This is consistent with estimates from a model of two-dimensional diffusive propagation in the superconductor [34, 42] for our sample parameters. The negative signal at low bias corresponds to a dominant contribution of CAR. Near the gap, positive peaks are observed, which may be attributed to charge imbalance.

4.3 Discussion

From the data obtained in hybrid structures with both highlytransparent point contacts and tunnel junctions, the following conclusions can be drawn: At low bias and temperature, non-local transport is governed by coherent subgap processes, i.e., crossed Andreev reflection and elastic cotunneling. At higher bias or temperature, charge imbalance dominates the signal. We have established a hierarchy of length scales, $\xi_S \ll \lambda_N \ll \lambda_{Q^*}$, which serves as an additional criterion to distinguish CAR/EC, spin accumulation, and charge imbalance. Spin-selection by ferromagnetic electrodes works as expected, favoring CAR for antiparallel alignment and EC for parallel alignment, respectively. For point contacts, the magnitude of the non-local signals agrees well with theory. EC dominates over CAR, as predicted theoretically [43, 44], and also observed in hybrid structures with normal-metal contacts [8]. For tunnel junctions, the situation appears to be more complicated. Dominating CAR has been observed not only for antiparallel but also for parallel magnetization alignment, which is not expected theoretically. Similar results have been obtained with normal-metal contacts [13], and a complex bias-dependent competition between CAR and EC may occur [7, 16]. Different attempts have been made to explain these results, including a non-local version of dynamical Coulomb blockade [45] and interference [44, 46]. In addition, for ferromagnetic contacts Andreev bound states induced by spin mixing may have an impact on non-local transport [47, 48]. At present, neither theory yields a comprehensive explanation of all experimental results on tunnel junctions, and further work is needed to clarify these issues.

References

1. J. Clarke, Phys. Rev. Lett. **28**, 1363 (1972)
2. M. Tinkham, J. Clarke, Phys. Rev. Lett. **28**, 1366 (1972)
3. J. Clarke, in *Nonequilibrium Superconductivity*, ed. by D.N. Langenberg, A.I. Larkin (North-Holland, 1986), p. 1
4. M. Johnson, R.H. Silsbee, Phys. Rev. Lett. **55**, 1790 (1985)
5. D. Beckmann, H.B. Weber, H.v. Löhneysen, Phys. Rev. Lett. **93**, 197003 (2004)
6. D. Beckmann, H.v. Löhneysen, Appl. Phys. A **89**, 603 (2007)
7. S. Russo, M. Kroug, T.M. Klapwijk, A.F. Morpurgo, Phys. Rev. Lett. **95**, 027002 (2005)
8. P. Cadden-Zimansky, V. Chandrasekhar, Phys. Rev. Lett. **97**, 237003 (2006)
9. P. Cadden-Zimansky, Z. Jiang, V. Chandrasekhar, New J. Phys. **9**, 116 (2007)
10. P. Cadden-Zimansky, J. Wei, V. Chandrasekhar, Nature Physics **5**, 393 (2009)
11. I. Asulin, O. Yuli, G. Koren, O. Millo, Phys. Rev. B **74**, 092501 (2006)
12. B. Almog, S. Hacohen-Gourgy, A. Tsukernik, G. Deutscher, Phys. Rev. B **80**, 220512 (2009)
13. A. Kleine, A. Baumgartner, J. Trbovic, C. Schönenberger, Europhys. Lett. **87**, 27011 (2009)
14. L. Hofstetter, S. Csonka, J. Nygård, C. Schönenberger, Nature **461**, 960 (2009)
15. L.G. Herrmann, F. Portier, P. Roche, A.L. Yeyati, T. Kontos, C. Strunk, Phys. Rev. Lett. **104**, 026801 (2010)
16. J. Brauer, F. Hübler, M. Smetanin, D. Beckmann, H.v. Löhneysen, Phys. Rev. B **81**, 024515 (2010)
17. J.M. Byers, M.E. Flatté, Phys. Rev. Lett. **74**, 306 (1995)
18. G. Deutscher, D. Feinberg, Appl. Phys. Lett. **76**, 487 (2000)
19. S. Dubois, L. Piraux, J.M. George, K. Ounadjela, J.L. Duvail, A. Fert, Phys. Rev. B **60**, 477 (1999)
20. F.J. Jedema, M.S. Nijboer, A.T. Filip, B.J. van Wees, Phys. Rev. B **67**, 085319 (2003)
21. G. Schmidt, D. Ferrand, L.W. Molenkamp, A.T. Filip, B.J. van Wees, Phys. Rev. B **62**, R4790 (2000)
22. S. Takahashi, S. Maekawa, Phys. Rev. B **67**, 052409 (2003)

23. R.J. Elliott, Phys. Rev. **96**, 266 (1954)
24. Y. Yafet, *Solid State Physics*, vol. 14 (Academic, New York, 1963)
25. T.Y. Hsiang, J. Clarke, Phys. Rev. B **21**, 945 (1980)
26. G.J. Dolan, L.D. Jackel, Phys. Rev. Lett. **39**, 1628 (1977)
27. C.J. Pethick, H. Smith, Ann. Phys. **119**, 133 (1979)
28. F. Hübler, J. Camirand Lemyre, D. Beckmann, H.v. Löhneysen, Phys. Rev. B **81**, 184524 (2010)
29. G.E. Blonder, M. Tinkham, T.M. Klapwijk, Phys. Rev. B **25**, 4515 (1982)
30. C.C. Chi, J. Clarke, Phys. Rev. B **21**, 333 (1980)
31. C.C. Chi, J. Clarke, Phys. Rev. B **19**, 4495 (1979)
32. M. Stuivinga, C.L.G. Ham, T.M. Klapwijk, J.E. Mooij, J. Low Temp. Phys. **53**, 633 (1983)
33. T.R. Lemberger, Phys. Rev. Lett. **52**, 1029 (1984)
34. D. Feinberg, Eur. Phys. J. B **36**, 419 (2003)
35. F. Pérez-Willard, J.C. Cuevas, C. Sürgers, P. Pfundstein, J. Kopu, M. Eschrig, H.v. Löhneysen, Phys. Rev. B **69**, 140502 (2004)
36. G.J. Strijkers, Y. Ji, F.Y. Yang, C.L. Chien, J.M. Byers, Phys. Rev. B **63**, 104510 (2001)
37. A.G.M. Jansen, A.P. van Gelder, P. Wyder, J. Phys. C: Solid St. Phys. **13**, 6073 (1980)
38. F.J. Jedema, B.J. van Wees, B.H. Hoving, A.T. Filip, T.M. Klapwijk, Phys. Rev. B **60**, 16549 (1999)
39. A. Kastalsky, A.W. Kleinsasser, L.H. Greene, R. Bhat, F.P. Milliken, J.P. Harbison, Phys. Rev. Lett. **67**, 3026 (1991)
40. E. Zhao, T. Löfwander, J.A. Sauls, Phys. Rev. B **70**, 134510 (2004)
41. B.P. Vodopyanov, JETP Lett. **87**, 328 (2008)
42. G. Bignon, M. Houzet, F. Pistolesi, F.W.J. Hekking, Europhys. Lett. **67**, 110 (2004)
43. R. Mélin, D. Feinberg, Phys. Rev. B **70**, 174509 (2004)
44. J.P. Morten, A. Brataas, W. Belzig, Phys. Rev. B **74**, 214510 (2006)
45. A. Levy Yeyati, F.S. Bergeret, A. Martin-Rodero, T.M. Klapwijk, Nat. Phys. **3**, 455 (2007)
46. D.S. Golubev, M.S. Kalenkov, A.D. Zaikin, Phys. Rev. Lett. **103**, 067006 (2009)
47. M.S. Kalenkov, A.D. Zaikin, Phys. Rev. B **76**, 224506 (2007)
48. G. Metalidis, M. Eschrig, R. Grein, G. Schön, arXiv:1006.3422

Chapter 5
Odd-Frequency Pairing in Superconducting Heterostructures

A.A. Golubov, Y. Tanaka, Y. Asano, and Y. Tanuma

Abstract We review the theory of odd-frequency pairing in superconducting heterostructures, where an odd-frequency pairing component is induced near interfaces. General description of superconducting proximity effect in a normal metal or a ferromagnet attached to an unconventional superconductor (S) is given within quasiclassical kinetic theory for various types of symmetry state in S. Various possible symmetry classes in a superconductor are considered, which are consistent with the Pauli principle: even-frequency spin-singlet even-parity (ESE) state, even-frequency spin-triplet odd-parity (ETO) state, odd-frequency spin-triplet even-parity (OTE) state and odd-frequency spin-singlet odd-parity (OSO) state. As an example, we consider junction between a diffusive normal metal (DN) and a p-wave superconductor (even-frequency spin-triplet odd-parity symmetry), where the pairing amplitude in DN belongs to an odd-frequency spin-triplet even-parity symmetry class. We also discuss the manifestation of odd-frequency pairing in conventional superconductor/normal (S/N) proximity systems and its relation to the classical McMillan-Rowell oscillations.

A.A. Golubov (✉)
Faculty of Science and Technology and MESA+ Institute of Nanotechnology,
University of Twente, 7500 AE, Enschede, The Netherlands
e-mail: A.A.Golubov@tnw.utwente.nl

Y. Tanaka
Department of Applied Physics, Nagoya University, Nagoya 464-8603, Japan

Y. Asano
Department of Applied Physics, Hokkaido University, Sapporo 060-8628, Japan

Y. Tanuma
Institute of Physics, Kanagawa University, 3-7-1, Rokkakubashi, Kanagawa-ku, Yokohama 221-8686, Japan

A. Sidorenko (ed.), *Fundamentals of Superconducting Nanoelectronics*,
NanoScience and Technology, DOI 10.1007/978-3-642-20158-5_5,

5.1 Introduction

It is well established that superconductivity is realized due to the formation of Cooper pairs consisting of two electrons. In accordance with the Pauli principle, it is customary to distinguish spin-singlet even-parity and spin-triplet odd-parity pairing states in superconductors, where odd (even) refer to the orbital part of the pair wave function. For example, s -wave and d-wave pairing states belong to the former case, while p-wave state belongs to the latter one [1]. In both cases, the pair amplitude is an even function of energy. However, the so-called odd-frequency pairing states when the pair amplitude is an odd function of energy can also exist. Then, the spin-singlet odd-parity and the spin-triplet even-parity pairing states are possible.

The possibility of realizing the odd-frequency pairing state was first proposed by Berezinskii in the context of ^{3}He, where the odd-frequency spin-triplet pairing was discussed [2]. The odd-frequency superconductivity was then discussed in the context of various mechanisms of superconductivity involving strong correlations [3–6]. The odd-frequency pairing state was recently proposed in ferromagnet/superconductor heterostructures with inhomogeneous magnetization or spin-mixing due to spin-dependent interface potential [7–16]. Manifestations of triplet pairs were recently observed experimentally as a long-range Josephson coupling across ferromagnets [17–20]. At the same time, the very important issue of the manifestation of the odd-frequency pairing in proximity systems without magnetic ordering received no attention yet. This question is addressed in this paper.

Note that the proximity effect involving spin-singlet even-frequency pairing state is realized in ferromagnet/superconductor heterostructures with homogeneous magnetization. Such effects are well understood in terms of oscillatory behavior of the order parameter [21,22] (Fulde-Ferrel-Larkin-Ovchinnikov state). This behavior manifests itself in oscillatory coupling in SFS Josephson junctions with transitions from 0- to π-state observed experimentally by Ryazanov et al. [23]. The physics of π-junctions is discussed in Chap. 9 of this book.

Coherent charge transport in structures involving diffusive normal metals (DN) and superconductors (S) was extensively studied during the past decade both experimentally and theoretically. However, almost all previous work was restricted to junctions based on conventional s-wave superconductors [24, 25]. Recently, new theoretical approach to study charge transport in junctions based on p-wave and d-wave superconductors was developed and applied to the even-frequency pairing state [26–31]. It is known that in the anisotropic paring state, due to the sign change of the pair potential on the Fermi surface, a so-called midgap Andreev resonant state (MARS) is formed at the interface [32–38]. As was found in [26–31], MARS competes with the proximity effect in contacts with spin-singlet superconductors, while it coexists with the proximity effect in junctions with spin-triplet superconductors. In the latter case, it was predicted that the induced pair amplitude in the DN has a peculiar energy dependence and the resulting local density of states (LDOS) has a zero energy peak (ZEP) [28–31]. Here we review a general theory of the proximity effect in the N/S junctions, both in the clean and in the dirty limit [39–42], applicable to any type of symmetry state in a superconductor forming the junction in the absence of spin-dependent electronic scattering at the N/S interface.

5.2 Junctions in the Dirty Limit

Let us first discuss the case of a diffusive normal metal attached to a superconductor (DN/S junction). Before proceeding with formal discussion, let us present qualitative arguments illustrating the main conclusions of the paper. Two constrains should be satisfied in the considered system: (1) only the s-wave even-parity state is possible in the DN due to isotropization by impurity scattering, (2) the spin structure of induced Cooper pairs in the DN is the same as in an attached superconductor. Then the Pauli principle provides the unique relations between the pairing symmetry in a superconductor and the resulting symmetry of the induced pairing state in the DN. Namely, for even-parity superconductors, even-frequency spin-singlet even-parity (ESE) and odd-frequency spin-triplet even-parity (OTE) states, the pairing symmetry in the DN should remain ESE and OTE. On the contrary, for odd-parity superconductors, even-frequency spin-triplet odd-parity (ETO) and odd-frequency spin-singlet odd-parity (OSO) states, the pairing symmetry in the DN should be OTE and ESE, respectively. The generation of the OTE state in the DN attached to the ETO p-wave superconductor is of particular interest. Similar OTE state can be generated in superconducting junctions with diffusive ferromagnets [7–12, 17] but due to different physical mechanism. Although the symmetry properties can be derived from the basic arguments given above, the quantitative model has to be considered to prove the existence of nontrivial solutions for the pairing amplitude in the DN in each of the above cases. These solutions and their main features will be discussed below.

Let us start with the general symmetry properties of the quasiclassical Green's functions in the considered system. The elements of retarded and advanced Nambu matrices $\widehat{g}^{R,A}$

$$\widehat{g}^{R,A} = \begin{pmatrix} g^{R,A} & f^{R,A} \\ \overline{f}^{R,A} & \overline{g}^{R,A} \end{pmatrix} \tag{5.1}$$

are composed of the normal $g^R_{\alpha,\beta}(\boldsymbol{r},\varepsilon,\boldsymbol{p})$ and anomalous $f^R_{\alpha,\beta}(\boldsymbol{r},\varepsilon,\boldsymbol{p})$ components with spin indices α and β. Here, $\boldsymbol{p} = \boldsymbol{p}_F / \mid \boldsymbol{p}_F \mid$, $\boldsymbol{p}_F$ is the Fermi momentum, $\boldsymbol{r}$ and ε denote coordinate and energy of a quasiparticle measured from the Fermi level.

The function f^R and the conjugated function $\bar{f}^R$ satisfy the following relation [43, 44]

$$\bar{f}^R_{\alpha,\beta}(\boldsymbol{r},\varepsilon,\boldsymbol{p}) = -[f^R_{\alpha,\beta}(\boldsymbol{r},-\varepsilon,-\boldsymbol{p})]^*. \tag{5.2}$$

The Pauli principle is formulated in terms of the retarded and the advanced Green's functions in the following way [43]

$$f^{A}_{\alpha,\beta}(\boldsymbol{r},\varepsilon,\boldsymbol{p}) = -f^{R}_{\beta,\alpha}(\boldsymbol{r},-\varepsilon,-\boldsymbol{p}). \tag{5.3}$$

By combining the two above equations, we obtain $\bar{f}^{R}_{\beta,\alpha}(\boldsymbol{r},\varepsilon,\boldsymbol{p}) = [f^{A}_{\alpha,\beta}(\boldsymbol{r},\varepsilon,\boldsymbol{p})]^{*}$. Further, the definitions of the even-frequency and the odd-frequency pairing are $f^{A}_{\alpha,\beta}(\boldsymbol{r},\varepsilon,\boldsymbol{p}) = f^{R}_{\alpha,\beta}(\boldsymbol{r},-\varepsilon,\boldsymbol{p})$ and $f^{A}_{\alpha,\beta}(\boldsymbol{r},\varepsilon,\boldsymbol{p}) = -f^{R}_{\alpha,\beta}(\boldsymbol{r},-\varepsilon,\boldsymbol{p})$, respectively. Finally, we get

$$\bar{f}^{R}_{\beta,\alpha}(\boldsymbol{r},\varepsilon,\boldsymbol{p}) = [f^{R}_{\alpha,\beta}(\boldsymbol{r},-\varepsilon,\boldsymbol{p})]^{*} \tag{5.4}$$

for the even-frequency pairing and

$$\bar{f}^{R}_{\beta,\alpha}(\boldsymbol{r},\varepsilon,\boldsymbol{p}) = -[f^{R}_{\alpha,\beta}(\boldsymbol{r},-\varepsilon,\boldsymbol{p})]^{*} \tag{5.5}$$

for the odd-frequency pairing. In the following, we focus on Cooper pairs with $S_z = 0$, remove the external phase of the pair potential in the superconductor and will concentrate on the retarded part of the Green's function.

We consider a junction consisting of a normal (N) and a superconducting reservoirs connected by a quasi-one-dimensional diffusive conductor (DN) with a length L much larger than the mean free path. The Green's function in the superconductor can be parameterized as $g_{\pm}(\varepsilon)\hat{\tau}_3 + f_{\pm}(\varepsilon)\hat{\tau}_2$ using Pauli matrices, where the suffix $+(-)$ denotes the right (left) going quasiparticles. $g_{\pm}(\varepsilon)$ and $f_{\pm}(\varepsilon)$ are given by $g_{+}(\varepsilon) = g^{R}_{\alpha,\beta}(\boldsymbol{r},\varepsilon,\boldsymbol{p})$ $g_{-}(\varepsilon) = g^{R}_{\alpha,\beta}(\boldsymbol{r},\varepsilon,\bar{\boldsymbol{p}})$ $f_{+}(\varepsilon) = f^{R}_{\alpha,\beta}(\boldsymbol{r},\varepsilon,\boldsymbol{p})$, and $f_{-}(\varepsilon) = f^{R}_{\alpha,\beta}(\boldsymbol{r},\varepsilon,\bar{\boldsymbol{p}})$, respectively, with $\bar{\boldsymbol{p}} = \bar{\boldsymbol{p}}_F / \mid \boldsymbol{p}_F \mid$ and $\bar{\boldsymbol{p}}_F = (-p_{\mathrm{Fx}}, p_{\mathrm{Fy}})$. Using the relations (5.4), (5.5), we obtain that $f_{\pm}(\varepsilon) = [f_{\pm}(\varepsilon)]^{*}$ for the even-frequency pairing and $f_{\pm}(\varepsilon) = -[f_{\pm}(-\varepsilon)]^{*}$ for the odd-frequency pairing, respectively, while $g_{\pm}(\varepsilon) = [g_{\pm}(-\varepsilon)]^{*}$ in both cases.

In the DN region only, the s-wave even-parity pairing state is allowed due to isotropization by impurity scattering. The resulting pair amplitude in the DN can be parameterized by $\cos\theta\hat{\tau}_3 + \sin\theta\hat{\tau}_2$ in a junction with an even-parity superconductor and by $\cos\theta\hat{\tau}_3 + \sin\theta\hat{\tau}_1$ in a junction with an odd-parity superconductor. The function θ satisfies the Usadel equation [45] with the corresponding boundary condition at the DN/S interface and at the N/DN interface [26, 27].

In the following, we will consider four possible symmetry classes of superconductor forming the junction and consistent with the Pauli principle: ESE, ETO, OTE, and OSO pairing states. We will use the fact that only the even-parity s-wave pairing is possible in the DN due to the impurity scattering and that the spin structure of pair amplitude in the DN is the same as in an attached superconductor.

1. Junction with ESE superconductor

   In this case, $f_{\pm}(\varepsilon) = f^{*}_{\pm}(-\varepsilon)$ and $g_{\pm}(\varepsilon) = g^{*}_{\pm}(-\varepsilon)$ are satisfied. Then, the Usadel equations and the boundary conditions are consistent with each other only when $\sin\theta^{*}(-\varepsilon) = \sin\theta(\varepsilon)$ and $\cos\theta^{*}(-\varepsilon) = \cos\theta(\varepsilon)$. Thus, the ESE state is formed in the DN, in accordance with the Pauli principle.
2. Junction with ETO superconductor

Now we have $f_\pm(\varepsilon) = f_\pm^*(-\varepsilon)$ and $g_\pm(\varepsilon) = g_\pm^*(-\varepsilon)$. Then $f_S(-\varepsilon) = -f_S^*(\varepsilon) = -f_S^*$ and $g_S(-\varepsilon) = g_S^*(\varepsilon) = g_S^*$. As a result, the boundary value problem is consistent if $\sin\theta^*(-\varepsilon) = -\sin\theta(\varepsilon)$ and $\cos\theta^*(-\varepsilon) = \cos\theta(\varepsilon)$. Thus, the OTE state is formed in the DN. Remarkably, the appearance of the OTE state is the only possibility to satisfy the Pauli principle, as we argued above. Interestingly, the OTE pairing state can also be realized in superconductor/ferromagnet junctions [7–12, 17], but the physical mechanism differs from the one considered here.

3. Junction with OTE superconductor

   In this case, $f_\pm(\varepsilon) = -f_\pm^*(-\varepsilon)$ and $g_\pm(\varepsilon) = g_\pm^*(-\varepsilon)$. Then $f_S(-\varepsilon) = -f_S^*(\varepsilon)$ and $g_S(-\varepsilon) = g_S^*(\varepsilon)$ and we obtain $\sin\theta^*(-\varepsilon) = -\sin\theta(\varepsilon)$ and $\cos\theta^*(-\varepsilon) = \cos\theta(\varepsilon)$. Due to the absence of the spin flip scattering, these relations mean that the OTE pairing state is induced in the DN.

4. Junction with OSO superconductor

   We have $f_\pm(\varepsilon) = -f_\pm^*(-\varepsilon)$, $g_\pm(\varepsilon) = g_\pm^*(-\varepsilon)$ and $f_S(-\varepsilon) = f_S^*(\varepsilon)$, $g_S(-\varepsilon) = g_S^*(\varepsilon)$. One can show that $\sin\theta^*(-\varepsilon) = \sin\theta(\varepsilon)$ and $\cos\theta^*(-\varepsilon) = \cos\theta(\varepsilon)$. Following the same lines as in case (1), we conclude that the ESE pairing state is induced in the DN.

The central conclusions are summarized in the table below.

	Symmetry of the pairing in superconductors	Symmetry of the pairing in the DN
(1)	Even-frequency spin-singlet even-parity (ESE)	ESE
(2)	Even-frequency spin-triplet odd-parity (ETO)	OTE
(3)	Odd-frequency spin-triplet even-parity (OTE)	OTE
(4)	Odd-frequency spin-singlet odd-parity (OSO)	ESE

Note that for even-parity superconductors the resulting symmetry of the induced pairing state in the DN is the same as that of a superconductor (the cases (1), (3)). On the contrary, for odd-parity superconductors, the induced pairing state in the DN has symmetry different from that of a superconductor (the cases (2), (4)).

To illustrate the main features of the proximity effect in all the above cases, we calculate the LDOS $\rho(\varepsilon) = \mathrm{Real}[\cos\theta]$ in the middle of the DN layer.

We start from junctions with ESE superconductors and choose the s-wave pair potential with $\Psi_\pm = 1$. The LDOS has a gap (Fig. 5.1a) and the Real(Imaginary) part of $f(\varepsilon)$ is an even(odd) function of ε consistent with the formation of the even-frequency pairing.

In junctions with ETO superconductors, we choose p_x-wave pair potential with $\Psi_+ = -\Psi_- = \cos\phi$ as a typical example. In this case, an unusual proximity effect is induced where the resulting LDOS has a zero energy peak (ZEP) [28–31]

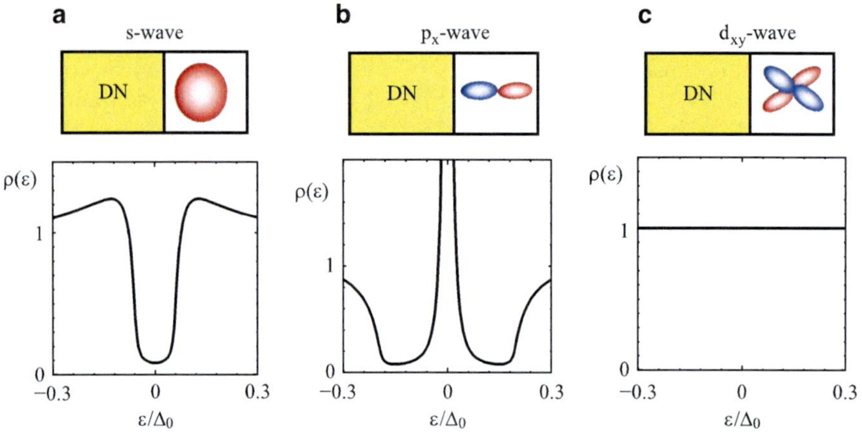

Fig. 5.1 Local density of states in the diffusive normal metal DN in a contact with s-wave superconductor (**a**), p_x-wave superconductor (**b**) and d_{xy}-wave superconductor (**c**)

as illustrated in Fig.5.1b. The resulting LDOS has a ZEP [28–31] since $f(\varepsilon = 0)$ becomes a purely imaginary number. This is consistent with $f(\varepsilon) = -f^*(-\varepsilon)$ and the formation of the OTE pairing in the DN.

It is instructive to compare the ETO state with p_x-wave pair potential and the ESE state with d_{xy}-wave pairing. In the latter case, as seen from Fig. 5.1c, there is no subgap structure at all in the LDOS in DN. This feature can be used to distinguish p_x-wave state from d_{xy}-wave one in tunneling experiments.

In summary, we considered four symmetry classes in a superconductor allowed by Pauli principle: (1) even-frequency spin-singlet even-parity (ESE), (2) even-frequency spin-triplet odd-parity (ETO), (3) odd-frequency spin-triplet even-parity (OTE) and (4) odd-frequency spin-singlet odd-parity (OSO). We have found that the resulting symmetry of the induced pairing state in the DN is (1) ESE (2) OTE (3) OTE and (4) ESE corresponding to the above four classes. When the even (odd) frequency pairing is induced in the DN, the resulting LDOS has a gap (peak) at zero energy.

5.3 Junctions in the Clean Limit

In this Section, we present the results of the theoretical study of the induced odd-frequency pairing state in ballistic normal metal/superconductor (N/S) junctions where a superconductor has even-frequency symmetry in the bulk and a normal metal layer has an arbitrary length.

We show that if a superconductor has even-parity pair potential (spin-singlet s-wave state), the odd-frequency pairing component with odd-parity is induced near the N/S interface, while in the case of odd-parity pair potential (spin-triplet p_x-

wave or spin-singlet d_{xy}-wave) the odd-frequency component with even-parity is generated. In conventional s-wave junctions, the amplitude of the odd-frequency pairing state is strongest in the case of fully transparent N/S interface and is enhanced at energies corresponding to the peaks in the local density of states (LDOS). In p_x- and d_{xy}-wave junctions, the amplitude of the odd-frequency component on the S side of the N/S interface is enhanced at zero energy, where the midgap Andreev resonant state (MARS) appears due to the sign change of the pair potential. The odd-frequency component extends into the N region and exceeds the even-frequency component at energies corresponding to the LDOS peak positions, including the MARS. At the edge of the N region, the odd-frequency component is non-zero while the even-frequency one vanishes.

In the following, we consider a N/S junction as the simplest example of non-uniform superconducting system without impurity scattering. Both cases of spin-triplet odd-parity and spin-singlet even-parity symmetries are considered in the superconductor. We assume a thin insulating barrier located at the N/S interface ($x = 0$) with N ($-L < x < 0$) and S ($x > 0$). The length of the normal region is L.

The quasiclassical Green's functions in a normal metal (N) and a superconductor (S) in the Matsubara frequency representation are parameterized as

$$\hat{g}_{\pm}^{(i)} = f_{1\pm}^{(i)}\hat{\tau}_1 + f_{2\pm}^{(i)}\hat{\tau}_2 + g_{\pm}^{(i)}\hat{\tau}_3, \quad (\hat{g}_{\pm}^{(i)})^2 = \hat{1}, \tag{5.6}$$

where the subscript $i\,(= N, S)$ refer to N and S, respectively. Here, $\hat{\tau}_j$ ($j = 1, 2, 3$) are Pauli matrices and $\hat{1}$ is a unit matrix. The subscript $+(-)$ denotes the left (right) going quasiparticles [43]. Functions $\hat{g}_{\pm}^{(i)}$ satisfy the Eilenberger equation [46]

$$\mathrm{i}v_{\mathrm{F}x}\hat{g}_{\pm}^{(i)} = \mp[\hat{H}_{\pm}, \hat{g}_{\pm}^{(i)}] \tag{5.7}$$

with

$$\hat{H}_{\pm} = \mathrm{i}\omega_n\tau_3 + \mathrm{i}\bar{\Delta}_{\pm}(x)\tau_2. \tag{5.8}$$

Here, $v_{\mathrm{F}x}$ is the x component of the Fermi velocity, $\omega_n = 2\pi T(n + 1/2)$ is the Matsubara frequency, n is an integer number and T is temperature. $\bar{\Delta}_{+}(x)$ ($\bar{\Delta}_{-}(x)$) is the effective pair potential for left (right) going quasiparticles. In the N region, $\bar{\Delta}_{\pm}(x)$ is set to zero due to the absence of a pairing interaction in the N metal. The above Green's functions can be expressed as

$$f_{1\pm}^{(i)} = \pm\mathrm{i}(F_{\pm}^{(i)} + D_{\pm}^{(i)})/(1 - D_{\pm}^{(i)}F_{\pm}^{(i)}), \tag{5.9}$$

$$f_{2\pm}^{(i)} = -(F_{\pm}^{(i)} - D_{\pm}^{(i)})/(1 - D_{\pm}^{(i)}F_{\pm}^{(i)}), \tag{5.10}$$

$$g_{\pm}^{(i)} = (1 + D_{\pm}^{(i)}F_{\pm}^{(i)})/(1 - D_{\pm}^{(i)}F_{\pm}^{(i)}).$$

Functions $D_{\pm}^{(i)}(x)$ and $F_{\pm}^{(i)}(x)$ satisfy the Ricatti equations [47–49] in the N and S regions, supplemented by the proper boundary conditions [39–42].

Here, we consider the situation without mixing of different symmetry channels for the pair potential. Then the pair potential $\bar{\Delta}_{\pm}(x)$ is expressed by

$$\bar{\Delta}_{\pm}(x) = \Delta(x)\Phi_{\pm}(\theta)\Theta(x) \tag{5.11}$$

with the form factor $\Phi_{\pm}(\theta)$ given by $\Phi_{\pm}(\theta) = 1, \pm\sin 2\theta$, and $\pm\cos\theta$ for s-wave, d_{xy}-wave, and p_x-wave superconductors, respectively. The pair potential $\Delta(x)$ is determined by the self-consistent equation

$$\Delta(x) = \frac{2T}{\log\frac{T}{T_C} + \sum_{n\geq 1}\frac{1}{n-\frac{1}{2}}}\sum_{n\geq 0}\int_{-\pi/2}^{\pi/2} \mathrm{d}\theta G(\theta) f_{2+} \tag{5.12}$$

with $G(\theta) = 1$ for s-wave case and $G(\theta) = 2\Phi(\theta)$ for other cases, respectively [50]. T_C is the transition temperature of the superconductor. The condition in the bulk is $\Delta(\infty) = \Delta_0$. Since the pair potential $\bar{\Delta}(x)$ is a real quantity, the resulting $f_{1\pm}$ is an imaginary quantity and $f_{2\pm}$ is a real one.

Before performing actual numerical calculations, we now discuss general properties of the pair amplitude. In the following, we explicitly write $f_{1\pm}^{(i)} = f_{1\pm}^{(i)}(\omega_n,\theta)$, $f_{2\pm}^{(i)} = f_{2\pm}^{(i)}(\omega_n,\theta)$, $F_{\pm}^{(i)} = F_{\pm}^{(i)}(\omega_n,\theta)$ and $D_{\pm}^{(i)} = D_{\pm}^{(i)}(\omega_n,\theta)$. For the limit $x=\infty$, we obtain

$$f_{1\pm}^{(S)}(\omega_n,\theta) = 0, \quad f_{2\pm}^{(S)}(\omega_n,\theta) = \frac{\Delta_0\Phi_{\pm}(\theta)}{\sqrt{\omega_n^2 + \Delta_0^2\Phi_{\pm}^2(\theta_{\pm})}}. \tag{5.13}$$

Note that $f_{1\pm}^{(i)}(\omega_n,\theta)$ becomes finite due to the spatial variation of the pair potential and it does not exist in the bulk. One can show that $D_{\pm}^{(i)}(-\omega_n,\theta) = 1/D_{\pm}^{(i)}(\omega_n,\theta)$ and $F_{\pm}^{(i)}(-\omega_n,\theta) = 1/F_{\pm}^{(i)}(\omega_n,\theta)$. After simple manipulation, we obtain

$$f_{1\pm}^{(i)}(\omega_n,\theta) = -f_{1\pm}^{(i)}(-\omega_n,\theta), \quad f_{2\pm}^{(i)}(\omega_n,\theta) = f_{2\pm}^{(i)}(-\omega_n,\theta), \tag{5.14}$$

for any x. It is remarkable that functions $f_{1\pm}^{(i)}(\omega_n,\theta)$ and $f_{2\pm}^{(i)}(\omega_n,\theta)$ correspond to odd-frequency and even-frequency components of the pair amplitude, respectively. Function $f_{1\pm}^{(1)}(\omega_n,\theta)$ describes the odd-frequency component of the pair amplitude penetrating from the superconductor.

Next, we discuss the parity of these pair amplitudes. The even-parity (odd-parity) pair amplitude should satisfy the following relation $f_{j\pm}^{(i)}(\omega_n,\theta) = f_{j\mp}^{(i)}(\omega_n,-\theta)$ $[f_{j\pm}^{(i)}(\omega_n,\theta) = -f_{j\mp}^{(i)}(\omega_n,-\theta)]$, with $j = 1,2$. For an even-parity (odd-parity) superconductor, $\Phi_{\pm}(-\theta) = \Phi_{\mp}(\theta)$ $[\Phi_{\pm}(-\theta) = -\Phi_{\mp}(\theta)]$. Then we can show that for the even-parity case

$$D_{\pm}^{(i)}(-\theta) = D_{\mp}^{(i)}(\theta), \quad F_{\pm}^{(i)}(-\theta) = F_{\mp}^{(i)}(\theta) \tag{5.15}$$

and for the odd-parity case

$$D_{\pm}^{(i)}(-\theta) = -D_{\mp}^{(i)}(\theta), \quad F_{\pm}^{(i)}(-\theta) = -F_{\mp}^{(i)}(\theta), \tag{5.16}$$

respectively.

The resulting $f_{1\pm}^{(i)}(\omega_n, \theta)$ and $f_{2\pm}^{(i)}(\omega_n, \theta)$ satisfy

$$\begin{aligned} f_{1\pm}^{(i)}(\omega_n, \theta) &= -f_{1\mp}^{(i)}(\omega_n, -\theta), \\ f_{2\pm}^{(i)}(\omega_n, \theta) &= f_{2\mp}^{(i)}(\omega_n, -\theta), \end{aligned} \tag{5.17}$$

for an even-parity superconductor and

$$\begin{aligned} f_{1\pm}^{(i)}(\omega_n, \theta) &= f_{1\mp}^{(i)}(\omega_n, -\theta), \\ f_{2\pm}^{(i)}(\omega_n, \theta) &= -f_{2\mp}^{(i)}(\omega_n, -\theta), \end{aligned} \tag{5.18}$$

for an odd-parity superconductor, respectively. Note that the parity of the odd-frequency component $f_{1\pm}^{(i)}(\omega_n, \theta)$ is always different from that in the bulk superconductor.

As shown above, the odd-frequency component $f_{1\pm}^{(i)}(\omega_n, \theta)$ is a purely imaginary quantity. The underlying physics behind this formal property is follows. Due to the breakdown of translational invariance near the N/S interface, the pair potential $\bar{\Delta}(x)$ acquires a spatial dependence, which leads to the coupling between even-parity and odd-parity states. Since the bulk pair potential has an even-frequency symmetry, the Fermi-Dirac statistics requires that the order parameter component induced near the interface should be odd in frequency. The phase of the induced pair amplitude undergoes a $\pi/2$ shift from that in the bulk S, thus removing internal phase shift between the even- and odd-frequency components and making the interface-induced state compatible with the time reversal invariance. As a result, function $f_{1\pm}^{(i)}(\omega_n, \theta)$ becomes a purely imaginary quantity.

Let us now focus on the values of the pair amplitudes at the edge of N region (at $x = -L$). We concentrate on two extreme cases with (I) $\Phi_+(\theta) = \Phi_-(\theta)$ and (II) $\Phi_+(\theta) = -\Phi_-(\theta)$. In the case (I), the MARS is absent since there is no sign change of the pair potential felt by the quasiparticle at the interface. Then the relation $D_+^{(N)} = D_-^{(N)}$ holds. However, in the case (II), the MARS is generated near the interface due to the sign change of the pair potential and the relation $D_+^{(N)} = -D_-^{(N)}$ is satisfied [32–34]. At the edge $x = -L$, it is easy to show that $F_{\pm}^{(N)} = -D_{\pm}^{(N)}$ for the former case and $F_{\pm}^{(N)} = D_{\pm}^{(N)}$ for the latter one. As a result, $f_{1\pm}^{(N)} = 0$ for the case (I) and $f_{2\pm}^{(N)} = 0$ for the case (II), respectively. Thus, we can conclude that in the absence of the MARS only the even-frequency pairing component exists at $x = -L$, while in the presence of the MARS only the odd-frequency one.

To understand the angular dependence of the pair amplitude in a more detail, we define $\hat{f}_1^{(i)}$ and $\hat{f}_2^{(i)}$ for $-\pi/2 < \theta < 3\pi/2$ with $\hat{f}_{1(2)}^{(i)} = f_{1(2)+}^{(i)}(\theta)$ for $-\pi/2 < \theta < \pi/2$ and $\hat{f}_{1(2)}^{(i)} = f_{1(2)-}^{(i)}(\pi - \theta)$ for $\pi/2 < \theta < 3\pi/2$. We decompose $\hat{f}_{1(2)}^{(i)}$ into various angular momentum component as follows,

$$\hat{f}_{1(2)}^{(i)} = \sum_m S_m^{(1(2))} \sin(m\theta) + \sum_m C_m^{(1(2))} \cos(m\theta) \tag{5.19}$$

with $m = 2l + 1$ for odd-parity case and $m = 2l$ for even-parity case with integer $l \geq 0$, where l is the quantum number of the angular momentum. Here, $C_m^{(1(2))}$ and $S_m^{(1(2))}$ are defined for all x. It is straightforward to show that the only nonzero components are (1) $C_{2l}^{(2)}$ and $C_{2l+1}^{(1)}$ for even-parity superconductor without sign change at the interface (i.e., s-wave or $d_{x^2-y^2}$-wave), (2) $S_{2l+2}^{(2)}$ and $S_{2l+1}^{(1)}$ for d_{xy}-wave, (3) $C_{2l+1}^{(2)}$ and $C_{2l}^{(1)}$ for p_x-wave, and (4) $S_{2l+1}^{(2)}$ and $S_{2l}^{(1)}$ for p_y-wave junctions, respectively. The allowed angular momenta for odd-frequency components are $2l + 1$, $2l + 1$, $2l$, and $2l + 2$ corresponding to each of the above four cases.

To get better insight into the spectral property of the odd-frequency pair amplitude, we perform an analytical continuation from the Matsubara frequency ω_n to the quasiparticle energy ε measured from the chemical potential. The retarded Green's function corresponding to (5.1) is defined as $\hat{g}_{\pm}^{(i)R} = f_{1\pm}^{(i)R}\hat{\tau}_1 + f_{2\pm}^{(i)R}\hat{\tau}_2 + g_{\pm}^{(i)R}\hat{\tau}_3$. One can show that $f_{1\pm}^{(i)R}(-\varepsilon) = -[f_{1\pm}^{(i)R}(\varepsilon)]^*$, $f_{2\pm}^{(i)R}(-\varepsilon) = [f_{2\pm}^{(i)R}(\varepsilon)]^*$, and $g_{\pm}^{(i)R}(-\varepsilon) = [g_{\pm}^{(i)R}(\varepsilon)]^*$. The LDOS $\rho(\varepsilon)$ at the N/S interface at $x = 0$ normalized to its value in the normal state is given by

$$\rho(\varepsilon) = \int_{-\pi/2}^{\pi/2} d\theta \mathrm{Real}\left(\frac{g_{+}^{(i)R}(\varepsilon) + g_{-}^{(i)R}(\varepsilon)}{2\pi}\right). \tag{5.20}$$

Let us discuss the case of s-wave superconductor junctions as shown in Fig. 5.2. By changing the length L of the N region and the transparency at the interface, we calculate the spatial dependence of the pair potential and the pair amplitudes in the Matsubara frequency representation. We only concentrate on the lowest angular momentum of the even-frequency pair amplitude $C_0^{(2)}$. As regards the odd-frequency pair amplitudes, we focus on the $C_1^{(1)}$, $C_3^{(1)}$ and $C_5^{(1)}$ components which all have odd-parity and depend on θ as $\cos\theta$, $\cos 3\theta$ and $\cos 5\theta$, respectively, and correspond to p_x-wave, f_1-wave and h_1-wave components shown in Fig. 5.1. In all cases, even-frequency component is constant in the S region far away from the interface and the corresponding odd-frequency components are absent. The s-wave pair potential is suppressed for the fully transparent case ($Z = 0$), while it is almost constant for low transparent case ($Z = 5$). It does not penetrate into the N region due to the absence of the attractive interaction in the N metal. On the

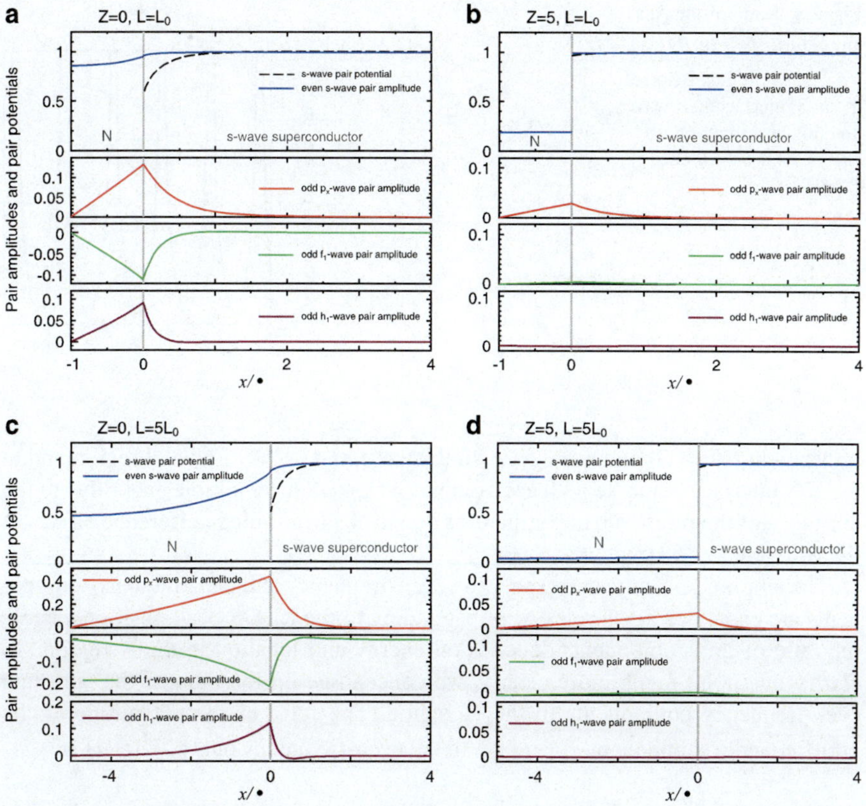

Fig. 5.2 Spatial dependence of the normalized pair potential, even-frequency pair amplitude and odd-frequency components of the pair amplitude for s-wave superconductor junctions. Here, we choose $\xi = v_F/\Delta_0$ in the S region ($x > 0$) and $\xi = L_0 = v_F/2\pi T_C$ in the N region. The pair amplitudes $C_0^{(2)}$, $C_1^{(1)}$, $C_3^{(1)}$, and $C_5^{(1)}$ are denoted as even s-wave, odd p_x-wave, odd f_1-wave, and odd h_1-wave pair amplitudes. (**a**) $Z = 0$, $L = L_0$, (**b**) $Z = 5$, $L = L_0$, (**c**) $Z = 0$, $L = 5L_0$, and (**d**) $Z = 5$, $L = 5L_0$, respectively

contrary, in all considered cases the spatial variation of the even-frequency s-wave pair amplitude is rather weak in the S region, while in the N region it is strong for $Z = 0$ and is reduced for $Z = 5$ since the proximity effect is weaker in the latter case. The odd-frequency component always vanishes at $x = -L$ and does not have a jump at the N/S interface even for nonzero Z. Its amplitude is strongly enhanced near the N/S interface especially for fully transparent junctions. Note that not only the p_x-wave but also f_1-wave and h_1-wave have sufficiently large magnitudes as shown in Fig. 5.2a, c. With the decrease in the transparency of the N/S interface, the odd-frequency components are suppressed as shown in Fig. 5.2b, d.

It is instructive to discuss the spectral properties of the induced pairing state in the N region. Here, we concentrate on the situation when the N/S/ interface is fully

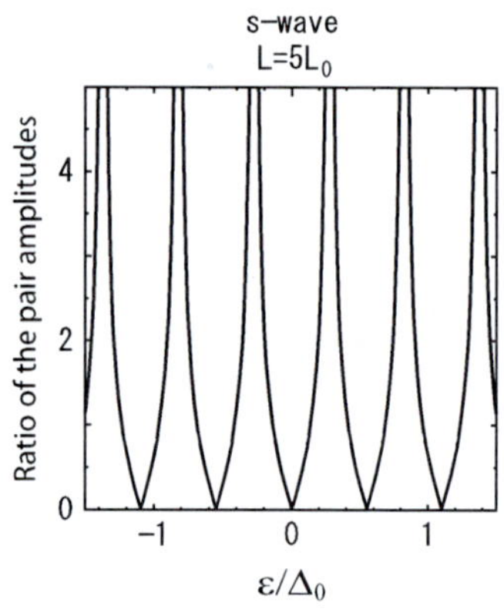

Fig. 5.3 Ratio of the pair amplitudes $f_{1+}^{(N)}(\varepsilon,\theta)/f_{2+}^{(N)}(\varepsilon,\theta)$ on the N-side of the N/S interface in s-wave junction as a function of energy ε for $\theta = 0$ and $L = 5L_0$

transparent ($Z = 0$) and $L = 5L_0$. In this case, the LDOS in the N region and at the N/S interface coincide with each other. The LDOS has multiple peaks due to the existence of the multi-sub gap structures due to electron-hole interference effects in the N region [52, 53].

The amplitudes of the corresponding even-frequency and odd-frequency components are enhanced at energies ε corresponding to the LDOS peak positions, while the ratio of this components depends on energy and location in the N region. To clarify this point much more clearly, we concentrate on the ratio of the odd- and even-frequency components in the N region. The ratio of the magnitude of the odd-frequency component $f_{1+}^{(N)}(\varepsilon,\theta)$ to the even-frequency one $f_{2+}^{(N)}(\varepsilon,\theta)$ is

$$\frac{\left|f_{1+}^{(N)}(\varepsilon,\theta)\right|}{\left|f_{2+}^{(N)}(\varepsilon,\theta)\right|} = \left|\tan\left(\frac{2\varepsilon}{v_{\mathrm{Fx}}}(L+x)\right)\right|. \tag{5.21}$$

At the edge of the N region, $x = -L$, the odd-frequency component vanishes at all energies. On the other hand, very interesting situation occurs at the N/S interface, $x = 0$ as will be shown below. In Fig. 5.3, we plot this ratio for $\theta = 0$ and $x = 0$.

It is remarkable that at some energies the amplitude of the odd-frequency pair amplitude exceeds that of the even-frequency one.

Let us clarify the relation between the positions of the bound states and the above ratio of the odd-to-even pair amplitude. In the limit $L >> L_0$, the bound states are determined by simple relation [52, 53]

$$\varepsilon_n = \frac{\pi v_{\mathrm{Fx}}}{2L}(n + 1/2), \quad n = 0, 1, 2, \ldots \tag{5.22}$$

That means that at the subgap peak energies the odd-frequency component dominates over the even-frequency one at the N/S interface. This is a remarkable property of the odd-frequency pairing, which makes it relevant to the classical McMillan-Rowell oscillations in the N/S geometry [52, 53]. To summarize, we have shown that the odd-frequency component is present even in the standard case of a ballistic N/S system, and it dominates at energies when the LDOS has subgap peaks.

5.4 Summary

Using the quasiclassical Green's function formalism, we have shown that the odd-frequency pairing state is ubiquitously generated in the normal metal/superconductor (N/S) ballistic junction system, where the length of the normal region is finite. It is shown that the even-parity (odd-parity) pair potential in the superconductor induces the odd-frequency pairing component with spin-singlet odd-parity (spin-triplet even-parity). Even for conventional s-wave junctions, the amplitude of the odd-frequency pairing state is enhanced at the N/S interface with fully transparent barrier. By analyzing the spectral properties of the pair amplitudes, we found that the magnitude of the resulting odd-frequency component at the interface can exceed that of the even-frequency one. For the case of p_x-wave and d_{xy}-wave junctions, the magnitude of the odd-frequency component at the S side of the N/S interface is significantly enhanced. The magnitude of the induced odd-frequency component is enhanced in the presence of the midgap Andreev resonant state due to the sign change of the anisotropic pair potential at the interface. The LDOS has a zero energy peak (ZEP) both at the interface and in the N region. At the edge of the N region, only the odd-frequency component is non-zero.

The underlying physics behind these phenomena is related to the breakdown of translational invariance near the N/S interface, where the pair potential $\bar{\Delta}(x)$ acquires a spatial dependence. As a result, an odd-frequency component is quite generally induced near the interface. The breakdown of translational invariance is the strongest when the pair potential changes sign upon reflection like in the case of p_x-wave and d_{xy}-wave junctions, then the magnitude of odd-frequency component is the largest. Moreover, the phase of the interface-induced odd-frequency component has a $\pi/2$ shift from that in the bulk of S. Therefore, as shown above, the odd-frequency component $f_{1\pm}^{(i)}(\omega_n, \theta)$ becomes purely imaginary quantity and the peak structure in the LDOS naturally follows from the normalization condition.

We have also shown that in the N/S junctions with s-wave superconductors the classical McMillan-Rowell oscillations [51–53] can also be reinterpreted in terms of odd-frequency pairing. At the energies corresponding to the subgap peaks in the N/S junction, the odd-frequency component dominates over the even-frequency one. This is remarkable application of the odd-frequency pairing concept.

Acknowledgements The authors express sincere gratitude to clarifying discussions with M. Eschrig, Ya.V. Fominov, Y. Fuseya, S. Kashiwaya, K. Miyake, K. Nagai, Yu. V. Nazarov, V. Ryazanov, M. Ueda and A.D. Zaikin. This work is partially supported by Grant-in-Aid for Scientific Research (Grant No. 17071007 and 17340106) from the Ministry of Education, Culture, Sports, Science and Technology of Japan, by Japan Society for the Promotion of Science (JSPS) and by NanoNed project TCS.7029.

References

1. M. Sigrist, K. Ueda, Rev. Mod. Phys. **63**, 239 (1991)
2. V.L. Berezinskii, JETP Lett. **20**, 287 (1974)
3. A. Balatsky, E. Abrahams, Phys. Rev. B **45**, 13125 (1992)
4. E. Abrahams, A. Balatsky, D.J. Scalapino, J.R. Schrieffer, Phys. Rev. B **52**, 1271 (1995)
5. M. Vojta, E. Dagotto, Phys. Rev. B **59**, R713 (1999)
6. Y. Fuseya, H. Kohno, K. Miyake, J. Phys. Soc. Jpn. **72**, 2914 (2003)
7. F.S. Bergeret, A.F. Volkov, K.B. Efetov, Phys. Rev. Lett. **86**, 4096 (2001)
8. F.S. Bergeret, A.F. Volkov, K.B. Efetov, Rev. Mod. Phys. **77**, 1321 (2005)
9. M.L. Kulic, M. Endres, Phys. Rev. B **62**, 11846 (2000)
10. A. Kadigrobov, R.I. Shekhter, M. Jonson, Eutophys. Lett. **90**, 394 (2001)
11. M. Eschrig, J. Kopu, J.C. Cuevas, G. Schön, Phys. Rev. Lett. **90**, 137003 (2003)
12. Ya.V. Fominov, A.A. Golubov, M.Yu. Kupriyanov, JETP Leet. **77**, 510 (2003)
13. M. Eschrig, T. Löfwander, Nat. Phys. **4**, 138 (2008)
14. V. Braude, Yu.V. Nazarov, Phys. Rev. Lett. **98**, 077003 (2007)
15. Y. Asano, Y. Tanaka, A.A. Golubov, Phys. Rev. Lett. **98**, 107002 (2007)
16. J. Linder, T. Yokoyama, A. Sudbø, M. Eschrig, Phys. Rev. Lett. **102**, 107008 (2009)
17. R.S. Keizer et al., Nature **439**, 825 (2006)
18. M.S. Anwar, F. Czeschka, M. Hesselberth, M. Porcue, J. Aarts J, Phys. Rev. B **82**, 100501(R) (2010)
19. T.S. Khaire, M.A. Khasawneh, W.P. Pratt Jr., N.O. Birge, Phys. Rev. Lett. **104**, 137002 (2010)
20. W.A. Robinson, J.D.S. Witt, M.G. Blamire, Science **329**, 59 (2010)
21. A. I. Buzdin, Rev. Mod. Phys. **77**, 935 (2005)
22. A.A.Golubov, M.Yu.Kupriyanov, E.Ilichev, Rev. Mod. Phys. **76**, 401 (2004)
23. V.V. Ryazanov, V.A. Oboznov, A.Yu. Rusanov, A.V. Veretennikov, A.A. Golubov, J. Aarts, Phys. Rev. Lett. **86**, 2427 (2001)
24. W. Belzig, F.K. Wilhelm, C. Bruder, G. Schön, A.D. Zaikin, Superlattice Microst. **25**, 1251 (1999)
25. A.A. Golubov, M.Yu. Kupriyanov, E, Il'ichev, Rev. Mod. Phys. **76**, 411 (2004)
26. Y. Tanaka, Yu.V. Nazarov, S. Kashiwaya, Phys. Rev. Lett. **90**, 167003 (2003)
27. Y. Tanaka, Yu.V. Nazarov, A.A. Golubov, S. Kashiwaya, Phys. Rev. B **69**, 144519 (2004)
28. Y. Tanaka, S. Kashiwaya, Phys. Rev. B **70** 012507 (2004)
29. Y. Tanaka, S. Kashiwaya, T. Yokoyama, Phys. Rev. B **71** 094513 (2005)
30. Y. Tanaka, Y. Asano, A.A. Golubov, S. Kashiwaya, Phys. Rev. B **72**, 140503(R) (2005)
31. Y. Asano, Y. Tanaka, S. Kashiwaya, Phys. Rev. Lett. **96** 097007 (2006)
32. Y. Tanaka, S. Kashiwaya, Phys. Rev. Lett. **74** 3451 (1995)
33. S. Kashiwaya, Y. Tanaka, Rep. Prog. Phys. **63**, 1641 (2000)
34. T. Löfwander, V.S. Shumeiko, G. Wendin, Supercond. Sci. Technol. **14**, R53 (2001)
35. L.J. Buchholtz, G. Zwicknagl, Phys. Rev. B **23** 5788 (1981)
36. J. Hara, K. Nagai, Prog. Theor. Phys. **74**, 1237 (1986)
37. C. Bruder, Phys. Rev. B **41**, 4017 (1990)
38. C.R. Hu, Phys. Rev. Lett. **72**, 1526 (1994)
39. Y. Tanaka, A.A. Golubov, Phys. Rev. Lett. **98**, 037003 (2007)
40. Y. Tanaka, A.A. Golubov, S. Kashiwaya, M. Ueda, Phys. Rev. Lett. **99**, 037005 (2007)
41. Y. Tanaka, Y. Tanuma, A.A. Golubov, Phys. Rev. B **76** 054522 (2007)
42. Y. Asano, Y. Tanaka, A.A. Golubov, Phys. Rev. Lett. **98**, 107002 (2007)
43. J.W. Serene, D. Rainer, Phys. Rep. **101**, 221 (1983)
44. M. Eschrig, Phys. Rev. B **61**, 9061 (2000)
45. K.D. Usadel, Phys. Rev. Lett. **25**, 507 (1970)
46. G. Eilenberger, Z. Phys. **214**, 195 (1968)
47. A. Shelankov, M. Ozana, Phys. Rev. B **61**, 7077 (2000)
48. N. Schopohl, K. Maki, Phys. Rev. B **52**, 490 (1995)

49. C. Iniotakis, S. Graser, T. Dahm, N. Schopohl, Phys. Rev. B **71** 214508 (2005)
50. M. Matsumoto, H. Shiba, J. Phys. Soc. Jpn. **64** 3384 (1995)
51. W.L. McMillan, Phys. Rev. **175** 537 (1968)
52. J.M. Rowell, W.L. McMillan, Phys. Rev. Lett. **16** 453 (1966)
53. J.M. Rowell, Phys. Rev. Lett. **30** 167 (1973)

Chapter 6
Ferromagnetic Josephson Junctions with Critical Current Density Artificially Modulated on a "Short" Scale

N.G. Pugach, M.Yu. Kupriyanov, E. Goldobin, D. Koelle, R. Kleiner, A.S. Sidorenko, and C. Lacroix

Abstract We study the Josephson effect in junctions with a ferromagnetic (F) barrier having its properties (interface transparency or the F-layer thickness) artificially modulated on a scale less than the Josephson penetration length. Within the framework of the quasiclassical Usadel equations, we describe SIFS and SIFNS (S is a superconductor, I is an insulator, N is a normal metal) structures with a step-like transparency of the FS or NS interface. The step-like change in parameters may lead to oscillations (including sign change) of the critical current density $J_C(y)$ along the junction in the vicinity of the step, resulting in the formation of a 0–π nano-junction near the step. Such structures exhibit an unusual behaviour in an external magnetic field H. The properties of arrays of nano-junctions with

N.G. Pugach (✉)
Faculty of Physics, M.V. Lomonosov Moscow State University,
119991 Leninskie Gory, GSP-2, Moscow, Russia
e-mail: pugach@magn.ru

M.Yu. Kupriyanov
Skobeltsyn Institute of Nuclear Physics, M.V. Lomonosov Moscow State University,
119991 Leninskie Gory, GSP-1, Moscow, Russia
e-mail: mkupr@pn.sinp.msu.ru

E. Goldobin · D. Koelle · R. Kleiner
Physikalisches Institut–Experimentalphysik II and Center for Collective Quantum Phenomena,
Universität Tübingen, Auf der Morgenstelle 14, D-72076 Tübingen, Germany
e-mail: gold@uni-tuebingen.de; koelle@uni-tuebingen.de; kleiner@uni-tuebingen.de

A.S. Sidorenko
Institute of Electronic Engineering and Nanotechnologies ASM, MD2028 Kishinev, Moldova,
Institute of Nanotechnology, Karlsruhe Institute of Technology (KIT), D-76021 Karlsruhe,
Germany
e-mail: anatoli.sidorenko@kit.edu

C. Lacroix
Institut N'eel, CNRS-UJF, BP 166, 38042 Grenoble Cedex 9, France
e-mail: claudine.lacroix@grenoble.cnrs.fr

A. Sidorenko (ed.), *Fundamentals of Superconducting Nanoelectronics*,
NanoScience and Technology, DOI 10.1007/978-3-642-20158-5_6,

several transparency steps are also investigated. We propose a method to realize a φ Josephson junction by combining alternating 0 and π parts made of "clean" SFS sub junctions with different F-layer thickness and an intrinsically non-sinusoidal current–phase relation (CPR). The latter can significantly enlarge the parameter range of the φ ground state and make the practical realization of φ Josephson junctions feasible. Such junctions may also have two different stable solutions, such as 0 and π, 0 and φ, or φ and π.

6.1 Introduction

The interest in Josephson junction (JJ) devices with a ferromagnetic barrier has been continuously increasing during the last years [1, 2]. Such a junction consists of two superconducting electrodes (S) separated by a ferromagnetic layer (F). It may include also a thin insulating tunnel barrier (I) and/or a normal metal layer (N), i.e. SFS, SIFS, or SIFNS multilayers may be considered. The critical current density J_c of an SFS junction exhibits damped oscillations as a function of the F-layer thickness d_F so that the Josephson phase ϕ can be 0 or π in the ground state [1, 2]. π junctions can be used as (non-dischargeable) on-chip π-phase batteries for self-biasing various electronic circuits in the classical and quantum domains, for example self-biased RSFQ logic [3] or flux qubits [4, 5]. In addition, for quantum circuits self-biasing also decouples the circuit from the environment and improves decoherence figures, for example in the quiet qubit. [4, 6, 7] In classical circuits, a phase battery may also substitute the conventional inductance and substantially reduce the size of an elementary cell [8]. Some of these proposals were already realized practically [3, 9].

Modern technology allows to manufacture not only 0 or π JJs, but also so-called 0–π Josephson junctions (see Fig. 6.1), i.e. junctions some parts of which behave as 0 junctions and other parts as π junctions [10]. In these structures, intensively studied experimentally, the different sign of J_C can be achieved by introducing a step-like change of the thickness of the F layer [11–15].

The interest in these 0–π junctions has been stimulated by the existence of unusual topological vortex solutions. A spontaneous Josephson vortex carrying a fraction of the magnetic flux quantum $\Phi_0 \approx 2.07 \times 10^{-15}$ Wb may appear at the 0–π boundary [10, 16, 17]. In the region where the phase ϕ changes from 0 to π there is a non-zero gradient $\partial\phi/\partial y$ of the Josephson phase along the junction that is proportional to the local magnetic field. In essence, this field is created by supercurrents $\sim \sin(\phi)$ circulating in this region. These currents are localized in a λ_J-vicinity of the 0–π boundary (λ_J is the Josephson penetration depth) and create a vortex of supercurrent with total magnetic flux equal to $\pm\Phi_0/2$, whereas a usual Josephson vortex carries $\pm\Phi_0$, provided that the junction length $L \gg \lambda_J$. In the case of $L \lesssim \lambda_J$ the spontaneous flux [10, 16, 18–20] $|\Phi| < \Phi_0/2$. It was shown theoretically [20, 21] and indicated in experiments [11, 14, 22] that for certain conditions the existence of a fractional Josephson vortex at the 0–π boundary is

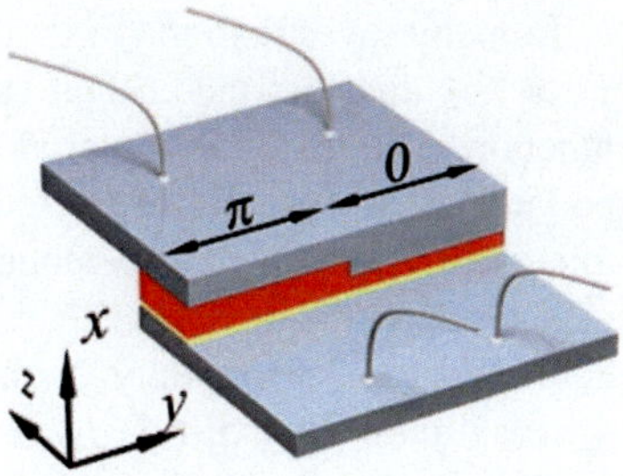

Fig. 6.1 Sketch of the 0–π ferromagnetic Josephson junction

energetically favorable in the ground state. The fractional vortex is pinned at the 0–π boundary and has two polarities that may be used for information storage and processing in the classical and quantum domains, for example to build JJ-based qubits [23]. We note that the fractional vortex described above is always pinned and is different from fractional Josephson vortices that are the solutions of a double sine-Gordon equation [24–26].

Not only single Josephson junctions, but also superconducting loops intersected by two JJs (dc SQUIDs) and JJ arrays may be used in applications. Such arrays consist of N Josephson junctions connected as a one-dimensional parallel chain in such a way that $N - 1$ individual superconducting loops are formed. Such an array exhibits an unusual dependence of its mean voltage on the magnetic field H for an overcritical applied bias current. If the loops are identical, the voltage response $V(H)$ is Φ_0-periodic. For JJ arrays with incommensurate loop areas the voltage response $V(H)$ is non-periodic, and can have a rather sharp dip at $H = 0$. This property may be used to create a sensitive absolute field magnetometer that is called superconducting quantum interference filter (SQIF) [27–31]. So far, these SQIFs are based on usual JJs. However, recently it was also suggested to realize 0–π SQUIFs, using constriction junctions in d-wave superconductors [32]. In the presented work we suggest SQIF-like structures of a new type based on 0–π s-wave JJs with a ferromagnetic barrier.

Recently, a set of SIFS JJs having a variety of different shapes was demonstrated experimentally: rectangular 0–π, 0–π–0, 20 × (0-π-); annular 0–π JJ and disk-shaped 0–π JJ, where the 0–π boundary forms a loop [33]. The Josephson supercurrent transport in all such structures was visualized [33] using Low Temperature Scanning Electron Microscopy (LTSEM) and current counterflow between 0 and π regions was confirmed. We note that among all 0–π JJ technologies available nowadays, only SIFS or SFS technology allows to fabricate JJs with a topology where the 0–π boundary forms a loop. Furthermore, the demonstration of a JJ consisting of 20 × (0-π-) periods was motivated by the possibility to produce a so-called φ JJ [24, 34], but this aim was not reached yet. Instead, in such structures (not only on SIFS, but also in Nb/cuprate ramp zigzag JJs [35, 36]) one observes experimentally an $I_c(H)$ dependence which is noticeably different from predicted theoretically [35]. It was recently found that such structures are extremely sensitive to non-uniformities of magnetic field (that arise as a result of field focusing by electrodes) and also to the misalignment angle [37].

In many investigations of 0–π JJs, it is assumed that the length of the junction $L \gg \lambda_J$ and that the critical current density is uniform along every part of the junction. The peculiarities arising on the nano-scale in the vicinity of the 0–π boundaries are usually neglected. In this work, we show in the framework of a microscopic theory of superconductivity that if some property of the JJ changes in a step-wise manner, its critical current density J_C may have a peculiar oscillatory dependence in the vicinity of the step. This leads to an unusual dependence of the maximum supercurrent $I_{max}(H)$ vs. the external magnetic field. Instead of the Fraunhofer pattern usually observed for a uniform JJ, I_{max} grows linearly with H on a comparatively large interval. This behaviour can be realized for certain junction parameters, for example for specific values of the ferromagnetic layer thickness d_F. We have obtained this result for SIFS JJs assuming the existence of a step-like non-uniformity in the transparency of the FS boundary, which should be sharp on the scale of the ferromagnetic coherence length ξ_F [38]. The Josephson junction with a ferromagnetic barrier has a complex coherence length

$$\xi^{-1} = \xi_1^{-1} + \mathrm{i}\xi_2^{-1} \tag{6.1}$$

In the "dirty" limit (electron mean free path $l \ll \xi_{1,2}$) if we neglect additional pair-breaking mechanisms one can obtain $\xi_1 = \xi_2 \equiv \xi_F$ [2]. The practical realization of the structure with a sharp SF boundary step on the scale ξ_F may be difficult as the typical value of $\xi_F \sim 1$ nm. One of the possible solutions is to introduce a normal metal (N) into the junction, i.e. to use an SIFNS structure. Then, the relevant length is $\xi_N \sim 100\,\text{nm} \gg \xi_F$, as a consequence of the proximity effect between F and N layers [39]. Fabrication of such steps should not pose technological problems. The next step would be to make an array of these junctions for example as in SQIFs.

It was shown that SFS JJs with periodically changing F-layer thickness at certain conditions can have not only 0 or π, but also an arbitrary Josephson phase φ ($0 < |\varphi| < \pi$) in the ground state – such a junction is called a φ JJ [26, 32, 40]. However, these conditions [26] are practically not achievable using contemporary technology. In addition to providing an arbitrary phase bias, such φ junctions have rather interesting physical properties such as two critical currents, non-Fraunhofer Josephson current dependence on an external magnetic field, half-integer Shapiro steps and an unusual behaviour when embedded in a SQUID loop. In *long* φ junctions, two types of mobile Josephson vortices carrying fractional magnetic flux $\Phi_1 < \Phi_0$ and $\Phi_2 = \Phi_0 - \Phi_1$ may exist, resulting in half integer zero field steps, two critical values of magnetic field penetration and other unusual properties [41].

The idea to realize a φ JJ was pushed forward by considering some practical aspects [42]. Following the original idea [24, 34] to achieve a φ ground state, one needs to control the size and critical current densities in 0 and π regions with unprecedented precision. It was shown that by using SFS JJs with a clean F-layer, one can obtain a substantial contribution from the intrinsic negative second harmonic in the CPR and substantially extend the regions of the φ ground state [42].

This article is organized as follows. In Sect. 6.2, we investigate ferromagnetic Josephson junctions with step-like interface transparency and its arrays.

Section 6.2.1 describes the model based on the linearized Usadel equation for an SIFS junction with a step-like transparency of the FS boundary. Its behaviour in an external magnetic field is investigated. Section 6.2.2 describes SIFNS structures with a step-like transparency of the NS boundary. The properties of asymmetric SIFNS junctions with few different steps of the boundary transparency are presented in Sect. 6.2.3. In Sect. 6.2.4, the magnetic properties of symmetric and asymmetric periodic arrays of these junctions are described. In Sect. 6.3, we propose a method for the reliable realization of a φ Josephson junction. In Sect. 6.3.1, we describe the model and derive the equations for the effective phase using an averaging procedure over the rapid oscillations of the phase. Section 6.3.2 contains the discussion and results of calculations done for a "clean" SFS junction. Section 6.4 concludes this work. The calculation details for SIFS and SIFNS Josephson junctions can be found in the appendices.

6.2 Ferromagnetic Josephson Junctions with Step-Like Interface Transparency

6.2.1 Model for SIFS Junction

6.2.1.1 Critical Current Density Distribution

We consider a Josephson junction of length L consisting of two semi-infinite superconducting electrodes, separated by a ferromagnetic layer of thickness d_F and a thin insulating film of thickness d_I, $d_I \ll d_F$ (see Fig. 6.2).

The IF interface coincides with the yz plane, and it is assumed that the structure is homogeneous in z-direction but not in y-direction. The suppression parameters γ_{B1} and γ_{B2} for the parts of the FS interface located at $0 \leq y \leq L/2$ and $L/2 \leq y \leq L$, respectively, are supposed to be large enough to neglect the suppression of superconductivity in the S part of the SF proximity system in the structure. $\gamma_{B1(2)} = R_{B1(2)}S/2\rho_F\xi_F$, where $R_{B1(2)}$ is the resistance of the corresponding part of the FS interface, S is the area of the junction , and ρ_F is the F metal resistivity. We assume that the dirty limit condition $l \ll \xi_{F,N,S}$ is fulfilled in the F and S layers and that the effective electron–phonon coupling constant is zero in the F metal.

Let either be the temperature T close to the critical temperature T_c of the superconducting electrodes or the suppression parameters at the FS interface large enough ($\gamma_{B1}, \gamma_{B2} \gg 1$) to permit the use of the linearized Usadel equations in the F film. Under the above restrictions, the problem of calculation of the critical current density in the structure reduces to the solution of the two-dimensional linearized Usadel equation:

$$\xi_F^2 \left\{ \frac{\partial^2}{\partial x^2} + \frac{\partial^2}{\partial y^2} \right\} \Theta_F - \frac{\widetilde{\omega}}{\pi T_c} \Theta_F = 0, \tag{6.2}$$

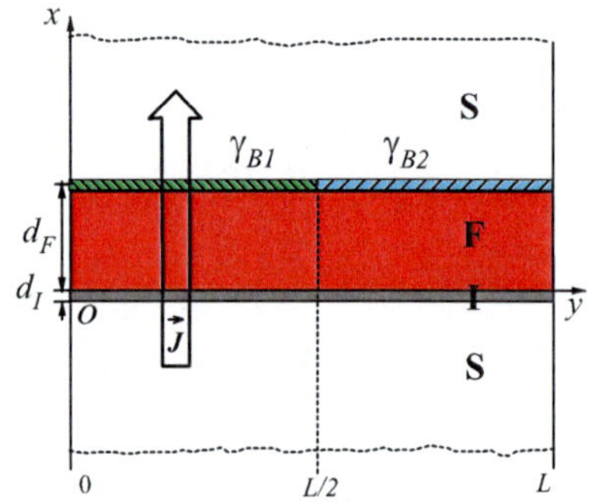

Fig. 6.2 Schematic view of the SIFS Josephson junction with a step-like change of the FS interface transparency ($\gamma_{B1} \neq \gamma_{B2}$)

where ω is the Matsubara frequency, $\widetilde{\omega} = |\omega| + \mathrm{i}E\,\mathrm{sign}(\omega)$, E is the exchange magnetic energy of the ferromagnetic material, its coherence length $\xi_F = (D_F/2\pi T_c)^{1/2}$, and D_F is the electron diffusion coefficient. $\Theta_F = \Theta_F(x, y, \omega)$ is the parameterized Usadel function, introduced by the expression $\Theta_F = \widetilde{\omega} F_F/G_F$, where F_F and G_F are Usadel functions for the ferromagnetic region. We use the units where the Plank and Boltzmann constants are $\hbar = 1$ and $k_\mathrm{B} = 1$.

Under the assumption $|\gamma_{B1} - \gamma_{B2}|/\gamma_{B1}\gamma_{B2} \ll 1$ the boundary conditions at the FS interfaces located at $x = d_F$ can be written in the form [43]:

$$\gamma_{B1}\frac{\xi_F}{\widetilde{\omega}}\frac{\partial}{\partial x}\Theta_F = \frac{\Delta \exp\left\{\mathrm{i}\frac{\phi}{2}\right\}}{\sqrt{\omega^2 + \Delta^2}}, \quad 0 \le y \le \frac{L}{2},$$

$$\gamma_{B2}\frac{\xi_F}{\widetilde{\omega}}\frac{\partial}{\partial x}\Theta_F = \frac{\Delta \exp\left\{\mathrm{i}\frac{\phi}{2}\right\}}{\sqrt{\omega^2 + \Delta^2}}, \quad \frac{L}{2} \le y \le L. \tag{6.3}$$

Here, Δ is the modulus of the order parameter of the superconducting electrodes, and the phase of the order parameter takes the values $\pm\phi/2$ on the two junction sides respectively. These conditions directly follow from Kupriyanov–Lukichev boundary conditions (see [43]) in the case of small interface transparency and had been intensively used for the analysis of a wide scope of problems in SF multilayers with small interface transparency [1, 2, 44].

The boundary condition at the interface covered by the insulating film ($x = 0$) is [43]:

$$\frac{\partial}{\partial x}\Theta_F = 0 \tag{6.4}$$

At the free ends of the junction located at $y = 0, L$ the boundary conditions correspond to zero current through these surfaces and have the form:

$$\frac{\partial}{\partial y}\Theta_F = 0. \tag{6.5}$$

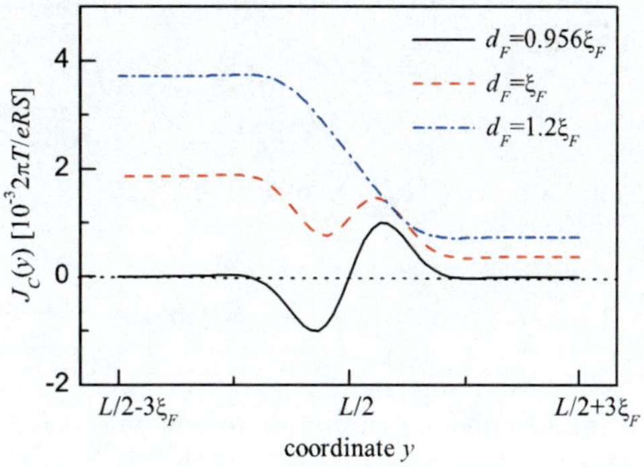

Fig. 6.3 Critical current density distribution (in units of $\pi T/eRS$) along the non-uniform SIFS Josephson junction. The temperature is $T = 0.1T_c$, the exchange magnetic energy is $E = 35\pi T_c$, $\gamma_{B1} = 2$, $\gamma_{B2} = 10$, $d_F = 0.956\xi_F$ corresponds to the point of the 0–π transition; for $d_F = 1.2\xi_F$ the junction is far from this transition

The final expressions for the solution of this two-dimensional boundary problem $\Theta_F(x, y, \omega)$ are given by (6.10) of Appendix 1. The Usadel function $\Theta_F(x, y, \omega)$ is substituted into the expression for the superconducting tunnel current density [45], that gives the sinusoidal current–phase relation

$$J(\phi) = J_C(y)\sin(\phi), \tag{6.6}$$

with the critical current density distribution of the following form:

$$J_C(y) = \frac{2\pi T}{eRS}\sum_{\omega=0}^{\infty}\frac{\Delta}{\sqrt{\Delta^2+\omega^2}}\,\mathrm{Re}\left[\frac{\Theta_F(0, y, \omega)}{\widetilde{\omega}}\right], \tag{6.7}$$

where R is the normal resistance of the tunnel SIFS junction. The calculation of $J_C(y)$ yields the following unexpected result: at some parameters (e.g., the ferromagnet thickness d_F [46]) when the uniform junction is close to a 0–π transition ($J_C = 0$), $J_C(y)$ (6.7) exhibits damped oscillations in the vicinity of the transparency step (Fig. 6.3). $J_C(y)$ changes sign so that the junction properties change from the 0 state to the π state on the scale of ξ_F. This means that a 0–π nano-junction with zero total critical current is formed inside the structure. Previously, a similar effect was predicted for an SFIFS-SNINS non-uniform junction [47]. Note that we call "0–π transition" the conditions when the uniform junction has zero critical current, and "0–π nano-junction" the region where the critical current density $J_C(y)$ is non-uniform and changes its sign near the junction non-uniformity under these conditions.

6.2.1.2 Maximum Josephson Current in an External Magnetic Field

The non-uniform distribution of the critical current density must lead to some peculiarities in the junction behaviour in an external magnetic field.

We start from the Ferrell–Prange type equation for an inhomogeneous Josephson junction

$$\lambda_{J0}^2 \frac{\partial^2}{\partial y^2}\phi(y) - \frac{J_C(y)}{J_{C0}} \sin\phi(y) = \frac{J}{J_{C0}}. \tag{6.8}$$

Here, $\lambda_{J0} = \lambda_J(J_{C0})$, and

$$\lambda_J(J_C) = \sqrt{\frac{\Phi_0}{2\mu_0\pi d'|J_C|}} \tag{6.9}$$

is the Josephson penetration depth, where J_{C0} is the maximum of the critical current density along the junction, J is the bias current density, $\mu_0 d'$ is the effective inductance per square of the junction electrodes, $d' \approx 2\lambda_L$ and λ_L is the London penetration depth of S regions. We assume that the ferromagnetic layer is thin enough to neglect its influence on λ_L.

Since the typical scale of the critical current oscillations is $\xi_F \ll \lambda_{J0}$, it is more interesting to examine a Josephson junction of intermediate length L in y-direction: $\xi_F, \lambda_L \ll L \ll \lambda_{J0}$. It is also assumed that the width of the junction in z direction exceeds the value λ_L. Under these conditions, it is possible to use the local equation (6.8) [48]. In this case, the solution of (6.8) can be found in the linear form $\phi(y) = \phi_0 + hy/\xi_F$, where $h = H/H_0$ is the normalized applied magnetic field in z-direction, and $H_0 = \Phi_0/2\pi\xi_F d'$.

The total current through the junction is calculated and the phase difference ϕ_0 providing the maximum of the total current at each value of h is determined. This yields the dependence of the maximum Josephson current through the junction on the external magnetic field $I_{max}(H)$, that is given by (6.11) in Appendix 1.

If the ferromagnet thickness d_F has a value such that the JJ is either in the 0 or π state, then $I_{max}(H)$ is mainly defined by the first term under the square root in the expression (6.11). This term describes the contribution of the average critical current density along the junction. The corresponding dependence $I_{max}(H)$ resembles a Fraunhofer pattern, that is typical for uniform Josephson junctions with $L \leq \lambda_{J0}$. With increasing non-uniformity $\gamma_{B2}/\gamma_{B1} \neq 1$, the oscillation period of the Fraunhofer-like pattern is doubled (compare the dash-dot and the dash-dot-dot lines in Fig. 6.4a). From calculation of the current distribution, we find that this is due to the fact that only one half of the junction actually conducts the current in this situation.

If the junction approaches the point of the 0–π transition, the 0–π nano-junction forms inside the structure (see $J_C(y)$ profile in Fig. 6.3) and the picture changes by the following way: The maximum Josephson current goes up if the magnetic flux through the JJ $\Phi = HS$ increases first keeping its oscillations (Fig. 6.4a). This increase continues up to a very large magnetic flux, much larger than Φ_0, while the Fraunhofer oscillations have a period of Φ_0. (Fig. 6.4b). $I_{max}(\Phi)$ achieves its maximum value when the magnetic flux through the nano-junction $\Phi'/\Phi_0 = 1$. Φ' depends on the length of the nano-junction that is defined by the critical current non-uniformity region $\sim \xi_F$ and does not depend on L. The magnetic flux Φ

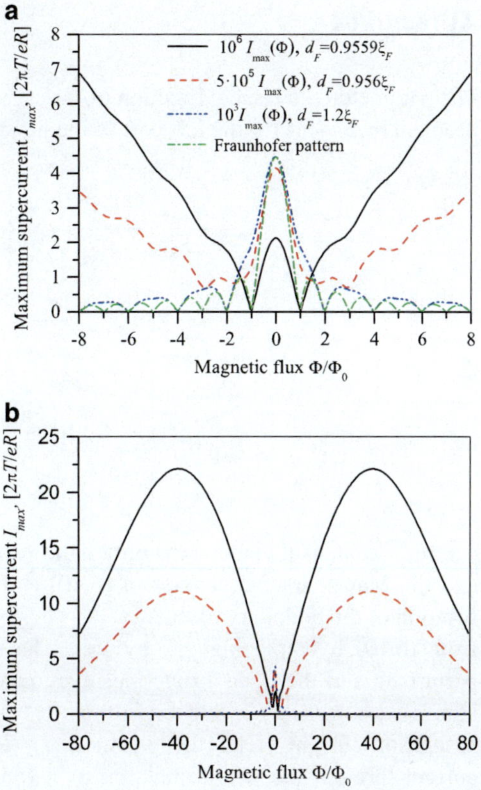

Fig. 6.4 Maximum total supercurrent (in units of $2\pi T/eR$) of an SIFS JJ as a function of the magnetic flux through the junction (in units of Φ_0) for different ferromagnetic layer thicknesses d_F (the 0–π transition is at $d_F = 0.9559\xi_F$). The junction length is $L = 50\xi_F$, the temperature is $T = 0.1T_c$, the exchange magnetic energy is $E = 35\pi T_c$, $\gamma_{B1} = 2$, $\gamma_{B2} = 10$. (**a**) also includes the Fraunhofer pattern (*dash-dot line*) for the homogeneous JJ ($\gamma_{B1} = \gamma_{B2} = 6$). (**b**) shows the same dependencies on a larger magnetic field scale

through the whole junction giving the maximum of $I_{\max}(\Phi)$ depends on the ratio L/ξ_F. The physics of this effect is rather transparent. The external field destroys the initial antisymmetric distribution of the supercurrent inside the nano-junction (Fig. 6.3). Since the external field increases, the antisymmetry vanishes, and the full current across the junction grows. A similar $I_{\max}(\Phi)$ dependence with a minimum current at zero field was obtained experimentally and described theoretically for 0–π JJs with a step in the F layer thickness [11,15,22]. In our case the extraordinary high value of the field is related to the very small size of the nano-junction $\sim \xi_F$.

To fabricate such structures one needs the junction to be close to the 0-π transition (d_F close to $d_F^{0\text{-}\pi}$) with high precision. Since $d_F^{0\text{-}\pi}$ depends on T this could be realized by changing T, as temperature induced 0–π transitions in SFS JJs were already observed [11,13,49].

To enhance this effect, one can create a periodic array of 0–π nano-junctions. The influence of the nano-junctions would be most significant if the length of every facet is comparable with the size $\sim \xi_F$ of $J_C(y)$ oscillations. This is difficult to realize technologically because the value of ξ_F is rather small $\sim$ 1nm. Moreover, it seems to be necessary to keep the layer thicknesses with very high precision along the entire array.

Appendix 1

The parameterized Usadel function $\Theta_F(x, y, \omega)$ found as a solution of the boundary problem (6.2)–(6.5) at the left part of the junction $0 \le y \le L/2$ can be written as:

$$\Theta_F = \frac{\Delta\widetilde{\omega}}{\sqrt{\Delta^2+\omega^2}} \frac{\exp\left\{i\frac{\phi}{2}\right\}}{\gamma_{B1}} \times \left\{ \frac{\cosh\left(\sqrt{\widetilde{\omega}}x/\xi_F\right)}{\sqrt{\widetilde{\omega}}\sinh\left(\sqrt{\widetilde{\omega}}d_F/\xi_F\right)} \right. \tag{6.10}$$
$$- \frac{\gamma_{B2}-\gamma_{B1}}{\gamma_{B2}} d_F\xi_F \sum_{k=0}^{\infty}{}' \frac{(-1)^k\cos(\pi k x/d_F)}{d_F^2\widetilde{\omega}+(\pi k\xi_F)^2}$$
$$\left. \times \exp\left[-\frac{|y-L/2|}{\xi_F}\sqrt{\widetilde{\omega}+(\pi k\xi_F/d_F)^2}\right]\right\},$$

where $\sum'$ means that at $k = 0$ only half of the term is taken. The order parameter and the Matsubara frequencies in (6.10) and below are normalized on πT_c. The solution of the boundary problem (6.2)–(6.5) at $L/2 \le y \le L$ can be reconstructed from (6.10) by replacing γ_{B1} by γ_{B2}. The second term in (6.10) describes the perturbation of the Usadel functions due to changing the FS interface transparency at $x = d_F$, $y = L/2$. Substitution of (6.10) into the standard expression for the Josephson current yields the $J_C(y)$ (6.7). To find out the maximum Josephson current through the whole junction as a function of an external magnetic field, $I_{\max}(H)$, we start from the Ferrell–Prange type equation for the non-uniform Josephson junction (6.8). In the practically interesting limit $\xi_F \ll L \ll \lambda_{J0}$, its solution $\phi(y)$ can be found in the linear form. To calculate the maximum value of the supercurrent $\phi(y)$ is substituted into (6.8) and (6.7) is integrated over y. Thus, the obtained total supercurrent through the junction is further maximized with respect to the phase difference ϕ_0. This procedure finally leads to

$$I_{\max}(h) = \frac{2\xi_F}{hL}\sqrt{[\Sigma_1\sin(hL/2\xi_F)]^2 + [\Sigma_2(h)]^2}, \tag{6.11}$$

where

$$\Sigma_1 = \frac{\gamma_{B1}+\gamma_{B2}}{\gamma_{B1}\gamma_{B2}} \frac{2\pi T}{eR} \sum_{\omega=0}^{\infty} \frac{\Delta^2}{\Delta^2+\omega^2} \operatorname{Re} \frac{\sqrt{\widetilde{\omega}}}{\sinh(\frac{d_F}{\xi_F}\sqrt{\widetilde{\omega}})}$$

$$\Sigma_2 = \frac{\gamma_{B2}-\gamma_{B1}}{\gamma_{B1}\gamma_{B2}} \frac{2\pi T}{eR} \sum_{\omega=0}^{\infty} \frac{\Delta^2\sqrt{\widetilde{\omega}}}{\Delta^2+\omega^2}$$
$$\times \operatorname{Re}\left[\frac{1}{\sqrt{\widetilde{\omega}+h^2}\sinh\left(\frac{d_F}{\xi_F}\sqrt{\widetilde{\omega}+h^2}\right)} - \frac{\cos(hL/2\xi_F)}{\sqrt{\widetilde{\omega}}\sinh\left(\frac{d_F}{\xi_F}\sqrt{\widetilde{\omega}}\right)}\right].$$

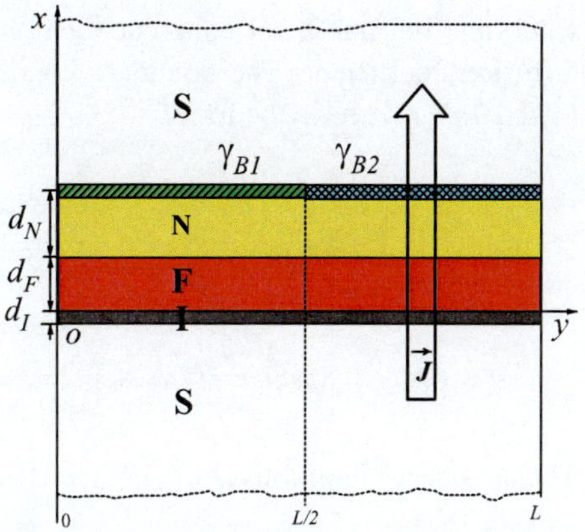

Fig. 6.5 Schematic view of a SIFNS Josephson junction with a step-like change of the NS interface transparency

The first term under the square root in the expression (6.11) corresponds to the contribution of uniform regions $|y - L/2| > \xi_F$ and has the Fraunhofer form, while the second term with Σ_2 contains a more complicated dependence on h and describes the non-uniformity of $J_C(y)$ and it is connected to peculiarities of $I_{\max}(h)$. It exceeds the first term in the vicinity of the 0–π transition. It is clearly seen from (6.11) that in the absence of the non-uniformity ($\gamma_{B1} = \gamma_{B2}$, thus $\Sigma_2 = 0$) the dependence of $I_{\max}(h)$ reduces to the well known Fraunhofer pattern.

6.2.2 *SIFNS and SINFS Structures*

6.2.2.1 The Boundary Problem

To overcome the difficulties mentioned above, we have considered a junction with an additional normal metal layer, i.e. non-magnetic ($E = 0$) and non-superconducting ($\Delta = 0$), with a thickness d_N (Fig. 6.5). The NS boundary has a step-like change in transparency. Now a non-uniform NS boundary is described by the suppression parameters γ_{B1} for $0 \le y \le L/2$ and γ_{B2} for $L/2 \le y \le L$, respectively, which are supposed to be large enough to neglect the suppression of superconductivity in the S electrode.

The corresponding boundary problem for the linearized Usadel equation differs from the previous one by the existence of the additional normal layer. Now the set of linearized Usadel equations includes also the equation for the N region that is written similarly to (6.2) as

$$\xi_N^2 \left\{ \frac{\partial^2}{\partial x^2} + \frac{\partial^2}{\partial y^2} \right\} \Theta_N - \frac{\omega}{\pi T_c} \Theta_N = 0, \tag{6.12}$$

where the normal metal coherence length $\xi_N = (D_N/2\pi T_c)^{1/2}$, and D_N is the diffusion coefficient. The boundary conditions for the non-uniform NS interface ($x = d_F + d_N$) have the form:

$$\gamma_{B1}\xi_N \frac{\partial}{\partial x}\Theta_N = \frac{\omega\Delta \exp\left\{\mathrm{i}\frac{\phi}{2}\right\}}{\sqrt{\omega^2+\Delta^2}}, \quad 0 \le y \le \frac{L}{2}, \tag{6.13}$$

$$\gamma_{B2}\xi_N \frac{\partial}{\partial x}\Theta_N = \frac{\omega\Delta \exp\left\{\mathrm{i}\frac{\phi}{2}\right\}}{\sqrt{\omega^2+\Delta^2}}, \quad \frac{L}{2} \le y \le L. \tag{6.14}$$

The boundary conditions at the FN interface located at $x = d_F$ can be written as

$$\frac{\xi_N}{|\omega|}\frac{\partial}{\partial x}\Theta_N = \gamma\frac{\xi_F}{\widetilde{\omega}}\frac{\partial}{\partial x}\Theta_F, \tag{6.15}$$

$$\Theta_F + \gamma_B\xi_F\frac{\partial}{\partial x}\Theta_F = \frac{\widetilde{\omega}}{|\omega|}\Theta_N, \tag{6.16}$$

$$\gamma_B = R_{\mathrm{BF}}2S/\rho_F\xi_F, \quad \gamma = \rho_N\xi_N/\rho_F\xi_F,$$

where R_{BF} is the resistance of the FN interface; $\rho_{N(F)}$ is the resistivity of the N or F layer, respectively.

There are also additional conditions at the free ends of the N layer located at $y = 0$ and $y = L$:

$$\frac{\partial}{\partial y}\Theta_N = 0. \tag{6.17}$$

This boundary problem (6.2), (6.3), (6.5), (6.12)–(6.17) was solved analytically. The solution is presented in Appendix 2 [(6.21), (6.22)]. Inserting (6.22) into the expression (6.7), it follows that $J_C(y)$ has changed in comparison with the SIFS junction, as can be seen in Fig. 6.6.

The current distribution keeps its form, but stretches along the y-axis proportionally to ξ_N. The reason is an interplay of the ferromagnetic and the normal coherence lengths due to the proximity effect. The Usadel function in the N region noticeably changes on the distance of the order of ξ_N from the step at $y = L/2$. Subsequently, the ferromagnet "feels" this change also on the distance $\sim \xi_N$ from the step and conducts this information to the other superconducting electrode. The FN interface parameters γ, γ_B and the ratio $s = \xi_N/\xi_F$ define the strength of this influence. This effect holds also when $d_N \ll \xi_N$. A thicker N layer slightly extends the length scale of the $J_C(y)$ oscillation, but also decreases the absolute value of J_C. The $J_C(y)$ distribution is shown in Fig. 6.6 for various values of d_N and s. While the d_N/s ratio stays constant, the 0–π transition occurs at the same conditions (d_F, T, γ, γ_B and so on).

Thus, the scale of the $J_C(y)$ changes in y-direction is $\propto \xi_N$ and can be hundred times larger than ξ_F. It is a very interesting manifestation of the FN proximity effect

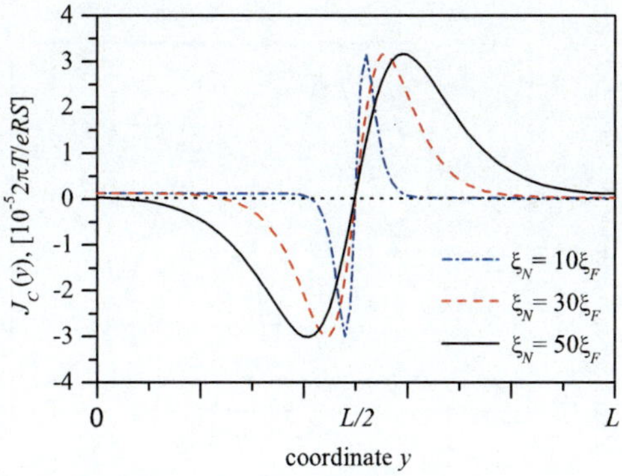

Fig. 6.6 Critical current density distribution (in units of $2\pi T/eRS$) along the non-uniform SIFNS Josephson junction of length $L = 2{,}000\xi_F$ at different N-layer thicknesses d_N and ξ_N so that $d_N/\xi_N = 5$. The temperature is $T = 0.1T_c$, the exchange magnetic energy is $E = 35\pi T_c$, the non-uniform NS boundary damping parameter is $\gamma_{B1} = 2$, $\gamma_{B2} = 10$, the FN boundary parameters are $\gamma = 0.1$, $\gamma_B = 0.2$, the ferromagnetic layer thickness $d_F = 0.987\xi_F$ corresponds to the point of the 0–π transition

that was already described for other geometries [50]. The critical current oscillation arises due to the presence of the ferromagnet in the structure, but the normal metal determines the period of J_C oscillations in y-direction. Thus, we have constructed a 0–π nano-junction on the scale $\xi_N \gg \xi_F$.

6.2.2.2 SIFNS Junction in an External Magnetic Field

It is clear that the size of the critical current non-uniformity influences the junction behaviour in an external magnetic field. The dependence $I_{\max}(H)$ for the junction near the 0–π transition and $I_{\max}(H)$ for the uniform JJ with an average interface transparency are shown in Fig. 6.7. Here, the 0–π nano-junction is wider than the one inside the SIFS junction, so that one needs much lower magnetic fields to destroy its asymmetry. Therefore, the width of the peaks becomes smaller and comparable to the Fraunhofer oscillation period when the junction length $L \sim \xi_N$. The details of these calculations are presented in the Appendix 2.

It is also interesting to investigate a tunnel JJ with reversed order of the F and N layers, namely a SINFS structure. The solution of the boundary problem is similar to (6.2), (6.3), (6.5), (6.12)–(6.17) for such a SINFS junction, and with the same NF boundary parameters γ, γ_B, and the same resistances R_{B1}, R_{B2} of the FS interface yields exactly the same expression (6.23) for $J_C(y)$. This means that the Josephson effect and the magnetic properties of these non-uniform tunnel structures do not depend on the order of the F and N layers. The length-scale of $J_C(y)$ oscillations is defined by the layer with the largest coherence length. This statement remains valid as long as one may use the linearized equations.

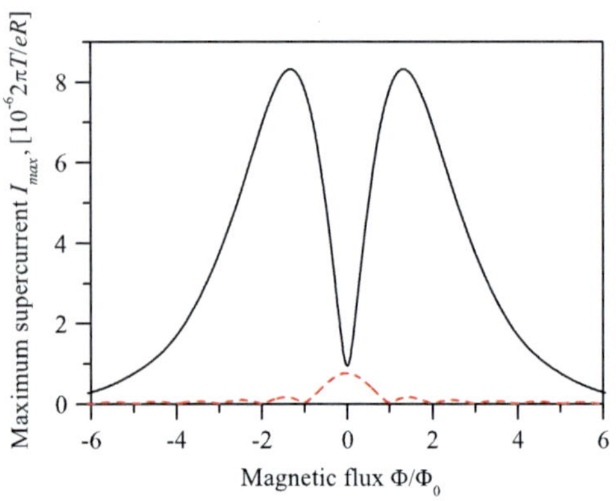

Fig. 6.7 $I_{\max}(\Phi)$ (in units of $2\pi T/eR$) of a SIFNS JJ and its Fraunhofer part given by the first term under the square root of (6.24) (*dashed line*). The conducting layer thicknesses are $d_F = 0.987\xi_F$, and $d_N = 5\xi_N$; $\xi_N = 50\xi_F$, the junction length is $L = 2{,}000\xi_F$, the temperature is $T = 0.1T_c$, the exchange magnetic energy is $E = 35\pi T_c$, the non-uniform NS boundary damping parameter is $\gamma_{B1} = 2$, $\gamma_{B2} = 10$, the FN boundary parameters are $\gamma = 0.1$, $\gamma_B = 0.2$

6.2.2.3 Micro-Vortex Forming

One can also consider the opposite problem, i.e. the influence of a direct current on the magnetic field distribution inside the SIFNS structure. An interesting question is: can the 0–π nano-junction include some fractional Josephson vortex as in long 0–π Josephson junctions? Naturally in the ground state when the current through the JJ is equal to zero, $I = 0$, the equilibrium phase distribution $\phi(y)$ remains constant along the junction due to the small size of this nano-junction $\sim \xi_N \ll \lambda_J$, as λ_J is the typical length-scale for phase changes.

Let us consider the case when the supercurrent through the junction is finite, but does not exceed its maximum value, i.e. $0 < I \leq I_{\max}$. Qualitatively, since one part of the non-uniform junction conducts the Josephson current much better than the other one (due to the difference in the NS boundary transparency and consequently in the value of J_C), the supercurrent redistributes within the junction area. This redistribution can be interpreted as a vortex of supercurrent. This vortex produces a magnetic field in z-direction, and is pinned around $y = L/2$, where the transparency changes step-wise. The distribution of the magnetic field and the corresponding magnetic flux have been calculated under the assumption that the local phase variation is much smaller than its average value.

The largest magnetic flux Φ is induced when the Josephson current takes its maximum value $I = I_{\max}$. Far from the 0–π transition, the magnetic field $H(y)$ increases almost linearly up to the non-uniformity point $y = L/2$ with a smooth extremum at $y = L/2$, as shown by the dashed line in Fig. 6.8. At the same time at $d_F = d_F^{0\text{-}\pi}$, the field distribution has a typical width (Fig. 6.8) comparable to the length of critical current non-uniformity (see Fig. 6.6). The corresponding magnetic flux $\Phi_{0\text{-}\pi}$ in this case does not depend on the junction length L contrary to the case

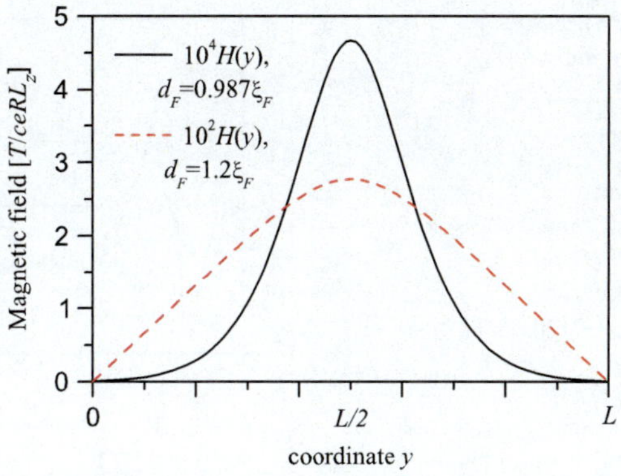

Fig. 6.8 Magnetic field distribution along a SIFNS junction at bias current $I = I_{\max}(0)$ for different thicknesses of the F-layer: $d_F = 0.987\xi_F$ corresponds to the point of the 0–π transition, with magnetic flux through the junction $\Phi_{0\text{-}\pi} = 0.135\mu_0 d_L T/eR$; far from the 0–$\pi$ transition at $d_F = 1.2\xi_N$, $\Phi = 15.8\mu_0 d_L T/eR$. The N-layer thickness is $d_N = 5\xi_N$, $\xi_N = 50\xi_F$, the junction length is $L = 2{,}000\xi_F$, the temperature is $T = 0.1T_c$, the exchange magnetic energy is $E = 35\pi T_c$, the non-uniform NS boundary damping parameter is $\gamma_{B1} = 2$, $\gamma_{B2} = 10$, the FN boundary parameters are $\gamma = 0.1$, $\gamma_B = 0.2$

of the junction far from the 0–π transition. The value of $\Phi_{0\text{-}\pi}$ is smaller than Φ due to the small length of the nano-junction and small value of the critical current, i.e.

$$\frac{\Phi}{\Phi_0} \sim \frac{L^2}{\lambda_{J0}^2} \frac{|\gamma_{B1} - \gamma_{B2}|}{\gamma_{B1} + \gamma_{B2}},$$
$$\frac{\Phi_{0\text{-}\pi}}{\Phi_0} \sim \frac{\xi_N^2}{\lambda_{J0}^2} \frac{|\gamma_{B1} - \gamma_{B2}|}{\gamma_{B1} + \gamma_{B2}} \tag{6.18}$$

Here, λ_{J0} is defined by (6.9).

6.2.3 SIFNS Junction with Few Steps of Boundary Transparency

Technology does not allow fabricating ferromagnetic JJs with ideally smooth interfaces. Moreover, one often uses a thin normal layer below the ferromagnet to improve the JJ properties [13, 51, 52]. The interlayer boundary non-uniformity may create peculiarities in the $J_C(y)$ distribution that have an effect on the junction behaviour in an external magnetic field. To describe this behaviour, as a first attempt it would be reasonable to consider a SIFNS Josephson junction having a few steps of the boundary transparency. At first, we take three steps with different lengths and FS boundary damping parameters γ_{B1}, γ_{B2} and γ_{B3} respectively. The length L_i of every step is assumed to be $L_i \geq \xi_N$, but the length of the whole junction

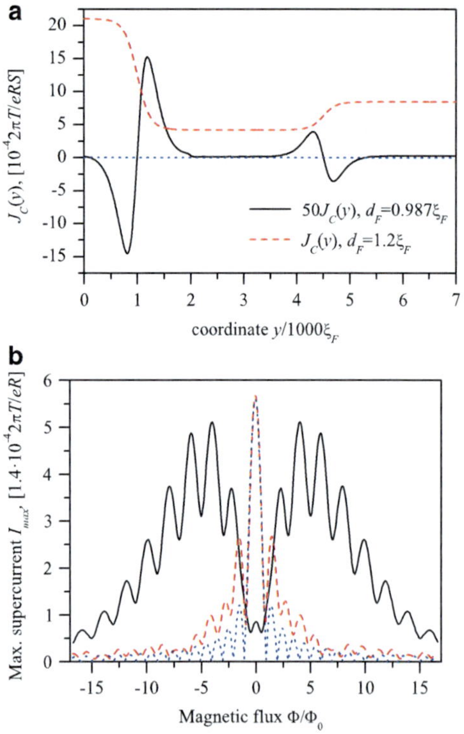

Fig. 6.9 Asymmetric SIFNS Josephson junction having three regions with NS boundary damping parameters $\gamma_{B1} = 2, \gamma_{B2} = 10$ and $\gamma_{B3} = 5$ and lengths $L_1 = 1{,}000\xi_F, L_2 = 3{,}500\xi_F$, and $L_3 = 2{,}500\xi_F$, respectively. (**a**) $J_C(y)$ and (**b**) $I_{\max}(\Phi)$ in the vicinity of the 0–π transition ($d_F = 0.987\xi_F$) and far from this transition ($d_F = 1.2\xi_F$). (**b**) also includes the Fraunhofer pattern (dash-dot line) for the homogeneous JJ ($\gamma_{B1} = \gamma_{B2} = \gamma_{B3} = 6$ and $d_F = 1.2\xi_F$). The N-layer thickness is $d_N = 5\xi_N$, $\xi_N = 50\xi_F$, the temperature is $T = 0.1T_c$, the exchange magnetic energy is $E = 35\pi T_c$, the FN boundary parameters are $\gamma = 0.1, \gamma_B = 0.2$

remains $L \ll \lambda_J$. Under these assumptions, the function $J_C(y) = J_{\mathrm{Ci}}$ becomes a constant far enough from the non-uniformities (J_{Ci} is the critical current density of the uniform JJ with the damping parameter γ_i). In this region, the condition $\partial J_C(y)/\partial y = 0$ is fulfilled as on a free end of the junction. Therefore, the solution of the corresponding boundary problem may be constructed from the solutions presented above:

$$J_C(y) = J_C(y, \gamma_{B1}, \gamma_{B2}), \text{ if } 0 \le y \le 2L_1$$

$$J_C(y) = J_C(y, \gamma_{B2}, \gamma_{B3}), \text{ if } 2L_1 \le y \le 2L_1 + 2L_3,$$

where $L_2 = L_1 + L_3$.

$J_C(y)$ at $d_F = d_F^{0\text{-}\pi}$ is neither symmetric, nor antisymmetric in y-direction, see Fig. 6.9a. Such a distribution $J_C(y)$ results in a rather complicated $J_{\max}(H)$ dependence, which remains non-Fraunhofer-like far from the 0–π transition, see Fig. 6.9b.

If the junction is symmetric $\gamma_{B1} = \gamma_{B3}$ and $L_1 = L_2/2 = L_3$, there are two different situations $\gamma_{B1} = \gamma_{B3} < \gamma_{B2}$ and $\gamma_{B1} = \gamma_{B3} > \gamma_{B2}$. The first one corresponds to one junction with a large J_C in the centre of the structure, and the second one corresponds to two junctions with a large J_C at the ends, which length is twice less than the length of the central section. Thus, in the second case and far from

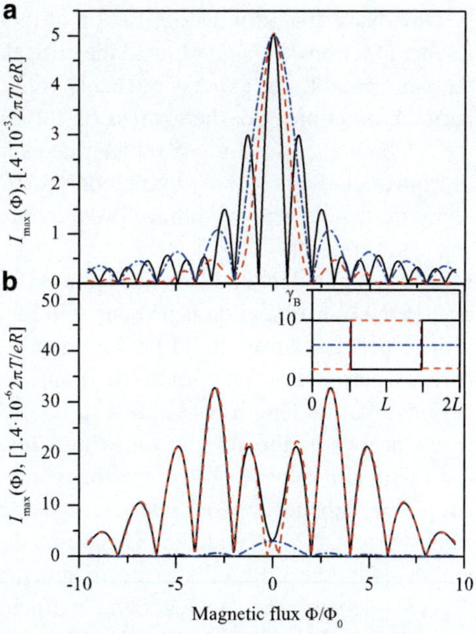

Fig. 6.10 $I_{\max}(\Phi)$ of the symmetric SIFNS Josephson junction having three regions of the NS boundary transparency (**a**) far from the 0–π transition $d_F = 1.2\xi_F$, and (**b**) in the vicinity of this transition $d_F = 0.987\xi_F$. NS interface damping parameters $\gamma_{B1} = \gamma_{B3}$ and γ_{B2} take the values 2 and 10, as it is shown on the inset. $L_1 = L_3 = L_2/2$, $L_1 + L_2 + L_3 = L = 3{,}500\xi_F$. The N-layer thickness is $d_N = 5\xi_N$, $\xi_N = 50\xi_F$, the temperature is $T = 0.1T_c$, the exchange magnetic energy is $E = 35\pi T_c$, the FN boundary parameters are $\gamma = 0.1$, $\gamma_B = 0.2$

the 0–π transition, the period of the $I_{\max}(H)$ dependence will be twice smaller than in the first case, see Fig. 6.10a. In the vicinity of the 0–π transition, the picture is practically the same for both cases, see Fig. 6.10b. The reason is that in this situation the Josephson current is defined mainly by two 0–π nano-junctions related to oscillations of $J_C(y)$ in the non-uniform areas.

Such a consideration may help to understand how the interlayer roughness or non-uniformity of the interfaces (of length $\sim \xi_N$) lead to deviations in $I_{\max}(H)$ from the Fraunhofer pattern obtained for the uniform structure with the averaged parameters. This deviation would depend not only on the value of the roughness but also on its distribution in the junction plane.

6.2.4 SIFNS Junctions Array

6.2.4.1 JJ Arrays in a Magnetic Field

The relatively long range of $J_C(y)$ non-uniformity in the SIFNS structure makes it possible to fabricate an array of nano-junctions with periodic step-like changes of the transparency of the NS interface. The transparency step must be sharp in comparison with ξ_N and it is assumed that the length L_i of every uniform part satisfies the conditions: $\xi_N < L_i \ll \lambda_J$.

Obviously, the periodic oscillation of the transparency leads to variations of the Usadel functions Θ_N and Θ_F and the critical current density along the junction with the same period. The expression (6.21), (6.22) for $\Theta_{N,F}$ has a form of a series of $2L_i$ periodic functions. So, the expression for the function of the periodic array, where the NS boundary damping parameter takes the alternating values γ_{B1} and γ_{B2}, may be constructed again as a simple continuation of the function $\Theta_{N,F}$ (6.21), (6.22) along the whole array. We have proven that this boundary problem does not have any other solution.

The $J_C(y)$ calculated from (6.7 and 6.23) for the periodic structure far from and near to the 0–π transition is presented in Fig. 6.11a.

The periodic array could be symmetric (contains an integer number of transparency periods) or not symmetric (contains semi-integer number of transparency periods), but its length is assumed to be much shorter than λ_{J0}. In the symmetric case, the expression for the maximum Josephson current $I_{\max}(H)$ has only the term with $\sin(\pi\Phi/\Phi_0)$ (Φ is the magnetic flux through the whole junction). The simple sin-dependence (6.25) leads to the value $\phi_0 = \pm\pi/2$ at which the maximum super-current $I_{\max}$ is realized. This value does not depend on H and on the array length as in the case of a uniform junction having the well-known Fraunhofer $I_{\max}(H)$ dependence. However, the coefficient in front of $\sin(\pi\Phi/\Phi_0)$ depends on the magnetic field due to the non-uniformity of the structure. It leads to deviations from the Fraunhofer pattern both far and near the 0–π transition, as shown in Fig. 6.11b,c.

If the structure is not symmetric, the expression $I_{\max}(H)$ (6.26) looks like a non-Fraunhofer function for a one-step structure. Both the symmetric and the non-symmetric arrays give similar complicated curves $I_{\max}(H)$. The applied magnetic field together with non-uniform current distribution along the junction yield the appearance of additional oscillations and sharp peaks in the dependence $I_{\max}(H)$ (Fig. 6.11b,c).

The short oscillation period is defined by the magnetic flux through the whole array. The long period of the additional oscillations depends on the length of every step (i.e. the semi-period of the structure non-uniformity), namely on the value of the magnetic flux through this step. Near the 0–π transition the appearing sharp maxima lay under the envelope curve which is nothing else but $I_{\max}(H)$ for a one step JJ of the same length. The distribution of the peak heights (the third oscillation period) depends on the $J_C(y)$ non-uniformity scale, that is defined by the ratio ξ_N/ξ_F and the value of d_N.

The sharp peaks in $I_{\max}(H)$ arise not only in the vicinity of the 0–π transition, but also for an arbitrary ferromagnetic layer thickness. This means that it is not necessary to keep high precision of the parameters along the whole structure. These sharp peaks have obviously a similar origin as the periodic peaks on the voltage-field dependence of the periodic array of usual SIS JJs described in [27, 29]. The JJ with periodically changing interface transparency also can be considered as an array of large J_C junctions inserted between junctions with small J_C.

Now, there are three scales: the length of the array L, the length of each region with constant transparency L_i and the width of the critical current non-uniformity

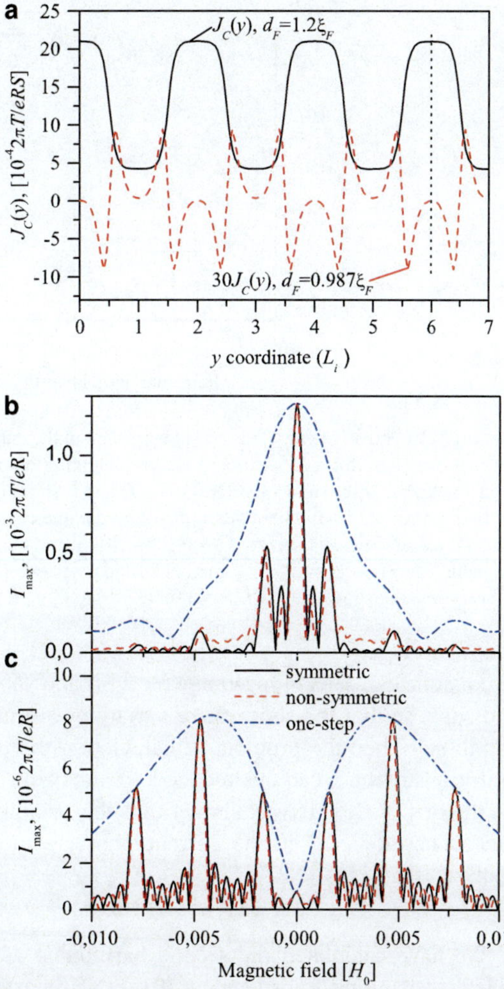

Fig. 6.11 SIFNS JJ with periodically changing damping parameter of the NS boundary between $\gamma_{B1} = 2$ and $\gamma_{B2} = 10$. (**a**) $J_C(y)$ and (**b**,**c**) $J_{\max}(H)$ for a JJ far from the 0–π transition ($d_F = 1.2\xi_F$) (**b**), and in the vicinity of the 0–π transition ($d_F = 0.987\xi_F$) (**c**). The N-layer thickness is $d_N = 5\xi_N$, $\xi_N = 50\xi_F$, the temperature is $T = 0.1T_c$, the exchange magnetic energy is $E = 35\pi T_c$, the FN boundary parameters are $\gamma = 0.1$, $\gamma_B = 0.2$, $L_i = 2{,}000\xi_F$. The length of the symmetric junction $L = 6L_i$ is indicated on the plot (**a**) by the *dashed line*. The dependence $I_{\max}(H)$ for the junction with only one NS boundary transparency step is shown on (**b**) and (**c**) for comparison

$\sim \xi_N$. All of them are assumed to be much smaller than λ_{J0}. Their interplay defines the behaviour of the structure in the external magnetic field and yields various unusual forms of the $I_{\max}(H)$ dependence. This allows to design JJ arrays with peculiar magnetic properties.

6.2.4.2 Averaging Over Short-Range Oscillations

Up to now it has been assumed that $L \ll \lambda_{J0}$. If the junction length becomes comparable to λ_{J0} [13, 15, 51], the linear phase ansatz cannot be used anymore. However, if the conditions $\xi_F, \xi_N \ll L_i \ll \lambda_{J0}$ remain valid for each transparency

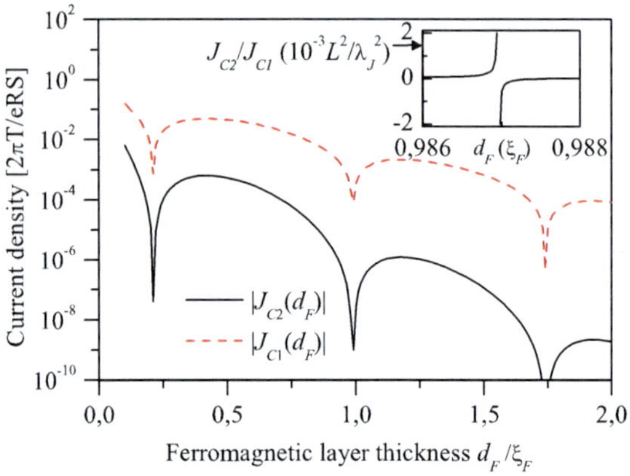

Fig. 6.12 Effectively generated second harmonic in the current–phase relation in comparison with the first harmonic for a JJ with periodically changing boundary transparency as a function of F-layer thickness. Inset shows the ratio J_{C2}/J_{C1} in a small region $0.986\xi_F < d_F < 0.988\xi_F$ near the 0–π transition. The NS boundary damping parameter takes the values $\gamma_{B1} = 2$ and $\gamma_{B2} = 10$, the N-layer thickness is $d_N = 5\xi_N$, $\xi_N = 50\xi_F$, the temperature is $T = 0.1T_c$, the exchange magnetic energy is $E = 35\pi T_c$, the FN boundary parameters are $\gamma = 0.1$, $\gamma_B = 0.2$, the length of every homogeneous area is $L_i = 2{,}000\xi_F$

region, one can select long-range ($\sim \lambda_{J0}$) and short range ($\sim L_i$) phase variations, and suppose that the short range variations are much smaller than the long range variations. Then the problem of calculating the phase distribution along the non-uniform structure can be reduced to the well-known problem of a non-linear oscillator [24, 26]. It was shown that the oscillating $J_C(y)$ leads to an effective second harmonic J_2 in the current–phase relation for the long-range (averaged) phase ψ, that is

$$J(\psi) = J_1 \sin(\psi) + J_2 \sin(2\psi). \tag{6.19}$$

We have calculated the second harmonics J_2 for the SIFNS array with periodically changing transparency of the NS interface. The expression obtained by averaging out the short-range oscillations is given by (6.27) in the Appendix 2. The second harmonic obtained in this way always has a negative value.

The dependence of the first and the second harmonics vs. d_F is presented in Fig. 6.12. The absolute value of J_2 is not large in comparison with J_1. When the junction has d_F far from the 0–π transition, J_2 increases and achieves its saturation as the length L_i increases. $J_C(y)$ does not change its sign along the junction, and the relation $|J_2/J_1| \sim L^2/2\pi\lambda_J^2$ is satisfied.

At $d_F \approx d_F^{0\text{-}\pi}$ where the junction is close to the 0–π transition and its first harmonic J_1 vanishes, the effective second harmonic J_2 is generated by the $J_C(y)$ non-uniformity. $J_C(y)$ changes its sign, but the regions where this occurs are very small (of the order of ξ_N, that defines the size of the $J_C(y)$ non-uniformity area), so that $|J_2/J_1| \sim \xi_N^2/2\pi\lambda_{J0}^2 \ll 1$. The ratio $|J_2/J_1|$ formally diverges at the 0–π transition point, where $J_1 \to 0$, see inset in Fig. 6.12. Thus, the question about the possibility of a φ-junction ($0 < \varphi < \pi$) realization arises here.

When in the JJ with current–phase relation (6.19), the conditions

$$\left|\frac{2J_2}{J_1}\right| > 1, \text{ and } J_2 < 0 \tag{6.20}$$

are satisfied, the φ-junction appears [26,41].

To arrive to the vicinity of the 0–π transition, one can prepare a high quality JJ with periodically changing interlayer boundary transparency and change the temperature during experiment. However, the conditions (6.20) are difficult to achieve due to the very small range of parameter values where they are satisfied.

A possibility to enlarge the domain of parameters where conditions (6.20) are satisfied is discussed in the next section.

Appendix 2

The boundary problem (6.2), (6.3), (6.5), (6.12)–(6.17) for the SIFNS junction has been solved analytically both for N and F layers. The derivatives on the free ends of the junction $\partial\Theta_{N,F}/\partial y = 0$ at $y = 0, L$ (6.5), (6.17); therefore, the solution in the normal metal layer is found in the form:

$$\Theta_N(x, y, \omega) = \exp\left\{\mathrm{i}\frac{\phi}{2}\right\}\left[A(x) + \sum_{n=0}^{\infty} A_n(x)\cos\left(\pi(2n+1)\frac{y}{L}\right)\right], \tag{6.21}$$

where the coefficients

$$A(x) = u\cosh\left(\frac{x - d_F - d_N}{\mu}\right) + \frac{\mu}{2\xi_N}\frac{\Delta}{\sqrt{\omega^2 + \Delta^2}}\frac{\gamma_{B1} + \gamma_{B2}}{\gamma_{B1}\gamma_{B2}}\sinh\left(\frac{x - d_F - d_N}{\mu}\right),$$

$$A_n(x) = u_n\cosh\left(\frac{x - d_F - d_N}{\mu_m}\right) - \frac{2(-1)^n\mu_n}{\pi(2n+1)\xi_N}\frac{\Delta}{\sqrt{\omega^2 + \Delta^2}}\frac{\gamma_{B1} - \gamma_{B2}}{\gamma_{B1}\gamma_{B2}}\sinh\left(\frac{x - d_F - d_N}{\mu_n}\right),$$

where

$$\mu = \frac{\xi_N}{\sqrt{\omega}}, \quad \mu_n = \frac{\xi_N}{\sqrt{\left[\frac{\xi_N}{L}\pi(2n+1)\right]^2 + \omega}}, \quad n = 0, 1, 2\ldots$$

$$u = \frac{\gamma_{B1}+\gamma_{B2}}{\gamma_{B1}\gamma_{B2}} \frac{\Delta}{\sqrt{\omega^2+\Delta^2}} \frac{\mu}{2\xi_N}$$

$$\times \frac{\left[\begin{array}{c}\xi_N \zeta \cosh\left(\frac{d_N}{\mu}\right)\cosh\left(\frac{d_F}{\zeta}\right) + \xi_F \gamma \mu \sinh\left(\frac{d_N}{\mu}\right)\\ \times \sinh\left(\frac{d_F}{\zeta}\right) + \gamma_B \xi_F \xi_N \cosh\left(\frac{d_N}{\mu}\right)\sinh\left(\frac{d_F}{\zeta}\right)\end{array}\right]}{\left[\begin{array}{c}\xi_F \mu \gamma \sinh\left(\frac{d_F}{\zeta}\right)\cosh\left(\frac{d_N}{\mu}\right) + \xi_N \zeta \sinh\left(\frac{d_N}{\mu}\right)\\ \times \cosh\left(\frac{d_F}{\zeta}\right) + \gamma_B \xi_N \xi_F \sinh\left(\frac{d_N}{\mu}\right)\sinh\left(\frac{d_F}{\zeta}\right)\end{array}\right]}$$

$$u_n = -\frac{\gamma_{B1}-\gamma_{B2}}{\gamma_{B1}\gamma_{B2}} \frac{\Delta}{\sqrt{\omega^2+\Delta^2}} \frac{2\mu_n(-1)^n}{\pi\xi_N(2n+1)}$$

$$\times \frac{\left[\begin{array}{c}\xi_N \zeta_n \cosh\left(\frac{d_N}{\mu_n}\right)\cosh\left(\frac{d_F}{\zeta_n}\right) + \xi_F \gamma \mu_n \sinh\left(\frac{d_N}{\mu_n}\right)\\ \times \sinh\left(\frac{d_F}{\zeta n}\right) + \gamma_B \xi_F \xi_N \cosh\left(\frac{d_N}{\mu_n}\right)\sinh\left(\frac{d_F}{\zeta_n}\right)\end{array}\right]}{\left[\begin{array}{c}\xi_N \zeta_n \sinh\left(\frac{d_N}{\mu_n}\right)\cosh\left(\frac{d_F}{\zeta_n}\right) + \xi_F \gamma \mu_n \sinh\left(\frac{d_F}{\zeta_n}\right)\\ \times \cosh\left(\frac{d_N}{\mu_n}\right) + \gamma_B \xi_N \xi_F \sinh\left(\frac{d_N}{\mu_n}\right)\sinh\left(\frac{d_F}{\zeta_n}\right)\end{array}\right]},$$

where the parameters corresponding to the ferromagnet:

$$\zeta = \frac{\xi_F}{\sqrt{\widetilde{\omega}}}, \quad \zeta_n = \frac{\xi_F}{\sqrt{\left[\frac{\xi_F}{L}\pi(2n+1)\right]^2 + \widetilde{\omega}}}, \qquad n = 0, 1, 2 \ldots$$

Taking into account the boundary conditions (6.4), (6.5), the solution for the ferromagnetic layer can be written as:

$$\Theta_F(x, y, \omega) = \frac{\widetilde{\omega}\Delta \exp\left\{ i\frac{\phi}{2}\right\}}{|\omega|\sqrt{\omega^2+\Delta^2}} \tag{6.22}$$

$$\times \left[\frac{\gamma_{B1}+\gamma_{B2}}{\gamma_{B1}\gamma_{B2}} v \cosh\left(\frac{x}{\zeta}\right) - \frac{\gamma_{B1}-\gamma_{B2}}{\gamma_{B1}\gamma_{B2}} \times \sum_{n=0}^{\infty} v_n \cosh\left(\frac{x}{\zeta_n}\right)\cos\left(\pi(2n+1)\frac{y}{L}\right)\right]$$

$$v = \frac{\mu\zeta/2}{\left[\begin{array}{l}\zeta\xi_N \sinh\left(\dfrac{d_N}{\mu}\right)\cosh\left(\dfrac{d_F}{\zeta}\right) + \mu\gamma\xi_F \sinh\left(\dfrac{d_F}{\zeta}\right)\\ \times\cosh\left(\dfrac{d_N}{\mu}\right) + \gamma_B\xi_N\xi_F \sinh\left(\dfrac{d_N}{\mu}\right)\sinh\left(\dfrac{d_F}{\zeta}\right)\end{array}\right]}$$

$$v_n = \frac{2(-1)^n}{\pi(2n+1)} \times \frac{\mu_n\zeta_n}{\left[\begin{array}{l}\zeta_n\xi_N \sinh\left(\dfrac{d_N}{\mu_n}\right)\cosh\left(\dfrac{d_F}{\zeta n}\right) + \gamma\mu_n\xi_F \cosh\left(\dfrac{d_N}{\mu_n}\right)\\ \times\sinh\left(\dfrac{d_F}{\zeta n}\right)\gamma_B\xi_F\xi_N + \sinh\left(\dfrac{d_N}{\mu_n}\right)\sinh\left(\dfrac{d_F}{\zeta n}\right)\end{array}\right]}$$

The critical current density given by the expression (6.7) turns out as follows:

$$J_C(y) = \frac{2\pi T}{eRS}\sum_{\omega=0}^{\infty}\frac{\Delta^2}{(\Delta^2+\omega^2)\,|\omega|} \times\left[\frac{\gamma_{B1}+\gamma_{B2}}{\gamma_{B1}\gamma_{B2}}\operatorname{Re} v - \frac{\gamma_{B1}-\gamma_{B2}}{\gamma_{B1}\gamma_{B2}}\operatorname{Re}\sum_{n=0}^{\infty} v_n \cos\left(\pi(2n+1)\frac{y}{L}\right)\right]. \quad (6.23)$$

Substituting the solution $\phi(y)$ in the linear form into the Ferrell-Prange equation (6.8), calculating the total current through the junction and maximizing it over ϕ_0, we arrive at the following $I_{\max}(h)$ dependence:

$$I_{\max}(h) = \frac{2\xi_F}{hL}\sqrt{B_0^2\sin^2\left(\frac{hL}{2\xi_F}\right) + B_1^2(h)\cos^2\left(\frac{hL}{2\xi_F}\right)}, \quad (6.24)$$

where

$$B_0 = \frac{\gamma_{B1}+\gamma_{B2}}{\gamma_{B1}\gamma_{B2}}\frac{2\pi T}{eR}\sum_{\omega=0}^{\infty}\frac{\Delta^2}{\Delta^2+\omega^2}\operatorname{Re}\frac{\sqrt{\varpi}}{\sinh(\frac{d}{\xi_F}\sqrt{\varpi})}v$$

$$B_1(h) = \frac{\gamma_{B2}-\gamma_{B1}}{\gamma_{B1}\gamma_{B2}}\frac{2\pi T}{eR}\sum_{\omega=0}^{\infty}\frac{\Delta^2}{\Delta^2+\omega^2}\sum_{n=0}^{\infty}\frac{\operatorname{Re} v_n}{\left[\frac{\pi(2n+1)\xi_F}{hL}\right]^2 - 1}$$

Here, B_0 describes the Fraunhofer contribution of the average $J_C(y)$, and B_1 responds to a non-Fraunhofer deviation due to the junction non-uniformity. The value $hL/2\xi_F = \pi\Phi/\Phi_0$ corresponds to the value of the magnetic flux Φ through the junction expressed in units of the magnetic flux quantum Φ_0.

Let us consider an array with a periodic variation of the boundary resistivity. The junction contains M resistivity steps and, therefore, $M + 1$ areas of constant resistivity, alternating between R_1 and R_2. The uniform areas have the lengths L_i each, and $L_i/2$ on the ends. The boundary problem has the same periodic solution described by the expressions (6.21), (6.22). The corresponding $J_C(y)$ is presented in Fig. 6.11a. It yields the dependence $I_{\max}(H)$ for a symmetric array $M = 2N$, (N is integer):

$$I_{\max}(h) = \frac{1}{f_s}\sin(f_s)\left|B_0 + B_2(h)\right| \tag{6.25}$$

here $f_s = M\,|h|\,L_i/2\xi_F = N\,|h|\,L_i/\xi_F = \pi\Phi/\Phi_0$ and

$$B_2(h) = \frac{\gamma_{B2} - \gamma_{B1}}{\gamma_{B1}\gamma_{B2}}\frac{2\pi T}{eR}\sum_{\omega=0}^{\infty}\frac{\Delta^2}{\Delta^2 + \omega^2}\sum_{n=0}^{\infty}\frac{(-1)^{N+1}\operatorname{Re}v_n}{\left[\frac{\pi(2n+1)\xi_F}{hL_i}\right]^2 - 1}.$$

The term $\sin(f_s)/f_s$ coincides with the Fraunhofer dependence, but the function $B_2(h)$ makes $I_{\max}(H)$ more complicated.

For a non-symmetric array containing $M = 2N + 1$ non-uniformities (see the corresponding $J_C(y)$ in Fig. 6.11a, the non-symmetric $J_C(y)$ yields the non-Fraunhofer pattern:

$$I_{\max}(h) = \frac{1}{f_{\text{as}}}\sqrt{B_0^2\sin^2 f_{\text{as}} + B_2^2(h)\cos^2 f_{\text{as}}}, \tag{6.26}$$

where $f_{\text{as}} = (2N + 1)\,|h|\,L_i/2\xi_F = \pi\Phi/\Phi_0$.

The dependencies (6.25) and (6.26) have a similar form with an additional oscillation period connected to the magnetic flux through every uniform part of the structure $|h|\,L_i/\xi_F$, while the main (the shorter) period depends on the value $f_{s,as} = \pi\Phi/\Phi_0$. When $|h| = (2n + 1)\pi\xi_F/L_i$, one of the terms of the sum $B_2(h)$ goes to infinity. It is not a real divergence but uncertainty of the type $\sin(x)/x$ at $x \to 0$. It gives a sharp maximum on the curve $I_{\max}(H)$, see Fig. 6.11b,c.

If the structure becomes long enough, its length compares to the Josephson penetration depth λ_{J0} but the length of each part remains $L_i \ll \lambda_{J0}$, and the linear approximation for the phase $\phi(y)$ is impossible to use. Then it is necessary to solve the sine-Gordon equation (6.8) [41]. It was shown [24, 26] that a second harmonic in the current–phase relation (6.19) for the long-range phase effectively generates in this situation. The first harmonic in the SIFNS array is the average critical current $J_1 = B_0/S$. The expression for the effective second harmonic obtained by the averaging procedure described in [24, 26] has the form

$$J_2 = -\frac{L_i^2}{4\pi^2\lambda_{J0}^2\,|J_{C0}|}\sum_{n=0}^{\infty}\frac{B_n^2}{(2n + 1)}, \tag{6.27}$$

where

$$B_n = \frac{\gamma_{B2} - \gamma_{B1}}{\gamma_{B1}\gamma_{B2}} \frac{2\pi T}{eRS} \sum_{\omega=0}^{\infty} \frac{\Delta^2}{\Delta^2 + \omega^2} \operatorname{Re} v_n.$$

It is necessary to note that the product $\lambda_{J0}^2 |J_{C0}|$ does not depend on the value J_{C0}: $\lambda_J^2 |J_{C0}| = \Phi_0 / 2\pi\mu_0(2\lambda_L)$ in accordance with the definition of λ_{J0} (6.9). The dependence of the effective first and second harmonic of d_F is presented in Fig. 6.12.

6.3 Method for the Reliable Realization of a φ Josephson Junction

6.3.1 *Phase Averaging of Rapid Oscillations with a Non-Sinusoidal CPR*

The φ junction is a JJ, where the Josephson energy density

$$E_J = \frac{\Phi_0}{2\pi} \int_0^{\phi} J(\phi')\mathrm{d}\phi' \tag{6.28}$$

has a local minimum at $\phi = \varphi$ $(0 < |\varphi| < \pi)$. To achieve this, the current–phase relation should be different from the usual sinusoidal one (6.6). It has been shown that a φ junction can be realized when the CPR has a second harmonic (6.19) and the conditions (6.20) are satisfied.

An intrinsic non-sinusoidal CPR $j(\phi) = j_1 \sin(\phi) + j_2 \sin(2\phi) + ...$ is not so exotic for Josephson junctions with a ferromagnetic barrier [2], as demonstrated recently in several experiments [53, 54]. Note that j_1 is not anymore the critical current density, but just the amplitude of the first harmonic and j_2 is the amplitude of the second harmonic. The critical current density J_C is then determined by local maxima of $j(\phi)$ as shown below. For example, in the simplest case of SFS junctions consisting of pure S and F metals, the CPR is strongly non-sinusoidal at a temperature $T \ll T_c$, where T_c is the critical temperature of the S metal. Different scattering mechanisms in the F-layer also influence the CPR: usual (non-pair breaking) scattering ("dirty" limit) [55–57]; spin-flip scattering [58] and scattering of electrons from the s to the d-band [5, 59]. In addition, the transparency of the interfaces or the presence of an extra insulating layer, like in SIFS or SIFIS junctions, influences the CPR as well [60–64]. However, the theoretical models [2, 5, 57–61] that take into account different scattering mechanisms give similar results: The amplitude of the second harmonic oscillates and decreases with d_F twice faster than the first one. To satisfy (6.20), one is tempted to choose d_F in the vicinity of the 0–π transition, where $|j_1| \ll |j_2|$. Unfortunately, there $j_2 > 0$ [5, 57–59], which excludes a φ ground state.

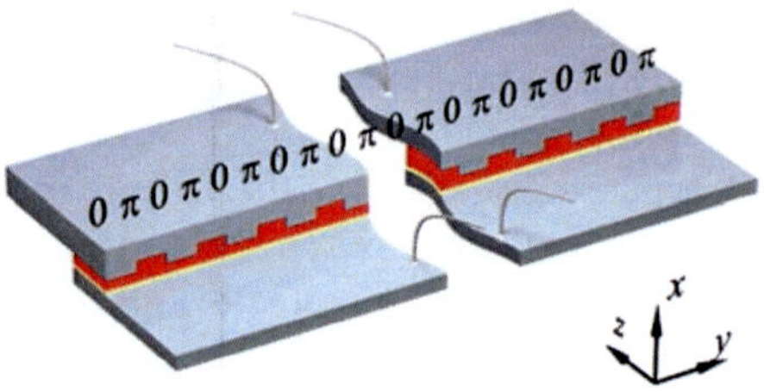

Fig. 6.13 Sketch of the 0–π multifacet ferromagnetic Josephson junction

A technique to create a negative second harmonics artificially was proposed recently [26, 65]. By using one of the available 0–π junction technologies [22, 35], one fabricates a Josephson junction with alternating short 0 ($j_1 > 0$) and π ($j_1 < 0$) regions (see Fig. 6.13), which all have the simple CPR like (6.6). Here, an *effective* second harmonic with negative amplitude is generated for the averaged phase, which is slowly varying on the scale of the facet length [26, 65]. Let the alternating 0 and π regions have the facet lengths L_a and L_b, respectively, so that $L_a \approx L_b$. It is assumed that $L_{a,b} < \lambda_J^a, \lambda_J^b$, where $\lambda_J^a = \lambda_J(j_{1a})$ and $\lambda_J^b = \lambda_J(j_{1b})$ are the Josephson penetration depth (6.9) in the corresponding parts.

The effective second harmonic has a maximum amplitude if $L_a = L_b \approx \lambda_J$, and if the values of the critical current densities in corresponding regions $j_{1a} \approx -j_{1b}$, see [26, 41]. The effective second harmonic amplitude is large enough only if j_{1a} and j_{1b} are very close by absolute values, demanding to choose thicknesses $d_{F,a}$ and $d_{F,b}$ with very high precision less than 1 Å for usual "dirty" SFS junctions. This precision is not achievable technologically. Therefore, a controllable φ-junction is quite hard to realize in this way.

However, ferromagnetic Josephson junctions already have some intrinsic second harmonic. Thus, the question is what will be the effective CPR in multifacet junctions if one takes into account such an intrinsic second harmonic. Usually, $j_2 < 0$ in the range of d_F where j_1 is rather large. The idea of the method used here is the following. We use two thicknesses $d_{F,a}$ and $d_{F,b}$ where $j_2(d_{F,a}) < 0$ and $j_1(d_{F,a}) \approx -j_1(d_{F,b})$. By making a step-like F-layer changing between $d_{F,a}$ and $d_{F,b}$, we effectively cancel the large first harmonic and create an effective negative second harmonic, which adds up with the intrinsic second harmonic.

Mints and coauthors [65–67] considered 1D junctions (along y direction in our notations) with a critical current density $j_1(y)$, which is a random or periodic function changing on a length-scale L_a, $L_b \ll \lambda_J^{a,b} = \lambda_J(j_{1a,b})$. Buzdin and Koshelev [26, 41] have found an exact solution of this problem if $J_C(y)$ in (6.6) alternates between the constant values j_{1a} and j_{1b}. Both groups assumed that $j_2 = 0$ within each region. Instead, here we assume that the CPR $j(d_F(y), \phi)$ is non-sinusoidal as a function of ϕ within each region, and alternates between $j(d_{F,a}, \phi)$ and $j(d_{F,b}, \phi)$ as a function of y. Then the problem of calculating the phase distribution along such a non-uniform structure can be reduced to the well-known problem of a non-linear oscillator [68].

The Josephson current in every region, as an odd function of the phase, can be expanded in a series of harmonics

$$j(d_F(y), \phi) = \sum_{n=1}^{\infty} j_n(d_F(y)) \sin n\phi, \tag{6.29}$$

and the problem is solved in a general form. Here, j_n denotes the amplitude of the *intrinsic* n-th harmonic of the current density.

We assume $\xi_F \ll L_{a,b} \leq \lambda_J^{a,b}$, were, ξ_F is the ferromagnetic coherence length, i.e. the characteristic length for the critical current density non-uniformity at the region of the step-like change of junction properties [38]. Let us consider one period of the structure $[-L_a; L_b)$ with F-layer thickness

$$d_F(y) = \begin{cases} d_{F,a}, \; y \in [-L_a, 0) \\ d_{F,b}, \; y \in [0, L_b) \end{cases}. \tag{6.30}$$

We rewrite (6.29) to separate the average and the oscillating parts of the Josephson current, that is,

$$j(y, \phi) = \sum_{n=1}^{\infty} \langle j_n \rangle \left[1 + g_n(y)\right] \sin n\phi, \tag{6.31}$$

where $j(y, \phi) = j(d_F(y), \phi)$. The average value of any function $f(y)$ is defined as

$$\langle f \rangle = \frac{1}{L_a + L_b} \int_{-L_a}^{L_b} f(y) \mathrm{d}y. \tag{6.32}$$

Then the averaged supercurrent of the n-th harmonic is

$$\langle j_n \rangle = \frac{L_a j_n(d_{F,a}) + L_b j_n(d_{F,b})}{L_a + L_b}, \tag{6.33}$$

and the corresponding oscillating part is

$$g_n(y) = (j_n(y) - \langle j_n \rangle) / \langle j_n \rangle, \tag{6.34}$$

so that $\langle g_n \rangle = 0$ by definition. Here $j_n(y, \phi) = j_n(d_F(y), \phi)$.

One can represent the Josephson phase $\phi(y)$ as a sum of a slow component $\psi(y)$, changing on a distance $\sim \Lambda_J = \lambda_J(\langle j_1 \rangle)$, see (6.9), and a rapid component $\zeta(y)$, changing on a distance $\sim L_a, L_b$, i.e.,

$$\phi(y) = \psi + \zeta(y). \tag{6.35}$$

Here, we assume that the junction is short ($l \ll \Lambda_J$) and do not write the y dependence for ψ. We also assume that the average of fast phase oscillations is vanishing, that is,

$$\langle \zeta \rangle = 0, \tag{6.36}$$

and their amplitude is small

$$\langle |\zeta| \rangle \ll 1. \tag{6.37}$$

The ground state of the Josephson junction is determined by the Ferrel–Prange equation (6.8), in this case

$$\Lambda_J^2 \frac{\partial^2 \phi}{\partial y^2} = \frac{j(y, \phi)}{|\langle j_1 \rangle|}. \tag{6.38}$$

Substituting (6.31) and (6.35) into (6.38) and keeping the terms up to first order in $\zeta(y)$, we can obtain equations for the rapid phase ζ:

$$\Lambda_J^2 \frac{\partial^2 \zeta}{\partial y^2} - \sum_{n=1}^{\infty} \beta_n g_n(y) \sin n\psi = 0. \tag{6.39}$$

To obtain the equation for the slow phase ψ, we average (6.38) over the length $(L_a + L_b) \ll \Lambda_J$ and get

$$\Lambda_J^2 \frac{\partial^2 \psi}{\partial y^2} - \sum_{n=1}^{N} [\beta_n \sin n\psi + n\beta_n \langle g_n \zeta \rangle \cos n\psi] = 0, \tag{6.40}$$

where $\beta_n = \langle j_n \rangle / |\langle j_1 \rangle|$. The number of harmonics N, that is reasonable to take into account within the given approximation, follows from the condition $N\,|\zeta(y)| \ll 1$. We have to find the function $\zeta(y)$ from (6.39), calculate average values $\langle g_n \zeta \rangle$, and substitute them into (6.40). Since for a step-like $d_F(y)$ the Josephson current (6.29) is a step-like function of y, (6.39) has the form $\partial^2 \zeta / \partial y^2 = const$ on every interval $[-L_a; 0)$ and $[0; L_b)$. Its solution is a parabolic segment. The function $\zeta(y)$ must be continuous at $y = 0$, and at the edges, that is it must satisfy the boundary condition $\zeta(-L_a) = \zeta(L_b)$. Moreover, it should satisfy (6.36).

It is convenient to expand the rapid phase $\zeta(y)$, as a solution of (6.39), into a series

$$\zeta(y) = \sum_{n=1}^{\infty} \zeta_n(y) \sin n\psi. \tag{6.41}$$

From this, the average values are calculated as

$$\beta_n \langle g_n \zeta_k \rangle = -\frac{2\alpha \delta j_n \delta j_k}{|\langle j_1 \rangle|}; \quad n, k = 1, 2... \tag{6.42}$$

where $\delta j_n \equiv j_n(d_{F,a}) - j_n(d_{F,b})$ and

$$\alpha \equiv \frac{L_a^2 L_b^2}{24\Lambda_J^2 (L_a + L_b)^2 \, |\langle j_1 \rangle|}. \tag{6.43}$$

By definition (6.9), $\Lambda_J^2 \, |\langle j_1 \rangle| = \lambda_J^2 \, |j_{1a}|$, where for brevity $j_{1a} \equiv j_1(d_{F,a}), \lambda_J \equiv \lambda_J(j_{1a})$.

The dependence of the Josephson current on the slow phase, which changes on a large distance of the order of Λ_J, follows from (6.40) as

$$J(\psi) = |\langle j_1 \rangle| \Big[\sum_n \beta_n \sin n\psi + \sum_{n,k} n\beta_n \, \langle g_n \zeta_k \rangle \cos n\psi \, \sin k\psi \Big] \tag{6.44}$$

The CPR (6.44) contains contributions of two types: intrinsic harmonics, that are given by the first term of (6.44) and the effectively generated ones, that are given by the second term. Intrinsic contributions are defined by the CPR (6.29) of the junction regions, while generated contributions effectively appear as a result of averaging over fast oscillations.

The amplitudes of the effectively generated harmonics, which are proportional to average values (6.42), are largest by absolute value if α and $|\delta j_n|$ reach their maximum. This happens if the lengths of a and b facets have the largest possible size, which still allows averaging, i.e,

$$L_a = L_b \approx \lambda_J \tag{6.45}$$

and

$$j_n(d_{F,a}) \approx -j_n(d_{F,b}) \tag{6.46}$$

Condition (6.46) ensures $|\langle j_1 \rangle| \ll |j_{1a}|$ and consequently $\Lambda_J \gg \lambda_J$. It was shown [26] that even if the equality (6.45) holds exactly, condition (6.37) is satisfied. In this case, $\alpha = 1/96 \, |j_{1a}|$ [c.f. (6.43)].

In all theoretical models developed up to now [2,5,57–61], the second harmonic was usually considered to be much smaller than the first one (except for the points of the 0–π transition, where $j_1 \to 0$), with an even smaller third harmonic. So, the expression (6.29) can be considered as a Tailor expansion in some small parameter. Then, keeping terms of the same order of this small parameter, we obtain $n + k = N + 1$ effective harmonics in the CPR of the multifacet 0–π junction.

The number N of harmonics that is reasonable to take into account follows from the estimate of the short-range phase $N \max |\zeta(y)| \ll 1$. As $\zeta(y)$ has a parabolic form, it takes its maximum absolute value at the center of every interval $-L_a/2$ and $L_b/2$. With the estimate

$$\sum_{n=1}^{\infty} \beta_n g_n(y) \sin n\psi \sim g_1 \tag{6.47}$$

one obtains

$$|\zeta(-L_a/2)| \sim \frac{L_a L_b (L_a + 2L_b)|\delta j_1|}{24\Lambda_J^2 (L_a + L_b)|\langle j_1 \rangle|}. \tag{6.48}$$

This is maximized if $L_a = L_b = \lambda_J$. Since $\max|\delta j_1| \approx 2|j_{1a}|$, we estimate $\max|\zeta(y)| \approx 1/8$. Thus, as a reasonable choice one can take $N = 3$. So, within the above approximations, it is reasonable to consider three intrinsic harmonics with amplitudes $j_{1..3}$ and four generated harmonics, i.e.

$$J(\psi) = \sum_{n=1}^{4} J_n \sin n\psi \tag{6.49}$$

with

$$\begin{aligned}
J_1 &= \langle j_1 \rangle + \alpha \delta j_1 \delta j_2 \\
J_2 &= \langle j_2 \rangle - \alpha \delta j_1^2 + 2\alpha \delta j_1 \delta j_3 \\
J_3 &= \langle j_3 \rangle - 3\alpha \delta j_1 \delta j_2 \\
J_4 &= -2\alpha \delta j_2^2 - 4\alpha \delta j_1 \delta j_3 \, .
\end{aligned} \tag{6.50}$$

Here, J_n denotes the total amplitude of the n-th harmonic in the CPR for the averaged phase. These expressions reduce to the earlier results [26,41] if one takes into account only the intrinsic first harmonic. The expression for every harmonic is a simple sum of the corresponding average intrinsic harmonic and the generated part.

6.3.2 Discussion of the φ Junction Conditions

We first address the question which type of ferromagnetic junction with 0–π facets can satisfy the conditions for a φ junction in the best way. The decay length ξ_1 (6.1) depends on l as well as on a pair-breaking scattering length in the ferromagnet (spin-flip scattering [58] or scattering into the d-band [69]; both influence the CPR). Pair-breaking scattering leads to [59, 70] $\xi_1 < \xi_2$ and the Josephson current decays so rapidly in the "dirty" limit that there are only very tiny regions of $d_{F,a}$ and $d_{F,b}$ on the dependence $j_1(d_F)$, where $j_1(d_{F,a}) \approx -j_1(d_{F,b})$. In the "clean" limit the Josephson current decays slower with increasing d_F than in dirty limit. The cleaner is the ferromagnet, the larger is l, and the slower is the decay; c.f. (23) and (24) with (7) from [57], or see the discussion in [69]. In the limit $l \gg (d_F, \xi_{1,2})$, the critical current density decreases as [55,56] $1/d_F$. Moreover, a "clean" SFS junction has a non-sinusoidal CPR, and its second harmonics $j_2 < 0$ in some regions of d_F far from 0–π transitions that can effectively help to satisfy the conditions for the realization of a φ junction. It was shown in different models that the second harmonic $j_2(d_F)$ decays and oscillates with d_F twice faster than $j_1(d_F)$. Therefore, it is reasonable to take an F-layer with $d_{F,a}$ and $d_{F,b}$ of the order of a few ξ_2. Usually, $\xi_2 = v_F/2E$, where v_F is the Fermi velocity and E is the exchange magnetic energy in the ferromagnet [57, 69, 71]. It was established experimentally, that for pure ferromagnetic metals ξ_2 is largest for Ni [72–75]. Higher harmonics also decay rapidly with increasing temperature [57, 76]. We have also checked this statement for models described in [55,56,59]. It is clear that by approaching T_c from below the superconducting gap $\Delta \to 0$, equations become linear and their simple exponential solutions yield only a sinusoidal CPR. Therefore, the most promising strategy for realizing a φ junction is to use a pure SFS junction with a thin Ni layer at low temperature ($\lesssim 0.1T_c$). The model describing the Josephson effect in such junctions was established long ago [55, 56]. It is based on the solution of the Eilenberger equations.

If the ferromagnetic film has a perpendicular magnetic anisotropy (i.e. if the ferromagnetic film has no in-plane magnetization in the ground state due to an easy axis of magnetization perpendicular to the junction plane), we can ignore the influence of the magnetization of the barrier on the phase difference, because the phase gradient is a consequence of the in-plane component of the magnetic field. We can neglect this effect also in the case, when the size l_d of magnetic domains in the F-layer $\xi_{1,2} \ll l_d \ll \lambda_J$. The corresponding phase variations would be much shorter in space than for the considered short-range phase, and their contribution would be of the order of $l_d^2/\lambda_J^2 \ll L_a L_b/\lambda_J^2$, by our assumption $L_a \sim L_b \sim \lambda_J$. Andreev bound states (in the "clean" limit) or a triplet pairing may arise in the area of domain walls and produce a long-range triplet component of the Josephson current [44, 77]. These triplet components have usually not a significant value for small values of $d_F \sim \xi_{1,2}$, where the usual singlet current is essential. The experiment [70] with Nb/Cu-Ni/Nb junctions demonstrated an absence of a significant influence of a domain structure on the CPR for multi-domain samples [78] (see the classical Fraunhofer dependence in Fig. 3 of [70]).

What is the measurable critical current density J_C (maximum supercurrent) of a Josephson junction with a non-sinusoidal CPR? It is not anymore $|j_1| = |j(\pi/2)|$ as follows from (6.6), but the local maximum of the expression (6.29) with respect to ϕ. Note that (6.29) may allow several local extrema in the interval $[0, 2\pi)$.

Figure 6.14a shows the three first harmonics vs F-layer thickness and Fig. 6.14b shows the corresponding $J_C(d_F)$ for the model of an SFS junction as described in [55, 56] (see also [57]). It is interesting to note, that near a 0–π transition, when the first harmonic is small, $j(\phi)$ may have two different local maxima, and the measurable $J_C(d_F)$ has two different values J_{C1}, $|J_{C2}|$ depending on the initial state of the junction [41] (see Fig. 6.14b). An example of a CPR $j(\phi)$ with two different maxima and the corresponding Josephson energy $E_J(\phi)$ (6.28) near a 0–π transition are presented in Fig. 6.14b(inset). Thus, even for a uniform Josephson junction (where the φ-ground state is impossible) close to a 0–π transition, two different values of critical current density could be realized if its CPR differs enough from the sinusoidal one.

In the framework of the clean SFS junction model [55, 56], we investigate a multifacet junction with $L_a = L_b = \lambda_J \ll \Lambda_J$. For λ_J, we take the value

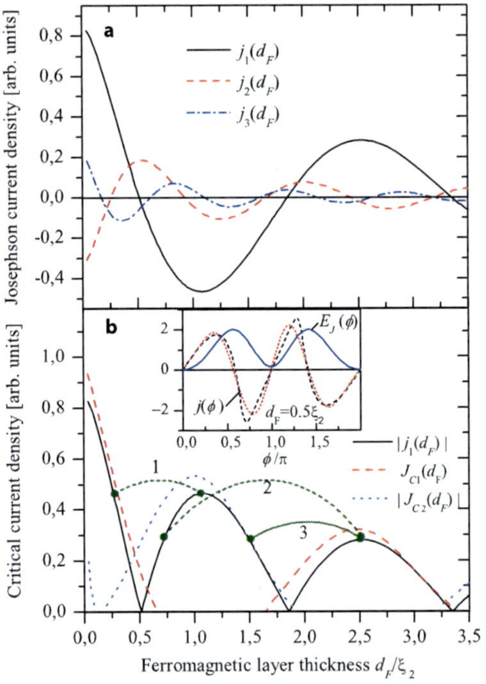

Fig. 6.14 Properties of a uniform SFS Josephson junction in the "clean" limit at $T = 0.1T_c$: (**a**) amplitudes of the first three harmonics j_1, j_2, j_3 of (6.29) as a function of the ferromagnet thickness d_F. (**b**) measurable critical current densities J_{C1} and $|J_{C2}|$ vs d_F, which are realized at the phase $0 < \phi < \pi$ and $\pi < \phi < 2\pi$, respectively. Inset shows for $d_F = 0.5\xi_2$ the CPR $j(\phi)$ from (3) in [55,56] (*dashed line*) and from the approximation by the first 3 harmonics (*dotted line*), with the corresponding $E_J(\phi)$ dependence (*solid line*). For comparison, in (**b**) the critical current $|j_1(d_F)|$ of a junction with only the first harmonic in the CPR is also shown. Here, pairs of points (connected by arcs), where $j_1(d_{F,a}) = -j_1(d_{F,b})$ correspond to a φ ground state of a multifacet SFS junction with alternating thicknesses $d_{F,a}$, $d_{F,b}$ (see areas 1,2,3 in Fig. 6.15a)

corresponding to the first minimum of j_1 at $d_F \approx \xi_2$, see Fig. 6.14a. For clean SFS junctions, this value is [74] $\approx 10\,\mathrm{kA/cm^2}$, which corresponds to $\lambda_J \approx 3\,\mu\mathrm{m}$ and which allows to realize $L_a \approx L_b$ with reasonable precision. Below we investigate the ground states in a Josephson junction with alternating regions of length L_a, L_b and F-layer thicknesses $d_{F,a}$ and $d_{F,b}$ varying from 0 to a few ξ_2.

The ground state corresponds to a local minimum of the energy (6.28) with the CPR (6.44). The junction has a stable static solution $\psi = 0$ (0-phase) if

$$\sum_{n=1}^{N} \left[n \langle j_n \rangle - \alpha \sum_{k=1}^{N+1-n} nk\delta j_n \delta j_k \right] > 0, \tag{6.51}$$

and the solution $\psi = \pi$ (π-phase) if

$$\sum_{n=1}^{N}\left[n\left\langle j_n\right\rangle-\alpha\sum_{k=1}^{N+1-n}(-1)^{n+k}nk\delta j_n\delta j_k\right]>0. \tag{6.52}$$

For the CPR (6.49), these conditions have the following form: For the 0-phase

$$J_1+2J_2+3J_3+4J_4>0, \tag{6.53}$$

and for the π-phase

$$J_1-2J_2+3J_3-4J_4<0. \tag{6.54}$$

Both solutions $\psi = 0$ and $\psi = \pi$ coexist when these conditions are satisfied simultaneously, i.e.

$$2J_2+4J_4>\left|J_1+3J_3\right|. \tag{6.55}$$

If both conditions (6.53) and (6.54) are not satisfied, that is

$$2J_2+4J_4<-\left|J_1+3J_3\right|, \tag{6.56}$$

only the φ ground state is possible. The conditions (6.53)–(6.56) coincide with the conditions (6.20) if $J_3 = J_4 = 0$. Generally, the φ junction is realized if the Josephson energy $E_J(\psi)$ (6.28) has a local minimum, i.e., if the CPR $J(\psi)$ (6.49) crosses $J = 0$ from a negative to a positive value at some point $\psi = \varphi \neq 0, \pi$. This is the case if the equation (obtained from $J(\psi) = 0$ in (6.49) and excluding $\sin\psi$ as it gives $\psi = 0, \pi$ not interesting now)

$$8J_4t^3+4J_3t^2+(2J_2-4J_4)t+J_1-J_3=0 \tag{6.57}$$

has at least one real solution $t = \cos\psi$, satisfying the conditions $|t| < 1$ (which gives $0 < |\psi| < \pi$), and

$$\begin{aligned}\frac{\partial J}{\partial\psi}\sim{}&32J_4t^4+12J_3t^3+(4J_2-32J_4)t^2\\&+(J_1-9J_3)t-2J_2+4J_4>0,\end{aligned} \tag{6.58}$$

(also obtained from (6.49)) that ensures the local minimum of $E_J(\psi)$. Starting from a pair of $d_{F,a}$, $d_{F,b}$ we calculate $J_{1\ldots4}$. Then from (6.53) to (6.58), the possible ground states are identified. The resulting phase diagram is shown in Fig. 6.15, where different ground states for each pair of $d_{F,a}$, $d_{F,b}$ are shown by different colors.

Figure 6.15a shows the results obtained if only the first intrinsic harmonic is taken into account. Here, the areas of 0 and π ground state phase are separated by slim regions of φ phase. It is clear that the phase diagram is symmetric with respect to the line $d_{F,a} = d_{F,b}$. Therefore, below, without losing generality, we focus on the case $d_{F,a} > d_{F,b}$. In the chosen interval of thicknesses, there are

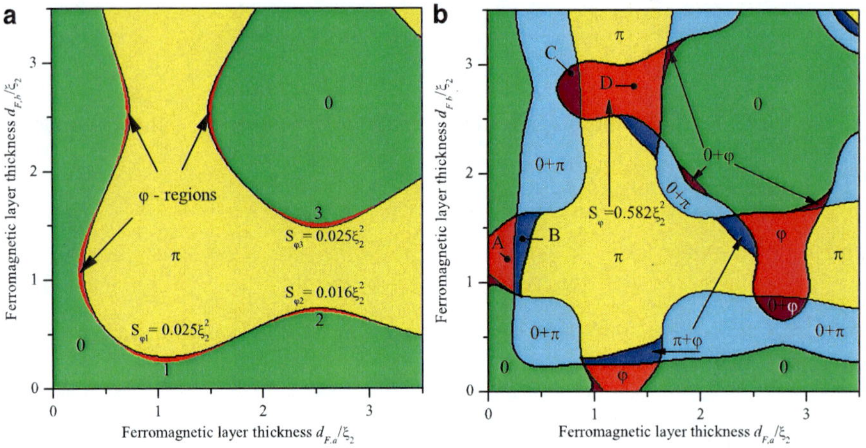

Fig. 6.15 Ground states scheme for the multifacet SFS junction with F-layer thickness periodically changing between $d_{F,a}$ and $d_{F,b}$ plotted on the $(d_{F,a}, d_{F,b})$ plane. The values of the ground state phase corresponding to each region are shown, and the areas S_φ for some of the different φ ground state regions are also indicated. In (**a**), the calculation includes only the first harmonics in every uniform part, that is, two generated harmonics. The φ – regions 1,2 and 3 in (**a**) correspond to the pairs of $d_{F,a}$ and $d_{F,b}$ that are shown by arcs in Fig. 6.14b. In (**b**), the first 3 harmonics and the corresponding 4 generated harmonics are taken into account, see (6.49) and (6.50). The dependence $E_J(\psi)$ for points A,B,C,D is shown in Fig. 6.16

3 areas of φ phase, marked as 1, 2 and 3. In these areas $j_1(d_{F,a}) \approx -j_1(d_{F,b})$, which corresponds to the pairs of $d_{F,a}$ and $d_{F,b}$ shown in Fig. 6.14b as 1, 2 and 3. If $d_{F,a}$ and $d_{F,b}$ are not well controllable, the area S_φ of φ regions is proportional to the probability of the φ junction realization. For all three different cases shown in Fig. 6.15a S_φ is rather small ($\leq 0.025\xi_2^2$). Hence, one has to control $d_{F,a}$ and $d_{F,b}$ extremely precisely to realize a φ junction. In the case of a "dirty" SFS junction, when $j_1(d_F)$ decays exponentially, the area S_φ is even smaller, and the probability to fabricate a φ-junction is vanishing.

However, if we take into account a non-sinusoidal CPR for the "clean" SFS junction, the areas of the φ ground state become much larger, as can be seen in Fig. 6.15b. This is a consequence of the fact that the intrinsic second harmonic $j_2(d_F) < 0$ in the corresponding regions, which efficiently helps to make the absolute value of the generated second harmonic large enough.

In particular, the areas 2 and 3 from Fig. 6.15a merge and form a compact φ ground state region with area $S_\varphi \approx 0.6\xi_2^2$ around $d_{F,a} \sim 2.8\xi_2$ and $d_{F,b} \sim 1.4\xi_2$. This region seems to be very well suited for the experimental realization of a φ junction, as it does not demand to produce an extremely thin F-layer and at the same time allows for reasonably large tolerances in sample fabrication. Taking the ferromagnetic coherence length $\xi_2 \sim 1.2$ nm from experiments [73, 74], the linear size of the φ region in Fig. 6.15b is $\Delta d_{F,a} \sim \Delta d_{F,b} \sim 0.5 \ldots 1$ nm. Modern technology allows the control of d_F with such precision [11, 79, 80].

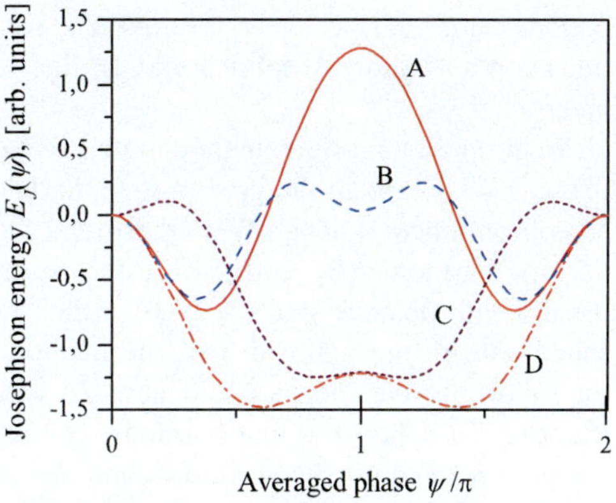

Fig. 6.16 Josephson energy $E_J(\psi)$ of multifacet SFS junctions for different combinations of F-layer thicknesses $d_{F,a}$ and $d_{F,b}$ as indicated by points A,B,C and D in Fig. 6.15b. A and D correspond to the φ ground state; B corresponds to two stable states π and φ; and C corresponds to two stable states 0 and φ

Finally, we note that the structure considered here may have simultaneously two different stable static solutions: 0 and π, 0 and φ, or π and φ, as can be seen in Fig. 6.15b. The corresponding $E_J(\psi)$ curves are presented in Fig. 6.16, where one can see several local minima of the Josephson energy as a function of the phase. Probably, taking into account the next harmonics, it is also possible to find two different φ ground states. However, we expect the corresponding regions within the $d_{F,a}, d_{F,b}$ plane to be so small that it will be quite hard to fabricate such a structure.

6.4 Conclusion

In the framework of the microscopic model based on the linearized Usadel equations, we have found that a step-like interface transparency inside SIFS and SIFNS JJs leads to oscillations of the critical current density $J_C(y)$ along the junction for d_F close to the 0–π transition value $d_F^{0\text{-}\pi}$. This, in turn, results in the formation of a non-uniform 0–π nano-junction with characteristic length of the order of the ferromagnet coherence length ξ_F for the SIFS JJ. For the SIFNS or SINFS structures, the nano-junction size is $\xi_N \gg \xi_F$.

The existence of this 0–π nano-junction inside the structure leads to an unusual dependence of the maximum supercurrent $I_{\max}$ on the external magnetic field H. If the junction is close to the 0–π transition, i.e. $d_F \approx d_F^{0\text{-}\pi}$, so that $J_C(y) \to 0$, $|y - L/2| \geq \max\{\xi_F, \xi_N\}$, then the $I_{\max}(H)$ dependence is very different from the usual Fraunhofer pattern. $I_{\max}$ increases with H up to a certain value which may be very large. This value depends on the size of the nano-junction (ξ_N or ξ_F).

The reasonably large size ($\sim \xi_N$) of the 0–π nano-junction in the SIFNS structure makes it possible to fabricate a JJ with periodic step-like changes of the NS interface transparency. $I_{\max}(H)$ in such a JJ may have additional oscillations and sharp peaks. Three relevant scales (the length of the array, the length of every region with constant transparency, and the length of the $J_C(y)$ non-uniformity) define three oscillation periods on $I_{\max}(H)$. These peculiarities are still present also far from the 0–π transition and allow to fabricate SQIF-like structures containing a ferromagnetic layer.

We propose a feasible method to realize a φ junction, that is with the Josephson phase $\varphi \neq 0$ or π in the ground state, based on SFS junctions. In a uniform SFS Josephson junction near a 0–π transition (with appropriate F-layer thickness d_F), the first harmonic $j_1 \to 0$, while the second harmonic j_2 dominates, however usually with positive sign ($j_2 > 0$), which excludes the formation of a φ ground state. Instead, the main idea of the method used here to make a φ junction is the following. We choose a thickness d_F where the second harmonic is large and negative. To cancel the first harmonic, we use a periodic step-like modulation of the F-layer thickness between $d_{F,a}$ and $d_{F,b}$. Here, $d_{F,a}$ and $d_{F,b}$ are chosen such that for both of them the second harmonic is negative, while the first harmonic is $j_1(d_{F,a}) \approx -j_1(d_{F,b})$. Periodic modulation not only cancels the first harmonic but also generates an additional negative second harmonic for the average phase. This effect can be made stronger if one works with a "clean" ferromagnetic barrier (with intrinsic negative second harmonic in the CPR) and uses alternating regions of equal length $L_a \approx L_b \sim \lambda_J$.

Different mechanisms, leading to a significantly non-sinusoidal current–phase relation of SFS junctions are analyzed. A CPR mostly different from the sinusoidal one is obtained for a "clean" SFS junction at low temperature. In this case, there are reasonably large regions of thicknesses $d_{F,a}$ and $d_{F,b}$ (in comparison with the description taking into account only the first harmonic), where the multifacet Josephson junction has a φ-ground state. Moreover, for some values of $d_{F,a}$ and $d_{F,b}$ such structures may have two different ground states (two local minima of the Josephson energy as a function of the phase): 0 and π, 0 and φ, or π and φ. Our analysis gives some practical recommendations for the fabrication of SFS junctions with arbitrary phase shifts φ in the ground state.

Acknowledgements This work was supported by the Russian Foundation for Basic Research (Grants 09-02-12176-ofi-m, 10-02-00569-a), by the German-Israeli Foundation (Grant G-967-126.14/2007) and by the Deutsche Forschungsgemeinschaft (DFG) via the SFB/TRR 21.

References

1. A.I. Buzdin, Rev. Mod. Phys. **77**(3), 935 (2005)
2. A. Golubov, M. Kupriyanov, E. Il'ichev, Rev. Mod. Phys. **76**, 411 (2004)
3. T. Ortlepp, Ariando, O. Mielke, C.J.M. Verwijs, K.F.K. Foo, H. Rogalla, F.H. Uhlmann, H. Hilgenkamp, Science **312**(5779), 1495 (2006)
4. L.B. Ioffe, V.B. Geshkenbein, M.V. Feigel'man, A.L. Faucheère, G. Blatter, Nature (London) **398**, 679 (1999)
5. N. Klenov, V. Kornev, A. Vedyayev, N. Ryzhanova, N. Pugach, T. Rumyantseva, J. Phys. Conf. Ser. **97**, 012037 (2008)

6. T. Yamashita, K. Tanikawa, S. Takahashi, S. Maekawa, Phys. Rev. Lett. **95**(9), 097001 (2005)
7. T. Yamashita, S. Takahashi, S. Maekawa, Appl. Phys. Lett. **88**(13), 132501 (2006)
8. A.V. Ustinov, V.K. Kaplunenko, J. Appl. Phys. **94**(8), 5405 (2003)
9. A.K. Feofanov, V.A. Oboznov, V. Bolginov, J. Lisenfeld, S. Poletto, V.V. Ryazanov, A.N. Rossolenko, M. Khabipov, D. Balashov, A.B. Zorin, P.N. Dmitriev, V.P. Koshelets, A.V. Ustinov, Implementation of superconductor-ferromagnet-superconductor π-shifters in superconducting digital and quantum circuits. Accepted to Nature Phys.
10. L.N. Bulaevskii, V.V. Kuzii, A.A. Sobyanin, Solid State Commun. **25**, 1053 (1978)
11. S. Frolov, D.V. Harlingen, V. Bolginov, V.A. Oboznov, V.V. Ryazanov, Phys. Rev. B **74**, 020503(R) (2006)
12. M.L.D. Rocca, M. Aprili, T. Kontos, A. Gomez, P. Spathis, Phys. Rev. Lett. **94**, 197003 (2005)
13. M. Weides, M. Kemmler, E. Goldobin, D. Koelle, R. Kleiner, H. Kohlstedt, A. Buzdin, Appl. Phys. Lett. **89**(12), 122511 (2006)
14. M. Weides, C. Schindler, H. Kohlstedt, J. Appl. Phys. **101**(6), 063902 (2007)
15. J. Pfeiffer, M. Kemmler, D. Koelle, R. Kleiner, E. Goldobin, M. Weides, A.K. Feofanov, J. Lisenfeld, A.V. Ustinov, Phys. Rev. B **77**(21), 214506 (2008)
16. J.H. Xu, J.H. Miller, C.S. Ting, Phys. Rev. B **51**, 11958 (1995)
17. E. Goldobin, D. Koelle, R. Kleiner, Phys. Rev. B **66**, 100508(R) (2002)
18. J.R. Kirtley, K.A. Moler, D.J. Scalapino, Phys. Rev. B **56**, 886 (1997)
19. T. Kato, M. Imada, J. Phys. Soc. Jpn. **66**(5), 1445 (1997)
20. E. Goldobin, D. Koelle, R. Kleiner, Phys. Rev. B **70**(17), 174519 (2004)
21. E. Goldobin, D. Koelle, R. Kleiner, Phys. Rev. B **67**, 224515 (2003)
22. M. Weides, M. Kemmler, H. Kohlstedt, R. Waser, D. Koelle, R. Kleiner, E. Goldobin, Phys. Rev. Lett. **97**, 247001 (2006)
23. E. Goldobin, K. Vogel, O. Crasser, R. Walser, W.P. Schleich, D. Koelle, R. Kleiner, Phys. Rev. B **72**(5), 054527 (2005)
24. R.G. Mints, Phys. Rev. B **57**(6), R3221 (1998)
25. R.G. Mints, I. Papiashvili, J.R. Kirtley, H. Hilgenkamp, G. Hammerl, J. Mannhart, Phys. Rev. Lett. **89**(6), 067004 (2002)
26. A. Buzdin, A. Koshelev, Phys. Rev. B **67**, 220504(R) (2003)
27. J. Oppenländer, C. Häussler, N. Schopohl, Phys. Rev. B **63**, 024511 (2000)
28. J. Oppenländer, T. Träubk, C. Häussler, N. Schopohl, Trans. Appl. Supercond. **11**, 1271 (2001)
29. C. Häussler, J. Oppenländer, N. Schopohl, J. Appl. Phys. **89**, 1875 (2001)
30. V. Schultze, R. Ijsselsteijn, H.G. Meyer, Supercond. Sci. Technol. **19**, S411 (2006)
31. V.K. Kornev, T.Y. Karminskaya, Y.V. Kislinskii, P.V. Komissinki, K.Y. Constantinian, G.A. Ovsyannikov, J. Phys. Conf. Ser. **43**, 1105 (2006)
32. A. Gumann, N. Schopohl, Phys. Rev. B **79**, 144505 (2009)
33. C. Gürlich, S. Scharinger, M. Weides, H. Kohlstedt, R.G. Mints, E. Goldobin, D. Koelle, R. Kleiner, Phys. Rev. B **81**(9), 094502 (2010)
34. A. Buzdin, A.E. Koshelev, Phys. Rev. B **67**, 220504(R) (2003)
35. H.J.H. Smilde, Ariando, D.H.A. Blank, G.J. Gerritsma, H. Hilgenkamp, H. Rogalla, Phys. Rev. Lett. **88**, 057004 (2002)
36. Ariando, D. Darminto, H.J.H. Smilde, V. Leca, D.H.A. Blank, H. Rogalla, H. Hilgenkamp, Phys. Rev. Lett. **94**(16), 167001 (2005)
37. S. Scharinger, C. Gürlich, R.G. Mints, M. Weides, H. Kohlstedt, E. Goldobin, D. Koelle, R. Kleiner, Phys. Rev. B **81**(17), 174535 (2010)
38. M. Kupriyanov, N. Pugach, M. Khapaev, A. Vedyayev, E. Goldobin, D. Koelle, R. Kleiner, JETP Lett **88**(1), 45 (2008). Pis'ma v ZhETF. **88**, 50 (2008)
39. N.G. Pugach, M.Y. Kupriyanov, A.V. Vedyayev, C. Lacroix, E. Goldobin, D. Koelle, R. Kleiner, A.S. Sidorenko, Phys. Rev. B **80**, 134516 (2009)
40. A. Zazunov, R. Egger, T. Jonckheere, T. Martin, Phys. Rev. Lett. **103**, 147004 (2009)
41. E. Goldobin, D. Koelle, R. Kleiner, A. Buzdin, Phys. Rev. B **76**, 224523 (2007)
42. N.G. Pugach, E. Goldobin, R. Kleiner, D. Koelle, Phys. Rev. B **81**(10), 104513 (2010)
43. M.Y. Kupriyanov, V.F. Lukichev, Sov. Phys. JETP **67**, 1163 (1988)

44. F. Bergeret, A. Volkov, K. Efetov, Rev. Mod. Phys. **77**, 1321 (2005)
45. V. Ambegaokar, A. Baratoff, Phys. Rev. Lett. **10**, 486 (1963)
46. A. Vasenko, A. Golubov, M. Kupriyanov, M. Weides, Phys. Rev. B **77**, 134507 (2008)
47. M.Y. Kupriyanov, A.A. Golubov, M. Siegel, *Nanoscale Devices – Fundamentals and Applications* (Springer, Amsterdam, 2006), chap. Josephson effect in composite junctions with ferromagnetic materials, p. 173
48. A. Abdumalikov, G.L.Alfimov, A. Malishevskii, Supercond. Sci. Technol. **22**, 023001 (2009)
49. V. Oboznov, V. Bolginov, A. Feofanov, V. Ryazanov, A. Buzdin, Phys. Rev. B (2005). ArXiv:cond-mat/0508573
50. T.Y. Karminskaya, A.A. Golubov, M.Y. Kupriyanov, A.S. Sidorenko, Phys. Rev. B **79**(21), 214509 (2009)
51. M. Weides, M. Kemmler, H. Kohlstedt, R. Waser, D. Koelle, R. Kleiner, E. Goldobin, Phys. Rev. Lett. **97**(24), 247001 (2006)
52. M. Weides, H. Kohlstedt, R. Waser, M. Kemmler, J. Pfeiffer, D. Koelle, R. Kleiner, E. Goldobin, Appl. Phys. A **89**(3), 613 (2007)
53. V.V. Ryazanov, V.A. Oboznov, V. Bolginov, A. Rossolenko, in *Proceedings of XII International Symphosium "Nanophysics and Nanoelectronics"*, vol. 1 (IFM RAS, Nizhniy Novgorod, 2008), vol. 1, p. 42. (in Russian)
54. H. Sellier, C. Baraduc, F. Lefloch, R. Calemczuk, Phys. Rev. Lett. **92**(25), 257005 (2004)
55. A.I. Buzdin, L. Bulaevskii, S. Panyukov, JETP Lett. **35**(4), 178 (1982).
56. A.I. Buzdin, L. Bulaevskii, S. Panyukov, Pis'ma v ZhETF **35**, 147 (1982)
57. F. Conschelle, J. Cayssol, A. Buzdin, Phys. Rev. B **78**, 134505 (2008)
58. A. Buzdin, Phys. Rev. B **72**(10), 100501 (2005)
59. A. Vedyayev, N. Ryzhanova, N. Pugach, J. Magn. Magn. Mat. **305**, 53 (2006)
60. A. Golubov, M. Kupriyanov, Pis'ma v ZhETF **81**(7), 419 (2005)
61. A. Golubov, M. Kupriyanov, JETP Lett. **81**, 335 (2005)
62. A. Golubov, M. Kupriyanov, Y.V. Fominov, JETP Lett. **75**(11), 588 (2002).
63. A. Golubov, M. Kupriyanov, Y.V. Fominov, Pis'ma v ZhETF **75**, 709 (2002)
64. Z. Radovic, N. Lazarides, N. Flytzanis, Phys. Rev. B **68**, 014501 (2003)
65. R. Mints, Phys. Rev. B **57**, R3221 (1998)
66. R.G. Mints, I. Papiashvili, Phys. Rev. B **62**(22), 15214 (2000)
67. R.G. Mints, I. Papiashvili, Phys. Rev. B **64**(13), 134501 (2001)
68. L. Landau, E. Lifshits, *Mechanics* (Pergamon Press, Oxford, 1960)
69. A. Vedyayev, C. Lacroix, N. Pugach, N. Ryzhanova, Europhys. Lett. **71**(4), 679 (2005)
70. V. Oboznov, V. Bolginov, A. Feofanov, V. Ryazanov, A. Buzdin, Phys. Rev. Lett. **96**, 197003 (2006)
71. F. Bergeret, A. Volkov, K. Efetov, Phys. Rev. B **64**, 134506 (2001)
72. J. Robinson, S. Piano, G. Burnell, C. Bell, M. Blamire, Phys. Rev. Lett. **97**, 177003 (2006)
73. J. Robinson, S. Piano, G. Burnell, C. Bell, M. Blamire, Phys. Rev. B **76**, 094522 (2007)
74. Y. Blum, A. Tsukernik, M. Karpovski, A. Palevski, Phys. Rev. Lett. **89**, 187004 (2002)
75. A.A. Bannykh, J. Pfeiffer, V.S. Stolyarov, I. Batov, V.V. Ryazanov, M. Weides, Phys. Rev. B **79**, 054501 (2009)
76. Z. Radović, L. Dobrosavljević-Grujić, B. Vujičić, Phys. Rev. B **63**(21), 214512 (2001)
77. Y. Fominov, A. Volkov, K. Efetov, Phys. Rev. B **75**, 104509 (2007)
78. I. Veschunov, V. Oboznov, A. Rossolenko, A. Prokofiev, L. Vinnikov, A. Rusanov, D. Matveev, Pis'ma v ZhETF **88**(11), 873 (2008). JETP Lett. **88**, 873 (2008)
79. M. Weides, K. Tillmann, H. Kohlstedt, Phys. C **437–438**, 349 (2006)
80. F. Born, M. Siegel, E.K. Hollmann, H. Braak, A.A. Golubov, D.Y. Gusakova, M.Y. Kupriyanov, Phys. Rev. B **74**(14), 140501 (2006)

Chapter 7
Josephson Effect in SFNS Josephson Junctions

T.Yu. Karminskaya, M.Yu. Kupriyanov, A.A. Golubov, and A.S. Sidorenko

Abstract The critical current, I_C, of Josephson junctions both in ramp-type (S-FN-S) and in overlap (SNF-FN-FNS, SN-FN-NS, SNF-N-FNS) geometries has been calculated in the frame of linearized Usadel equations (S–superconductor, F–ferromagnetic, N–normal metal). For the ramp-type structures, in which S electrodes contact directly the end walls of FN bilayer, it is shown that I_C may exhibit damping oscillations as a function of both the distance L between superconductors and thicknesses $d_{F,N}$ of ferromagnetic and normal layers. The conditions have been determined under which the decay length and period of oscillation of $I_C(L)$ at fixed d_F are of the order of decay length of superconducting correlations in the N metal, ξ_N, that is much larger than in F film. In overlap configurations, in which S films are placed on the top of NF bilayer, the studied junctions have complex SNF or SN electrodes (N or NF bilayer are situated under a superconductor). We demonstrate that in these geometries the critical current can exceed that in ramp-type junctions. Based on these results, the choice of the most practically applicable geometry is discussed.

7.1 Introduction

The existence of the oscillatory dependence of the critical current on the distance between superconducting electrodes reliably confirmed in a number of experiments

T.Yu. Karminskaya · M.Yu. Kupriyanov (✉)
Nuclear Physics Institute, Moscow State University, 119992 Moscow, Russia
e-mail: mkupr@pn.sinp.msu.ru

A.A. Golubov
Faculty of Science and Technology and MESA+ Institute of Nanotechnology,
University of Twente, 7500 AE, Enschede, The Netherlands

A.S. Sidorenko
Institute of Nanotechnology, Karlsruhe Institute of Technology, D-76021 Karlsruhe, Germany
and
Institute of Electronic Engineering and Nanotechnologies Chisinau, Moldova

A. Sidorenko (ed.), *Fundamentals of Superconducting Nanoelectronics*,
NanoScience and Technology, DOI 10.1007/978-3-642-20158-5_7,

using a variety of ferromagnetic materials and the types of Josephson junctions [1–16]. Promising use of π transitions, for which the critical current has a negative value, has been discussed in [17–21] for the implementation of qubits and for superconducting electronics. However, all these structures have some significant drawbacks, limiting their application.

The first of them is the smallness of the characteristic scale penetration of superconductivity in a ferromagnet. Indeed, analysis existing experimental data [1–16] shows that the value of exchange energy, H in ferromagnetic materials scales in between 850 and 2,300 K. Such large values of H lead to effective decay length, $\xi_{F1} \approx 1.2$–4.6 nm, and period of oscillations, $\xi_{F2} \approx 0.3$–2 nm, of thickness dependence of an SFS junction critical current, I_C. These values turned out to be much smaller compared to the decay length, $\xi_N \approx 10$–100 nm, in similar SNS structures. This fact makes it difficult to fabricate SFS junctions with reproducible parameters. It also leads to suppression of $I_C R_N$ product, thus limiting the cutoff frequency of the junctions. Since a search of exotic ferromagnetic materials with smaller value of H is challenging problem [16], one has to seek for another solutions.

Possible way to increase the decay length in a ferromagnetic barrier is the use of long-range proximity effect due to induced spin-triplet superconductivity [26–53] in structures with nonuniform magnetization. If magnetization of a ferromagnetic barrier is homogeneous, then only singlet component and triplet component with projection $S_z = 0$ of the total Cooper-pair spin are induced in the F region. These superconducting correlations are short-ranged, that is they extend into the F layer over a short distance of the order of $\xi_{F1} = \sqrt{D_F/H}$ in the diffusive case. However, in the case of inhomogeneous magnetization, for example in the presence of magnetic domain walls or in SF multilayer with noncollinear directions of magnetization of different F layers, a long-range triplet component (LRTC) with $S_z = \pm 1$ may appear. It decays into F region over distance $\xi_F = \sqrt{D_F/2\pi T_C}$ (here T_C is the critical temperature of S layer), which is by the factor $\sqrt{H/2\pi T_c}$ larger than ξ_{F1}. The latter property might lead to the long-range effects observed in some experiments [27, 28].

The transformation of decay length from ξ_{F1} to ξ_F might also take place in a vicinity of a domain wall even without generation of an odd triplet component [29–37]. This enhancement depends on an effective exchange field, which is determined by thicknesses and exchange fields of the neighboring domains. If a sharp domain wall is parallel [33, 36] or perpendicular to SF interface [37] and the thickness of ferromagnetic layers, $d_f \lesssim \xi_{F1}$, then for antiparallel direction of magnetization the exchange field effectively averages out, and the decay length of superconducting correlations becomes close to that of a single nonmagnetic N metal $\xi_F = \sqrt{D_F/2\pi T_C}$. It should be mentioned that for typical ferromagnetic materials ξ_F is still small compared to decay length $\xi_N \gtrsim 100$ nm of high conductivity metals such as Au, Cu, or Ag. This difference can be understood if one takes into account at least two factors. The first of them is that typical values of Fermi velocities in ferromagnetic materials (see, e.g., the analysis of experimental data done in [13, 14]) are of the order of 2×10^5 m/s, about an order of magnitude smaller than in high conductivity metals. The second factor is rather small electron mean free path in ferromagnets, especially in alloys such as CuNi, PtNi, etc.

In this work, research has focused on finding solutions to eliminate the above-stated deficiencies in SFS Josephson junctions with traditional geometry. To this end, new types of SFS Josephson junctions were suggested in which the weak links composed from NF multilayer structure. This work was aimed at carrying out theoretical studies of processes in these structures and the proof of the fundamental features such as extending the period of oscillation and the scale of the decay of the critical current to values of about ξ_N.

7.2 Effective Decrease in the Exchange Energy in S-(FN)S Josephson Structures

7.2.1 Structure of S-FN-S Junction and its Mathematical Description

We consider S-FN-S Josephson junction (Fig. 7.1) that consists of two massive superconducting electrodes connected to end-walls of a bilayer NF structure. The width of F layer is d_F and of N layer is d_N. It is suggested that the dirty limit conditions are satisfied in the N and F materials, and exchange energy $H = 0$ in the normal metal. We will consider that NF interface has finite transparency and that N and F metals have different transport parameters. The origin of the coordinate system is in the middle of the structure and the x- and y-axes are perpendicular and parallel to the NF interface, respectively (Fig. 7.1).

It is suggested that the structure is completely symmetric and the suppression parameters $\gamma_{\mathrm{BN}} = R_{\mathrm{BN}}\mathcal{A}_{\mathrm{BN}}/\rho_N\xi_N$ and $\gamma_{\mathrm{BF}} = R_{\mathrm{BF}}\mathcal{A}_{\mathrm{BF}}/\rho_F\xi_F$, characterizing the NS and FS interfaces, respectively, are large:

$$\gamma_{\mathrm{BN}} \gg \max\left\{1, \frac{\rho_S\xi_S}{\rho_N\xi_N}\right\}, \quad \gamma_{\mathrm{BF}} \gg \max\left\{1, \frac{\rho_S\xi_S}{\rho_F\xi_F}\right\},$$

so that the suppression of superconductivity in S electrodes can be disregarded. Here, $R_{\mathrm{BN}}, R_{\mathrm{BF}}$ and $\mathcal{A}_{\mathrm{BN}}, \mathcal{A}_{\mathrm{BF}}$ are the resistance and area of the SN (SF) interface, respectively, ρ_S, ρ_F, ρ_N and $\xi_S = (D_S/2\pi T_C)^{1/2}$, $\xi_F = (D_F/2\pi T_C)^{1/2}$, $\xi_N = (D_N/2\pi T_C)^{1/2}$ are the resistivities and coherence lengths of the materials,

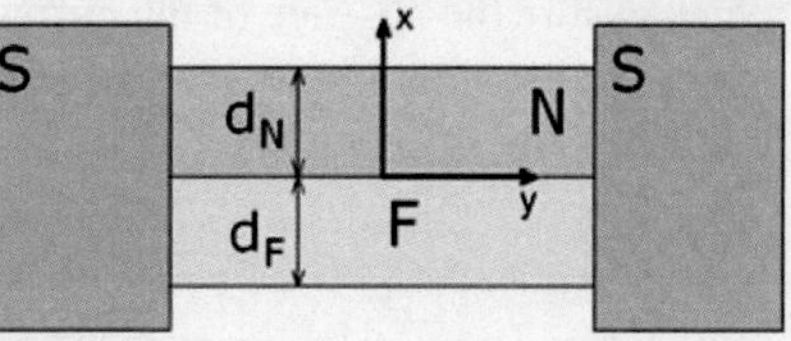

Fig. 7.1 Structure of S-FN-S Josephson junction

respectively; $D_{S,N,F}$ are the diffusion coefficients of the respective materials; and T_C is the critical temperature of the superconducting electrodes.

Under the above assumptions, it can be suggested that the Green's functions G_S and Φ_S in superconducting electrodes are equal to their equilibrium values $G_S = \omega/\sqrt{\omega^2+\Delta^2}$, $\Phi_S = \Delta \exp\{\pm i\varphi/2\}$, where Δ and φ are the absolute value and phase difference of the order parameters of the superconducting electrodes. The properties of the weak-link region can be described by using the linearized Usadel equation [38]. In the Φ – parameterization, they are represented in the form:

$$\xi_N^2 \left\{ \frac{\partial^2}{\partial x^2} + \frac{\partial^2}{\partial y^2} \right\} \Phi_N - \frac{|\omega|}{\pi T_c} \Phi_N = 0, \tag{7.1}$$

$$\xi_F^2 \left\{ \frac{\partial^2}{\partial x^2} + \frac{\partial^2}{\partial y^2} \right\} \Phi_F - \frac{\widetilde{\omega}}{\pi T_c} \Phi_F = 0, \tag{7.2}$$

where $\omega = T\pi(2n+1)$ are the Matsubara frequencies and $(n = 0, \pm 1, \pm 2 ...)$, $\widetilde{\omega} = |\omega| + iH \operatorname{sgn} \omega$.

The boundary conditions at the SN and SF interfaces (for $y = \pm L/2$) have the form cite [39,40]

$$\gamma_{\mathrm{BN}} \xi_N \frac{\partial}{\partial y} \Phi_N = \pm G_S \Delta \exp\{\pm i\varphi/2\}, \tag{7.3}$$

$$\gamma_{\mathrm{BF}} \xi_F \frac{\partial}{\partial y} \Phi_F = \pm \frac{\widetilde{\omega}}{|\omega|} G_S \Delta \exp\{\pm i\varphi/2\}. \tag{7.4}$$

The boundary conditions at the FN interface (for $x = 0$) have the form [39,40]:

$$\frac{\xi_N}{|\omega|} \frac{\partial}{\partial x} \Phi_N = \gamma \frac{\xi_F}{\widetilde{\omega}} \frac{\partial}{\partial x} \Phi_F, \tag{7.5}$$

$$\gamma_B \xi_F \frac{\partial}{\partial x} \Phi_F + \Phi_F = \frac{\widetilde{\omega}}{|\omega|} \Phi_N, \tag{7.6}$$

$$\gamma_B = R_B \mathcal{A}_B / \rho_F \xi_F, \quad \gamma = \rho_N \xi_N / \rho_F \xi_F,$$

where R_B and $\mathcal{A}_B$ are the resistance and area of the NF interface, respectively. The conditions at the free boundaries of the weak-link region at $x = d_N$ and $x = -d_F$ reduce to the equations

$$\frac{\partial \Phi_F}{\partial x} = 0, \quad \frac{\partial \Phi_N}{\partial x} = 0, \tag{7.7}$$

which ensure the absence of the current through these boundaries.

For further simplification of the problem, the thicknesses of the F and N films are assumed to be sufficiently small:

$$d_N \ll \xi_N, \quad d_F \ll \xi_F \tag{7.8}$$

and the solution of the boundary value problem given by (7.1)–(7.7) is sought in the form of the expansion in small parameters (d_N/ξ_N) and (d_F/ξ_F). This assumption allows one to simplify the problem and to reduce two-dimensional problem to one dimensional. In the first approximation, the functions Φ_N and Φ_F

$$\Phi_N(x, y) = A(y), \Phi_F(x, y) = B(y) \tag{7.9}$$

are independent of the coordinate x. In the next approximation, taking into account (7.7) we arrive at the expressions

$$\Phi_N = A(y) + \left\{ \frac{|\omega|}{\pi T_c \xi_N^2} A(y) - \frac{\partial^2}{\partial y^2} A(y) \right\} \frac{(x - d_N)^2}{2}, \tag{7.10}$$

$$\Phi_F = B(y) + \left\{ \frac{\widetilde{\omega}}{\pi T_c \xi_F^2} B(y) - \frac{\partial^2}{\partial y^2} B(y) \right\} \frac{(x + d_F)^2}{2}. \tag{7.11}$$

The substitution of (7.10), (7.11) into boundary conditions (7.5), (7.6) yields the following system of two equations for the functions $A(y)$ and $B(y)$:

$$\left[\zeta_F^2 \frac{\partial^2}{\partial y^2} - \left(\gamma_F \frac{\widetilde{\omega}}{\pi T_c} + 1 \right) \right] B(y) + \frac{\widetilde{\omega}}{|\omega|} A(y) = 0, \tag{7.12}$$

$$B(y) \frac{|\omega|}{\widetilde{\omega}} + \left[\zeta_N^2 \frac{\partial^2}{\partial y^2} - \left(\gamma_N \frac{|\omega|}{\pi T_c} + 1 \right) \right] A(y) = 0, \tag{7.13}$$

where

$$\zeta_F = \sqrt{\gamma_F} \xi_F, \quad \zeta_N = \sqrt{\gamma_N} \xi_N, \tag{7.14}$$

$$\gamma_F = \gamma_B \frac{d_F}{\xi_F}, \quad \gamma_N = \frac{\gamma_B}{\gamma} \frac{d_N}{\xi_N}. \tag{7.15}$$

The solution of this system of equations is represented in the form

$$A(y) = A_1 \cosh q_1 y + A_2 \sinh q_1 y + A_3 \cosh q_2 y + A_4 \sinh q_2 y,$$

$$B(y) = B_1 \cosh q_1 y + B_2 \sinh q_1 y + B_3 \cosh q_2 y + B_4 \sinh q_2 y,$$

where coefficients are related as follows :

$$B_1 = -\frac{1}{\zeta_F^2} \beta \frac{\widetilde{\omega}}{|\omega|} A_1, \; B_2 = -\frac{1}{\zeta_F^2} \beta \frac{\widetilde{\omega}}{|\omega|} A_2, \tag{7.16}$$

$$B_3 = \zeta_N^2 \frac{1}{\beta} \frac{\widetilde{\omega}}{|\omega|} A_3, \; B_4 = \zeta_N^2 \frac{1}{\beta} \frac{\widetilde{\omega}}{|\omega|} A_4.$$

Here, the inverse coherence lengths q_1 and q_2 are the roots of the characteristic equation

$$q_{1,2}^2 = \frac{1}{2}\left[u^2 + v^2 \pm \sqrt{(u^2 - v^2)^2 + 4\zeta_F^{-2}\zeta_N^{-2}}\right], \tag{7.17}$$

$$u^2 = \frac{1}{\zeta_N^2} + \frac{\Omega}{\xi_N^2}, \quad v^2 = \frac{1}{\zeta_F^2} + \frac{\Omega}{\xi_F^2} + \mathrm{i}\frac{h}{\xi_F^2}, \tag{7.18}$$

and parameter β is:

$$\beta = \frac{2}{\left[u^2 - v^2 + \sqrt{(u^2 - v^2)^2 + 4\zeta_F^{-2}\zeta_N^{-2}}\right]}, \tag{7.19}$$

and $\Omega = |\omega| \backslash \pi T_C$, $\widetilde{\Omega} = \widetilde{\omega}\backslash \pi T_C$, $h = H/\pi T_C \operatorname{sgn}(\omega)$.

The integration constants A_1, A_2, A_3, A_4 are determined from boundary conditions (7.3), (7.4):

$$A_1 = \frac{1 - s\beta\frac{\xi_N}{\xi_F}\zeta_N^{-2}}{\gamma_{\mathrm{BN}}(1 + \kappa^2)}\frac{G_S\Delta\sin(\varphi/2)}{\xi_N q_1 \cosh q_1\frac{L}{2}}, \tag{7.20}$$

$$A_2 = \mathrm{i}\frac{1 - s\beta\frac{\xi_N}{\xi_F}\zeta_N^{-2}}{\gamma_{\mathrm{BN}}(1 + \kappa^2)}\frac{G_S\Delta\cos(\varphi/2)}{\xi_N q_1 \sinh q_1\frac{L}{2}}, \tag{7.21}$$

$$A_3 = \frac{1 + \beta\frac{\xi_F}{\xi_N\zeta_F^2 s}}{\gamma_{\mathrm{BF}}(1 + \kappa^2)}\frac{G_S\Delta\cos(\varphi/2)}{\xi_F q_2 \sinh q_2\frac{L}{2}}, \tag{7.22}$$

$$A_2 = \mathrm{i}\frac{1 + \beta\frac{\xi_F}{\xi_N\zeta_F^2 s}}{\gamma_{\mathrm{BF}}(1 + \kappa^2)}\frac{G_S\Delta\sin(\varphi/2)}{\xi_F q_2 \cosh q_2\frac{L}{2}}, \tag{7.23}$$

here $s = \gamma_{\mathrm{BN}}/\gamma_{\mathrm{BF}}$, $\kappa = \beta(\zeta_F\zeta_N)^{-1}$.

The substitution of the solution obtained in the form of into the expression J_S for the superconducting current

$$J_S = \frac{\mathrm{i}\pi T \mathcal{A}_{\mathrm{BF}}}{2e\rho_F}\sum_{\omega=-\infty}^{\infty}\frac{1}{\widetilde{\omega}^2}\left[B_\omega\frac{\partial}{\partial y}B_{-\omega}^* - B_{-\omega}^*\frac{\partial}{\partial y}B_\omega\right]$$
$$+\frac{\mathrm{i}\pi T \mathcal{A}_{\mathrm{BN}}}{2e\rho_N}\sum_{\omega=-\infty}^{\infty}\frac{1}{\omega^2}\left[A_\omega\frac{\partial}{\partial y}A_{-\omega}^* - A_{-\omega}^*\frac{\partial}{\partial y}A_\omega\right] \tag{7.24}$$

yields the sinusoidal dependence $J_S = I_C \sin\varphi$. It is convenient to represent the critical current $I_C = I_{C1} + I_{C2}$ as the sum of two terms:

$$I_{C2} = \frac{2\pi T}{eR_{\mathrm{BF}}\gamma_{\mathrm{BF}}}\operatorname{Re}\sum_{\omega>0}^{\infty}\frac{G_S^2\Delta^2\omega^{-2}(1 + \beta s^{-1}\frac{\xi_F}{\xi_N\zeta_F^2})^2}{(1 + \kappa^2)\,\xi_F q_2 \sinh Lq_2}, \tag{7.25}$$

$$I_{C1} = \frac{2\pi T}{eR_{\text{BN}}\gamma_{\text{BN}}} \operatorname{Re} \sum_{\omega>0}^{\infty} \frac{G_S^2 \Delta^2 \omega^{-2} (1 - \beta s \frac{\xi_N}{\xi_F \zeta_N^2})^2}{(1+\kappa^2)\, \xi_N q_1 \sinh L q_1}. \tag{7.26}$$

Expressions (7.25), (7.26) specify a general expression for the critical current of the S(FN)S Josephson junctions under investigation. According to these relations, by complete analogy with oscillatory systems with two degrees of freedom, the S–(FN)–S structure under consideration can be characterized in terms of the partial inverse coherence lengths u, v and the proper inverse coherence lengths q_1, q_2.

The parameters ζ_F^{-1} and ζ_N^{-1} are the coupling constants. It is easy to see that the amplitude distribution coefficients at the proper coherence lengths, which are proportional to β are determined only by the material constants of the structure and are independent of the boundary conditions at the SN and SF interfaces, respectively. The way of current injection in the weak- link region (through the ratio $\gamma_{\text{BN}}/\gamma_{\text{BF}}$) is taken into account by the coefficient s, and the subsequent redistribution of the injected current between the F and N films is determined by the ratio γ_F/γ_N.

7.2.2 *Analysis of Inverse Coherence Lengths and Critical Current*

Analysis of expressions (7.25), (7.26) for the critical current components and inverse coherence lengths are simplified for a number of limiting cases.

In the limit of a high resistance of the FN weak-link interface

$$\zeta_N \gg \xi_N, \quad \zeta_F \gg \xi_F \tag{7.27}$$

the coupling constants between the F and N films are small. In the first approximation in ζ_N^{-1} and ζ_F^{-1}, the supercurrent in the structure flows through two independent channels and formulas for I_{C1} and I_{C2} are transformed to the expressions for the critical currents [39,41,42] that were previously obtained for two-barrier SIFIS and SINIS junctions:

$$\frac{eR_{\text{BF}} I_{C2}}{2\pi T_C} = \frac{T}{\gamma_{\text{BF}} T_C} \sum_{\omega>0}^{\infty} \operatorname{Re} \left\{ \frac{G_S^2 \Delta^2}{\omega^2 \xi_F q_2 \sinh L q_2} \right\}, \tag{7.28}$$

$$\frac{eR_{\text{BN}} I_{C1}}{2\pi T_C} = \frac{T}{\gamma_{\text{BN}} T_C} \sum_{\omega>0}^{\infty} \frac{G_S^2 \Delta^2}{\omega^2 \xi_N q_1 \sinh L q_1}, \tag{7.29}$$

where

$$q_2^2 = q_{20}^2 = \frac{\Omega}{\xi_F^2} + \mathrm{i} \frac{h}{\xi_F^2}, q_1^2 = q_{10}^2 = \frac{\Omega}{\xi_N^2}. \tag{7.30}$$

In the next approximation, the proper inverse coherence lengths are easily expressed as

$$q_1^2 = q_{10}^2 + \frac{1}{\zeta_N^2} + \frac{\xi_F^2(\Omega + ih)}{\zeta_F^2\zeta_N^2\,(h^2 + \Omega^2)}, \tag{7.31}$$

$$q_2^2 = q_{20}^2 + \frac{1}{\zeta_F^2} - \frac{\xi_F^2(\Omega + ih)}{\zeta_F^2\zeta_N^2\,(h^2 + \Omega^2)}. \tag{7.32}$$

According to (7.31), (7.32) the proximity effect between the N and F films leads to a small decrease in the effective exchange energy in the F film. The physical meaning of these changes is obvious. An electron for a certain time can be in the N part of the FN film of the structure. This is equivalent to the subjection of electrons to the effective exchange energy averaged over the thickness of the FN film, which is obviously lower than the exchange energy in the ferromagnetic part of the structure. Changes in the damping of the superconductivity in the N film are more significant. In this case, the exponential decrease law changes to damping oscillations. However, their period in this approximation is much larger than ξ_N

$$\Lambda = 4\pi\xi_N \frac{\sqrt{\Omega}\zeta_F^2\zeta_N^2\,(h^2 + \Omega^2)}{\xi_N^2\xi_F^2 h} \gg \xi_N. \tag{7.33}$$

It increases infinitely for $h \to 0$ and is proportional to h for $h \gg \Omega$. This means that the term I_{C2} in this case in the expression for the critical current is negligibly small and $I_C \approx I_{C1}$. In contrast to similar SNS junctions without F films, the dependence $I_{C1}(L)$ has the form of damping oscillations. This effect is a consequence of the double proximity effect because the superposition between the superconducting correlations induced from superconductors and spin ordering from the ferromagnet occurs in the N film. However, the oscillation period is very large; for this reason, the experimental observation of the transition to the π-state is complicated in the case considered above.

In the opposite limiting case $\zeta_F \ll \xi_F$ and $\zeta_N \ll \xi_N$, strong coupling between the F and N films occurs in the weak-link region. In this case, the inverse proper coherence lengths are easily obtained in the form

$$q_{1+\theta(\zeta_N-\zeta_F)} = \frac{\sqrt{\zeta_N^2 + \zeta_F^2}}{\zeta_N\zeta_F} + \frac{\Omega\zeta_N\zeta_F}{2(\zeta_F^2 + \zeta_N^2)^{3/2}}\left(\frac{\zeta_F^2}{\xi_N^2} + \frac{\zeta_N^2}{\xi_F^2} + ih\frac{\zeta_N^2}{\xi_F^2}\right), \tag{7.34}$$

$$q_{2-\theta(\zeta_N-\zeta_F)} = \frac{1}{\sqrt{\zeta_N^2 + \zeta_F^2}}\sqrt{\left(\frac{\zeta_F^2}{\xi_F^2} + \frac{\zeta_N^2}{\xi_N^2}\right)\Omega + ih\frac{\zeta_F^2}{\xi_F^2}}, \tag{7.35}$$

where θ – Hevecide function. From (7.34), (7.35) it follows that the ferromagnetic film in the limit $\zeta_F \gg \zeta_N$ additionally suppresses superconductivity induced in the N region, so that

$$q_1 = \frac{1}{\zeta_N} + \frac{\Omega \zeta_N}{2\zeta_F^2}\left(\frac{\zeta_F^2}{\xi_N^2} + \frac{\zeta_N^2}{\xi_F^2}\right) + \mathrm{i}h\frac{\Omega \zeta_N^3}{2\zeta_F^2 \xi_F^2}, \tag{7.36}$$

$$q_2 = \sqrt{\frac{\Omega + \mathrm{i}h}{\xi_F^2} + \frac{\zeta_N^2 \Omega}{\zeta_F^2 \xi_N^2}}. \tag{7.37}$$

It is seen that the coherence length and oscillation period of the term I_{C2} in this case coincides in the first approximation with the respective quantities for the SFS junctions, whereas the term I_C in I_{C1} damps at lengths $(\mathrm{Re}(q_1))^{-1} \approx \zeta_N \ll \xi_N$.

In the limit $\zeta_N \gg \zeta_F$, the processes in the N film are determining, so that

$$q_2 = \frac{1}{\zeta_F} + \frac{\Omega \zeta_F}{2\xi_F^2} + \mathrm{i}h\frac{\Omega \zeta_F}{2\xi_F^2}, \tag{7.38}$$

$$q_1 = \sqrt{\frac{\Omega}{\xi_N^2}\left(1 + \frac{\zeta_F^2 \xi_N^2}{\zeta_N^2 \xi_F^2}\right) + \mathrm{i}\frac{h}{\xi_F^2}\frac{\zeta_F^2}{\zeta_N^2}}. \tag{7.39}$$

Therefore, the term I_{C2} in the critical current decreases more sharply than I_{C1}. In particular, the typical damping scale for superconducting correlations is approximately equal to $\xi_{N\Omega} = \xi_N/\sqrt{\Omega}$, whereas the effective exchange energy decreases by a factor of $\zeta_F^2/\zeta_N^2 \ll 1$.

Thus, when $\zeta_F \ll \zeta_N \ll \xi_N$ both the damping scale and oscillation period of $I_C(L)$ in the S–(FN)–S structures under consideration are much larger than the respective values in similar SFS junctions, where the normal film is absent. This statement is illustrated by the numerical calculation results shown in Figs. 7.2–7.6.

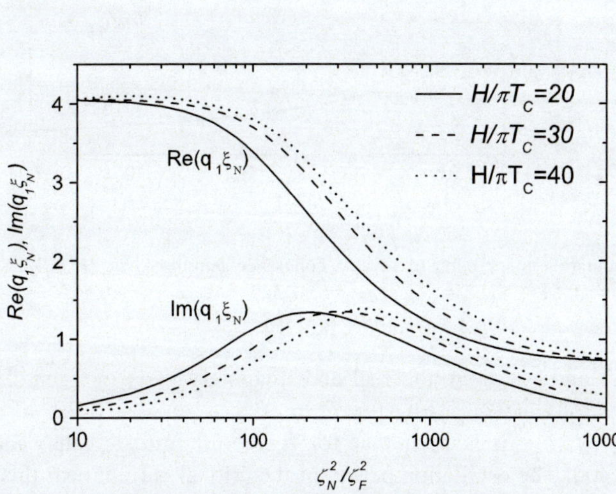

Fig. 7.2 Real and imaginary parts of inverse coherence length q_2 versus the parameter $z = (\zeta_N/\zeta_F)^2$ at $\xi_N/\zeta_N = 4$, $\xi_N/\xi_F = 10$, $T = 0.5T_C$, and $h = 20, 30, 40$

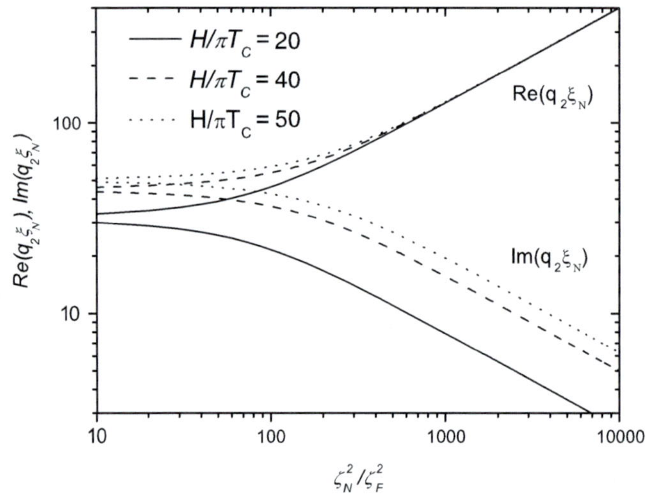

Fig. 7.3 Real and imaginary parts of inverse coherence length q_1 versus the parameter $z = (\zeta_N/\zeta_F)^2$ at $\xi_N/\zeta_N = 4$, $\xi_N/\xi_F = 10$, $T = 0.5T_C$, and $h = 20, 40, 50$

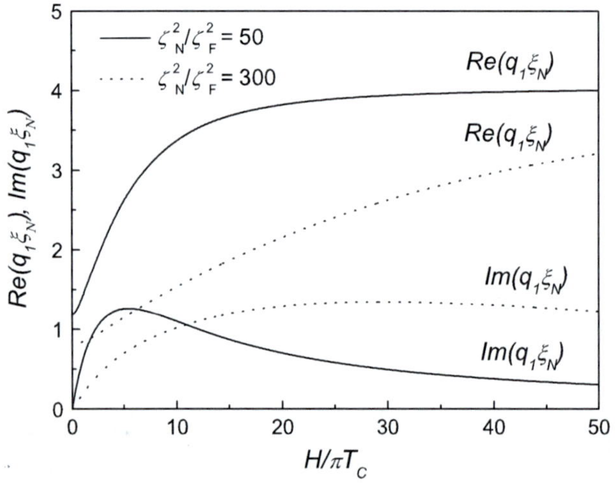

Fig. 7.4 Real and imaginary parts of inverse coherence lengths q_1 at $H/\pi T_C$, $\xi_N/\zeta_N = 4$, $\xi_N/\xi_F = 10$, $T = 0.5T_C$, and $z = (\zeta_N/\zeta_F)^2 = 50, 300$

Figures 7.2 and 7.3 show the real and imaginary parts of q_2 and q_1, respectively, as functions of $(\zeta_N/\zeta_F)^2$, for $T = 0.5T_C$, $h = 20, 30, 40$ and $\xi_N = 10\xi_F$ and $\xi_N = 4\zeta_N$. it is seen that for $h = 30$ $\mathrm{Im}(q_2\xi_N)$ has maximum at $(\zeta_N/\zeta_F)^2 \approx 300$. The oscillation period of the critical current near this maximum is $\Lambda = 2\pi(\mathrm{Im}(q_2)^{-1} \approx 1.5\pi\xi_N$, and its damping length is $(\mathrm{Re}(q_2))^{-1} \approx 0.4\xi_N$. The damping scale of the second term in the expression for the critical current is $(\mathrm{Re}(q_2))^{-1} \approx 0.014\xi_N$, which is two orders of magnitude smaller. Such strong

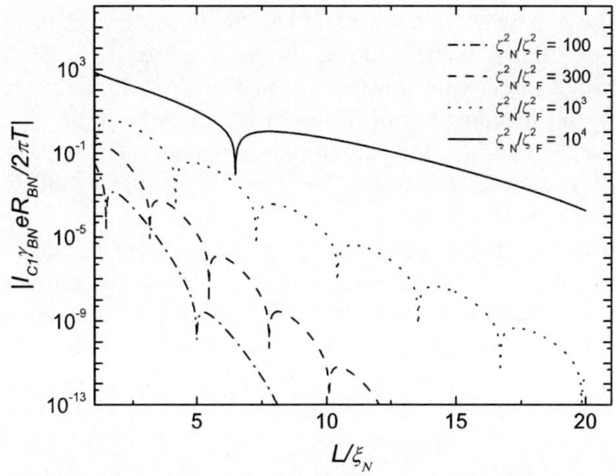

Fig. 7.5 Normalized value of part of critical current I_{C1} versus the distance L/ξ_N between the superconducting electrodes for $h = 30$, $\xi_N/\zeta_N = 4$, $\xi_N/\xi_F = 10$, $T = 0.5T_C$, $s = 1$, and $z = (\zeta_N/\zeta_F)^2 = 100, 300, 1{,}000, 10{,}000$

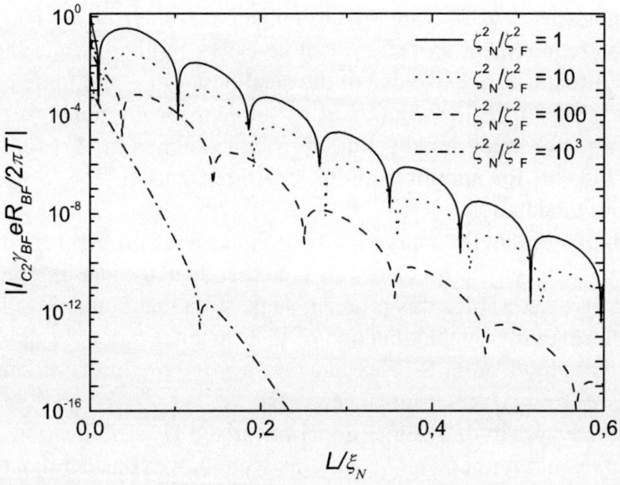

Fig. 7.6 Normalized value of part of critical current I_{C2} of the critical current versus the distance L/ξ_N between the superconducting electrodes for $h = 30$, $\xi_N/\zeta_N = 4$, $\xi_N/\xi_F = 10$, $T = 0.5T_C$, $s = 1$, and $z = (\zeta_N/\zeta_F)^2 = 1, 10, 100, 1{,}000$

difference between the damping lengths allows observation of the transition to the π state in the structures, where the distance between the electrodes is an order of magnitude larger than that in the available structures.

Figure 7.4 shows dependences of real and imaginary parts of q_1 on exchange energy for parameter $z = 50, 300$. At small H critical current decrease without oscillations. With H increase, the period of oscillations is decreased. It is seen that the imaginary part of q_1 has a maximum as function exchange energy and, with increase of z this maximum moves toward large values of H. The value of this maximum increases with growth of z, and at $z \sim 50$ leaves on saturation. Simultaneously, the damping length decreases with increase of H.

Figures 7.5 and 7.6 show the critical current components I_{C1} and I_{C2} as functions of the distance L between electrodes for $T = 0.5T_C$ and various values of the parameter $z = (\zeta_N/\zeta_F)^2$.

It is easy to see that the component I_{C2} at the given parameters decreases sharply with an increase in L and, in agreement with expectations, its contribution to I_C is negligibly small already at $L \approx 0.5$, that is, long before the appearance of the first minimum in I_{C2}. It is interesting that the oscillation period in $I_C \approx I_{C2}$ is a nonmonotonic function of the parameter z. It has the minimum at $z \approx 300$.

7.3 Josephson Effect in S-FN-S Structures with Arbitrary Thickness of Ferromagnetic and Normal Layers

The results obtained in previous section are essentially based on the assumption that the thicknesses, d_N, d_F, are small compared to their decay lengths. In real structures, the requirement $d_N \ll \xi_N$ can be easily fulfilled, while the inequality $d_F \ll \xi_F$ is difficult to achieve due to the smallness of ξ_F and finite roughness of NF interfaces. Therefore, the solution of two-dimensional problem is needed. The structures with two-dimensional geometry were examined in [37] for two-domain junction, in [43, 44] for multidomain SF structures, and in [45] for junction with helicoidal spin modulation.

It is worth to note that the solution of two-dimensional problem arising in the "in plane" geometry, when the domain wall is perpendicular to SF interface [32, 37], is simplified by a natural for this problem suggestion that domain walls consist of materials differing only by the direction of their magnetization. In this section, we will discuss properties of an S-FN-S junction beyond the limits of small F and N film thicknesses for two dimensional geometry.

We will solve system of 2-dimensional linearized Usadel equations (7.1), (7.2) with boundary conditions (7.3)–(7.6). In this section, we consider that the origin of coordinate axis y is on left interface between S electrode and F and N film. It is convenient to write the general solution of the boundary value problem in the form:

$$\Phi_N(x, y) = \Phi_N(y) + \sum_{n=-\infty}^{\infty} A_n(x) \cos\frac{\pi n(y-L)}{L}, \tag{7.40}$$

$$\Phi_F(x, y) = \Phi_F(y) + \sum_{n=-\infty}^{\infty} B_n(x) \cos\frac{\pi n(y-L)}{L}, \tag{7.41}$$

where $\Phi_N(y)$ and $\Phi_F(y)$ are asymptotic solutions of (7.40), (7.41) at the distance far from FN interface

$$\Phi_N(y) = \frac{G_S\Delta}{\sqrt{\Omega}\gamma_{\mathrm{BN}}}\left(\frac{\cos\frac{\varphi}{2}\cosh\frac{L-2y}{2\xi_{N\Omega}}}{\sinh\frac{L}{2\xi_{N\Omega}}} - \frac{\mathrm{i}\sin\frac{\varphi}{2}\sinh\frac{L-2y}{2\xi_{N\Omega}}}{\cosh\frac{L}{2\xi_{N\Omega}}}\right), \tag{7.42}$$

$$\Phi_F(y) = \frac{\sqrt{\widetilde{\Omega}}G_S\Delta}{\Omega\gamma_{\mathrm{BF}}}\left(\frac{\cos\frac{\varphi}{2}\cosh\frac{L-2y}{2\xi_{F\Omega}}}{\sinh\frac{L}{2\xi_{F\Omega}}} - \frac{\mathrm{i}\sin\frac{\varphi}{2}\sinh\frac{L-2y}{2\xi_{F\Omega}}}{\cosh\frac{L}{2\xi_{F\Omega}}}\right), \tag{7.43}$$

where $\xi_{N\Omega} = \xi_N/\sqrt{\Omega}$, $\xi_{F\Omega} = \xi_F/\sqrt{\widetilde{\Omega}}$, while functions $A_n(x)$ and $B_n(x)$ are solutions of appropriate one-dimensional boundary problem. The details of $A_n(x)$ and $B_n(x)$ determination are given in Appendix.

Substitution of the expressions (7.40), (7.41) into formula for the supercurrent, I_S, after routine calculations and several simplifications presented in Appendix in the most interesting from the practical point of view situation is when

$$H \gg \pi T_C,\ \xi_F \ll \xi_N, \tag{7.44}$$

results in sinusoidal dependence $I_S = I_C \sin\varphi$, where

$$I_C = \frac{2\pi T}{e}\frac{d_N}{\xi_N}\frac{W}{\gamma_{\mathrm{BN}}^2\rho_N}\mathrm{Re}\sum_{\omega>0}^{\infty}\frac{G_S^2\Delta^2}{\omega^2 q\xi_N\sinh(qL)}, \tag{7.45}$$

and inverse coherence length is given by

$$q = \frac{1}{\xi_N}\sqrt{\frac{\xi_N}{d_N}\frac{\gamma\sqrt{\widetilde{\Omega}}}{\gamma_B\sqrt{\widetilde{\Omega}} + \coth\left\{\frac{d_F}{\xi_F}\sqrt{\widetilde{\Omega}}\right\}} + \Omega}. \tag{7.46}$$

It is necessary to note that in general the expression for the critical current except summation over ω contains also summation over infinite number of inverse coherence length, which are the eigenvalues of boundary problem. As it is followed from estimations given in Appendix under the restrictions on the distance L between superconductors

$$L >> \mathrm{Re}\left(\frac{1}{q}\,\mathrm{arctanh}\,\frac{1}{\gamma_{\mathrm{BN}}q\xi_N}\right) \tag{7.47}$$

and on thickness of the N layer

$$\frac{\xi_F^2}{\eta h\xi_N^2} << \frac{d_N}{\xi_N} << \eta, \tag{7.48}$$

where

$$\eta = \begin{cases} \frac{1}{\gamma}\sqrt{\gamma_B^2 + \gamma_B\sqrt{2h^{-1}} + h^{-1}}, & \frac{d_F}{\xi_F} >> 1/\sqrt{h}, \\ \frac{1}{\gamma}\sqrt{\gamma_B^2 + 2\gamma_B\frac{\xi_F}{d_F}\frac{\Omega}{h^2} + \frac{\xi_F^2}{d_F^2h^2}}, & \frac{d_F}{\xi_F} << 1/\sqrt{h} \end{cases} \tag{7.49}$$

and $h = H/\pi T_C$ the main contribution to I_C comes from the item corresponding to the first eigenvalue, q, which is given by expression (7.46) and for I_C one can get formula (7.45).

It is necessary to point out that the experimentally studied parameters such as decay length of I_C as a function of L and period of I_C oscillations should be mainly controlled by the real and imaginary parts of inverse coherence length q calculated at $\omega = \pi T$.

Below here will be detailed examination of behavior of inverse coherence length q as a functions of geometrical and transport parameters of weak link. The calculated dependencies do not only provide the knowledge, which is necessary to take into account for design of S-FNF-S structures with input properties, but also will be useful for understanding the features of $I_C(L, d_F)$ dependencies.

7.3.1 *Properties of Inverse Coherence Length q*

In the limit of thin F film $d_F/\xi_F << 1/\sqrt{h}$ expression (7.46) for q transforms into result obtained in previous Section in the limit $\zeta_N >> \zeta_F$:

$$q^2 = \frac{\Omega}{\xi_N^2} + \frac{(h^2 + \Omega^2)\zeta_F^2 + \Omega\xi_F^2 + ih\xi_F^2}{\zeta_N^2\zeta_F^2(h^2 + (\xi_F^2\zeta_F^{-2} + \Omega)^2)}, \tag{7.50}$$

where $\zeta_F^2 = \gamma_B d_F/\xi_F$, $\zeta_N^2 = \gamma_B d_N/(\xi_N\gamma)$.

In the opposite case $d_F/\xi_F >> 1/\sqrt{h}$ from (7.46), we get

$$q = \frac{1}{\xi_N}\sqrt{\Omega + \frac{\xi_N}{d_N}\gamma\sqrt{\frac{h}{2}}\frac{i + 1 + \sqrt{2}\gamma_B\sqrt{h}}{\sqrt{h}\gamma_B(\sqrt{2} + \gamma_B\sqrt{h}) + 1}}. \tag{7.51}$$

At temperature $T = 0.5T_C$, the main contribution into the critical current is provided by the term corresponding to the first Matsubara frequency ($n = 0$). For this reason here, we will study the properties of inverse coherence lengths q for $\Omega = 0.5$ that is the value of Ω for ($n = 0$) and $T/T_C = 0.5$.

In Figs. 7.7 and 7.8, solid curves show the real and imaginary parts of inverse coherence lengths q as a function of d_F/ξ_F calculated from (7.46) for two ratios of normal film thickness $d_N/\xi_N = 0.01$ and $d_N/\xi_N = 0.1$.

The dotted lines in these figures are the same dependencies, which are followed from asymptotic formula (7.50). All calculations were done for a set of parameters $\xi_N/\xi_F = 10$, $\gamma = 0.03$, $\gamma_B = 0.2$, $h = 30$, which provide the maximum value for imaginary part of q at $d_F/\xi_F = 0.1$, $d_N/\xi_N = 0.01$ and $d_N/\xi_N = 0.1$.

It is clearly seen that the solid and dotted curves are in close agreement for $d_F/\xi_F \lesssim 0.1$. At $d_F/\xi_F \gtrsim 0.4$, the inverse coherence length starts to be practically independent of d_F reaching the value determined by (7.51). This fact is very important from practical point of view. It says that the parameters of

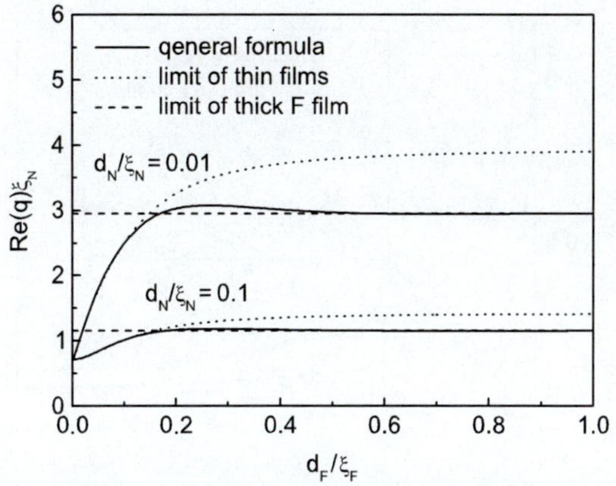

Fig. 7.7 Real part of q versus the thickness of F film d_F/ξ_F for two values of $d_N/\xi_N = 0.1, 0.01$ (*solid lines* for q calculated from (7.46), *dotted lines* for q calculated from formula for thin films, *dashed lines* correspond to the limit oh thick F film (7.51)) at $\xi_N/\xi_F = 10$, $\Omega = 0.5$, $h = 30$, $\gamma_B = 0.2$, $\gamma = 0.03$

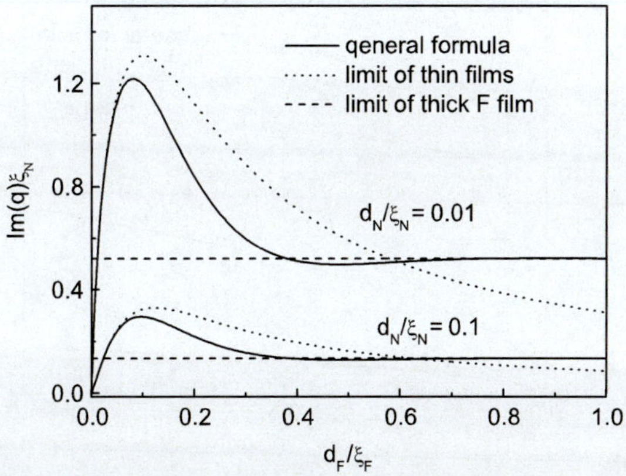

Fig. 7.8 Imaginary part of q versus the thickness of F film d_F/ξ_F for two values of $d_N/\xi_N = 0.1, 0.01$ (*solid lines* for q calculated from (7.46), *dotted lines* for q calculated from formula for thin films, *dashed lines* correspond to the limit oh thick F film (7.51)) at $\xi_N/\xi_F = 10$, $\Omega = 0.5$, $h = 30, \gamma_B = 0.2$, $\gamma = 0.03$

S-FN-S junctions do not deteriorate with increase of d_F (see the dashed lines in Figs. 7.7, 7.8). Moreover, imaginary parts of q become very robust against the fluctuation of ferromagnetic film thickness in the practically important interval $d_F/\xi_F \gtrsim 0.4$. From Figs. 7.7, 7.8, one can also see that taking into account finite value of thicknesses of films leads to some increase of decay length.

In Fig. 7.9, solid and dotted lines show the real and imaginary parts of inverse coherence length q as function of d_N/ξ_N calculated from (7.46) and (7.50), respectively, for two ratio of ferromagnetic film thickness $d_F/\xi_F = 0.1$ and

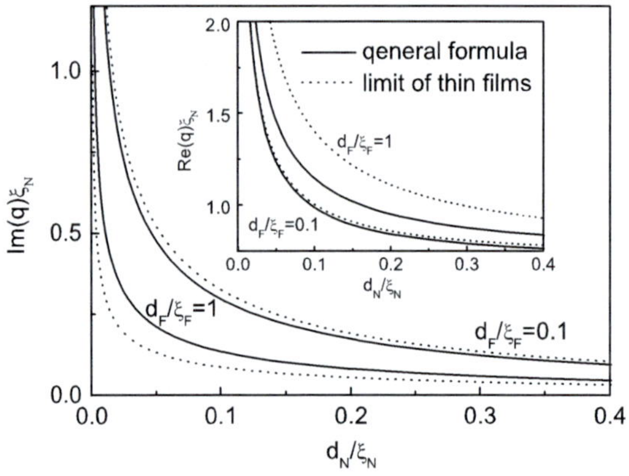

Fig. 7.9 Imaginary part of q versus the thickness of N film d_N/ξ_N for two values of $d_F/\xi_F = 0.1, 1$ (*solid lines* for q calculated from (7.46) and *dotted lines* for q calculated from formula for thin films [48]) at $\xi_N/\xi_F = 10$, $\Omega = 0.5$, $h = 30$, $\gamma_B = 0.2$, $\gamma = 0.03$. Inset shows real part of q versus the thickness of N film d_N/ξ_N at the same parameters

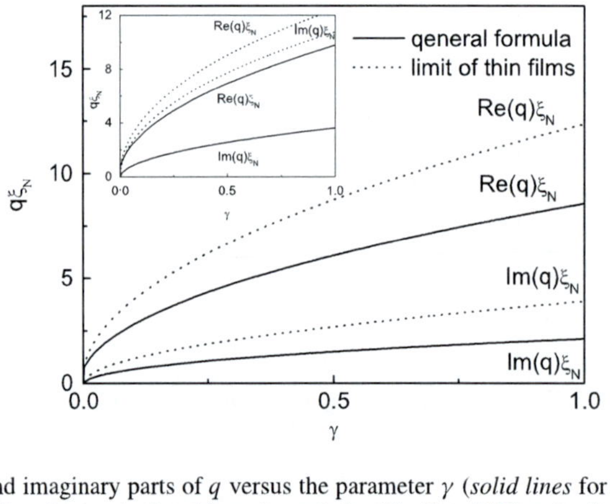

Fig. 7.10 Real and imaginary parts of q versus the parameter γ (*solid lines* for q calculated from (7.46) and *dotted lines* for q calculated from formula for thin films [48]) at $\xi_N/\xi_F = 10$, $\Omega = 0.5$, $h = 30$, $d_F/\xi_F = 0.5$, $d_N/\xi_N = 0.05$ and $\gamma_B = 0.1$. Inset shows the same dependence at $\gamma_B = 0.01$ and the same other parameters

$d_F/\xi_F = 1$ and at the same set of other parameters ($\xi_N/\xi_F = 10, \gamma = 0.03, \gamma_B = 0.2, \Omega = 0.5, h = 30$).

From the data, it follows that an increase of thickness of normal film leads to an increase of period of oscillations, which tends to infinity at large d_N. The decay length also increases with d_N and for $d_F/\xi_F = 0.1$ it practically approaches the value $\xi_N/\sqrt{\Omega}$, that is the decay length of the single normal film.

Figures 7.10 and 7.11 show the real and imaginary parts of inverse coherence lengths q as a function of γ and γ_B calculated at $d_N/\xi_N = 0.05$ for $\xi_N/\xi_F = 10$,

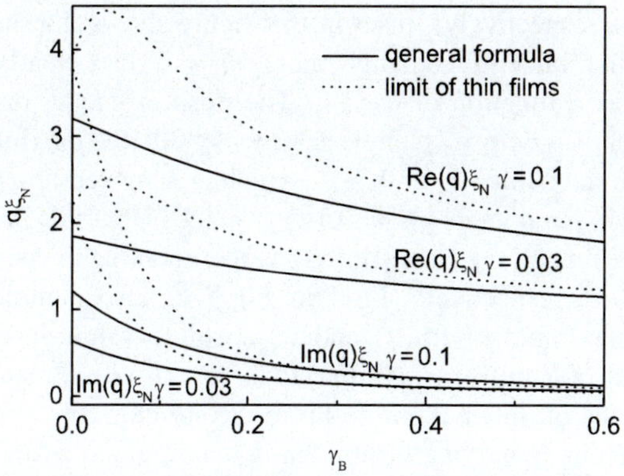

Fig. 7.11 Real and imaginary parts of q versus the parameter γ_B (*solid lines* for q calculated from (7.46) and *dotted lines* for q calculated from formula for thin films) at $\xi_N/\xi_F = 10$, $\Omega = 0.5$, $h = 30$, $d_F/\xi_F = 0.5$, $d_N/\xi_N = 0.05$ and two values of $\gamma = 0.1, 0.03$

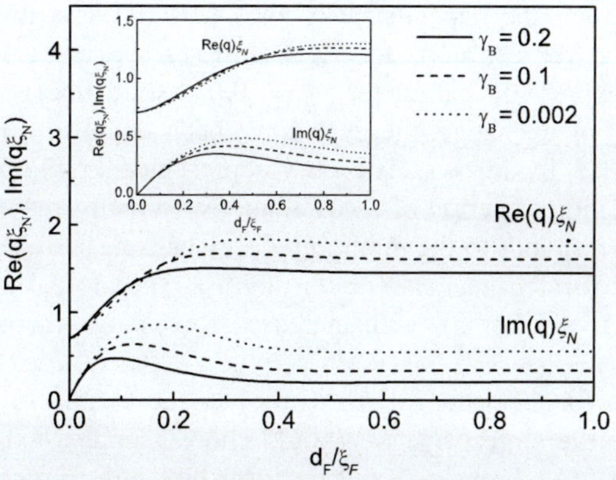

Fig. 7.12 Real and imaginary parts of q versus the thickness of F film d_F/ξ_F for $\gamma_B = 0.2, 0.1, 0.01$ (*solid, dashed, and dotted lines*) at $\xi_N/\xi_F = 10$, $\Omega = 0.5$, $h = 30$, $\gamma = 0.03$, $d_N/\xi_N = 0.05$. Inset shows the same dependence for $h = 5$

$\Omega = 0.5$, $h = 30$. In Fig. 7.10, $\gamma_B = 0.1$ and inset shows the same dependencies calculated for $\gamma_B = 0.01$. In Fig. 7.11, $\gamma = 0.1, 0.03$.

There is significant discrepancy between the curves calculated from the general expression (7.46) for q and from asymptotic dependence (7.50) for q. This discrepancy the larger the smaller is the suppression parameter γ_B. This result is obvious since the expression (7.50) is not valid at small γ_B. From direct comparison of the curves, we can conclude that in practically important range of $\gamma \gtrsim 0.1$ within the accuracy of 20% we may use the results of previous Section.

Figure 7.12 shows the dependencies of real and imaginary parts of q upon thickness of F film calculated for $\xi_N/\xi_F = 10$, $\Omega = 0.5$, $h = 30$, $\gamma = 0.03$, $d_N/\xi_N = 0.05$ and the set of parameters $\gamma_B = 0.2, 0.1, 0.01$ (solid lines, dashed lines, dotted lines, respectively). Inset in this figure shows the same dependencies obtained for smaller value of exchange energy $h = 5$. It is clearly seen that $\mathrm{Im}(q)$ has a maximum as a function of d_F/ξ_F. The position of the maximum shifts to larger F layer thickness with γ_B decrease. At $\gamma_B = 0.2$, the maximum of imaginary part has the value $\max(\mathrm{Im}(q)) \approx 0.5\xi_N$, which is achieved at $d_F/\xi_F \approx 0.1$. For smaller suppression parameters $\gamma_B = 0.1$ ($\gamma_B = 0.01$), the maximum of imaginary part equals to $\mathrm{Im}(q) \approx 2/3\xi_F$ ($\max(\mathrm{Im}(q)) \approx \xi_F$) and is achieved at $d_F/\xi_F \approx 0.13$ and $d_F/\xi_F \approx 0.2$, respectively. Inset in Fig. 7.12 also demonstrates that both position of the maximum of $\mathrm{Im}(q)$ and its absolute value depend on exchange energy h. Decrease of h shifts $\max(\mathrm{Im}(q))$ to larger ratio d_F/ξ_F and simultaneously suppresses the value of this maximum. From the structure of expression (7.46) for q, it follows that its imaginary part $\mathrm{Im}(q)$ has a maximum as a function of exchange energy h. Indeed, at $h \to 0$ the period of $I_C(L)$ oscillation tends to infinity, which is equivalent to $\mathrm{Im}(q) \to 0$. At large h, the imaginary part $\mathrm{Im}(q) \propto h^{-1/2}$, that is also goes to zero with h increase. At $d_F/\xi_F \gtrsim 0.4$ both $\mathrm{Im}(q)$ and $\mathrm{Re}(q)$ saturate and practically become independent on F layer thickness.

Figure 7.13 shows the dependence of $\mathrm{Im}(q)/\mathrm{Re}(q)$ as a function of d_F/ξ_F. The calculations have been done for $\gamma_B = 0.01$, $\xi_N/\xi_F = 10$, $\Omega = 0.5$, and for two values of suppression parameter $\gamma = 0.03$ (solid line), $\gamma = 0.1$ (dotted line). The values of exchange energy h have been equal to 10 and 30, as it is marked in Fig. 7.13 by arrows. The curves presented in Fig. 7.13 can be also used for minimization of period of I_C oscillations. Actually, the maximum of ratio $\mathrm{Im}(q)/\mathrm{Re}(q)$ corresponds to the minimum decay per one period. It is obvious that this maximum is located near maximum of $\mathrm{Im}\, q$. Therefore, the position of this maximum shifts to smaller d_F with increase of exchange energy or suppression parameter γ.

If we want to fix the value of $\max(\mathrm{Im}\, q)$ in the vicinity of ξ_N and to shift this maximum to the largest d_F, we should choose suppression parameter γ_B in the range of 0.01 (for large d_F) and perform the fitting procedure to estimate suppression parameter γ and exchange energy h. For instance, we may find that

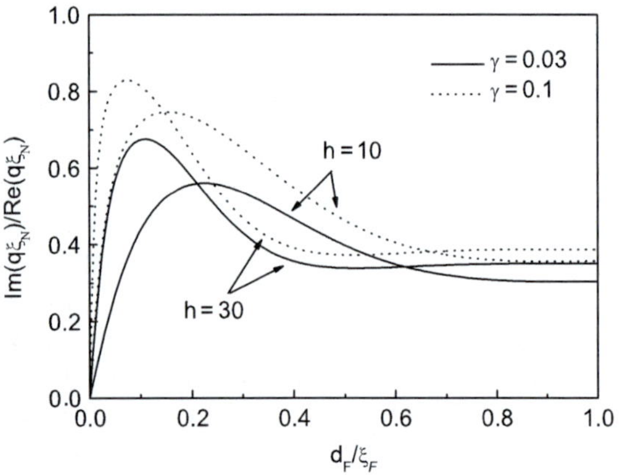

Fig. 7.13 Ratio of imaginary part of q to its real part versus the thickness of F film d_F/ξ_F for two values of $\gamma = 0.03, 0.1$ (*solid and dotted lines*) at $\xi_N/\xi_F = 10, \gamma_B = 0.01$, $d_N/\xi_N = 0.05$, $\Omega = 0.5$, and $h = 30, 10$

for $h = 30$ maximum of $\mathrm{Im}(q)$ is of the order of ξ_N and it is achieved for $\gamma = 0.03$ at $d_F/\xi_F \approx 0.18$, while for $h = 10$, $\gamma = 0.03$ the maximum is shifted to $d_F/\xi_F \approx 0.3$. The smaller the exchange energy the thicker should be the thickness of F film to get $\mathrm{Im}(q)$ in the range of ξ_N. Thus for $h = 5$, $\gamma = 0.09$, the maximum is achieved at $d_F/\xi_F = 0.45$. From the data presented in

Figures 7.12 and 7.13, it follows that for all mentioned above sets of parameters the ratio $\mathrm{Im}(q)/\mathrm{Re}(q) \approx 0.6$ and does not exceed this value.

For thick F film $d_F/\xi_F >> 1/\sqrt{h}$, the maximum of $Im(q)/Re(q)$ is achieved at small $\gamma_B \to 0$, $h >> \Omega$ and $\gamma\xi_N/d_N\sqrt{h/2} >> \Omega$ and this maximum relation is $Im(q)/Re(q) \sim 0.4$.

7.3.2 Thickness Dependence of the Critical Current

The critical current (7.45) of the studied S-FN-S Josephson junction (see Fig. 7.14) is a function of two arguments. They are the distance between superconducting electrodes L/ξ_N and the thickness of ferromagnetic film d_F/ξ_F. The dependence of $I_C(L/\xi_N, d_F/\xi_F)$ has shown in Fig. 7.14. It has been calculated from (7.45), (7.46) for $h = 10$, $d_N/\xi_N = 0.05$, $\gamma_B = 0.01$, $\gamma = 0.03$, $\gamma_{\mathrm{BF}}/\gamma_{\mathrm{BN}} = 1$ and $T = 0.5T_C$.

In the limit $d_F \to 0$ period of critical current oscillations tends to infinity (see Fig. 7.14) and I_C decays monotonically with L, as it must be for SNS Josephson junctions. With d_F, increase (see Fig. 7.15) the dependence of critical current as a function of L/ξ_N has the form of damped oscillations. The decay length of

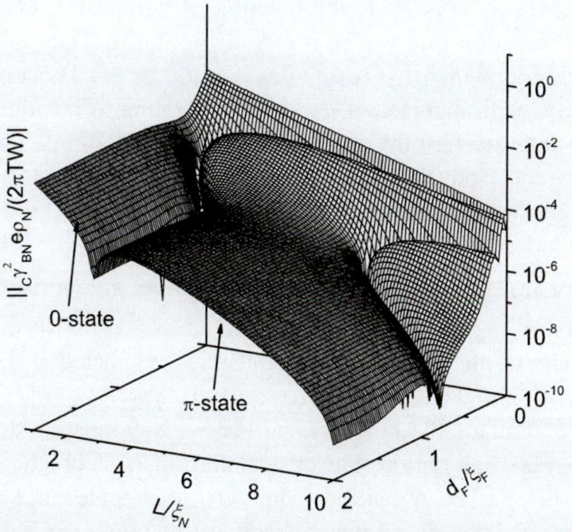

Fig. 7.14 Normalized absolute value of critical current versus the thickness of F film d_F/ξ_F and distance between the superconducting electrodes L/ξ_N for $\gamma = 0.03, \gamma_B = 0.01, \xi_N/\xi_F = 10, d_N/\xi_N = 0.05, h = 10, \gamma_{\mathrm{BN}}/\gamma_{\mathrm{BF}} = 1, T = 0.5T_C$

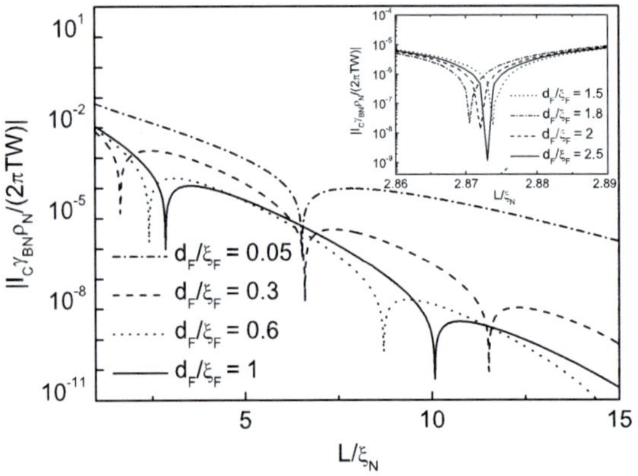

Fig. 7.15 Normalized absolute value of critical current versus the distance between electrodes L/ξ_N for $\gamma = 0.03, \gamma_B = 0.01, \xi_N/\xi_F = 10, d_N/\xi_N = 0.05, h = 10, \gamma_{\mathrm{BN}}/\gamma_{\mathrm{BF}} = 1, T = 0.5T_C$ for several thicknesses of F film $d_F/\xi_F = 0.05, 0.3, 0.6, 1$. Insert shows the same dependence for $d_F/\xi_F = 1.5, 1.8, 2, 2.2$.

these oscillations is different for different thickness of F film. The most intensive suppression is localized in the vicinity of $d_F/\xi_F \approx 0.6$ since Re(q) has maximum at this thickness of F layer. It is seen that the suppression of I_C is smaller for thicker ($d_F > 0.6\xi_F$) and thinner ($d_F < 0.6\xi_F$) F films. Period of I_C oscillations decreases with d_F achieving the smallest value at $d_F/\xi_F \approx 0.3$. Further increase of d_F results in increase of this period. Finally in the range of thickness $d_F/\xi_F > 0.5$ both period of I_C oscillations and decay length are nearly constant. In the interval of F layer thickness $d_F/\xi_F \gtrsim 1$, the position of zeros of $I_C(L, d_F)$ undergoes oscillations as a function of d_F (see insert in Fig. 7.15). They take place around values $L = L_n$, under which $I_C(L_n, d_F) = 0$ at $d_F \gg \xi_F$. The amplitude of these oscillations decays with increase of d_F. It is interesting to mention that the larger is L_n, the more intensive are the amplitudes of the oscillations. This behavior can be easily understood from the form of $q(d_F)$ dependence (7.46). In the vicinity of $L = L_n$, the critical current is small due to the 0-π transition of I_C as a function of L. Under this condition, any small variations of q, which occur due to factor coth $\left(d_F\sqrt{\widetilde{\Omega}}/\xi_F\right)$ in (7.46) start to be important giving rise to the discussed $I_C(L, d_F)$ behavior.

Figure 7.16 shows the $I_C(L, d_F)$ dependencies calculated at fixed values of L under $h = 10$, $d_N/\xi_N = 0.05$, $\gamma_B = 0.01$, $\gamma = 0.03$, $\gamma_{\mathrm{BF}}/\gamma_{\mathrm{BN}} = 1$, and $T = 0.5T_C$.

In the small L domain, the properties of the S-FN-S junction do not depend on the structure of weak link region. The critical current I_C is practically independent on d_F, so that there is no transition from zero to π state on $I_C(d_F)$. With the increase of L the $I_C(d_F)$ dependence becomes apparent (see Fig. 7.14) resulting in suppression of I_C. This suppression is different for different thicknesses of F film. The strongest suppression is realized in the vicinity of $d_F/\xi_F \approx 0.3$. This fact is illustrated in Fig. 7.16a by the line corresponding to the ratio $L/\xi_N = 1.3$.

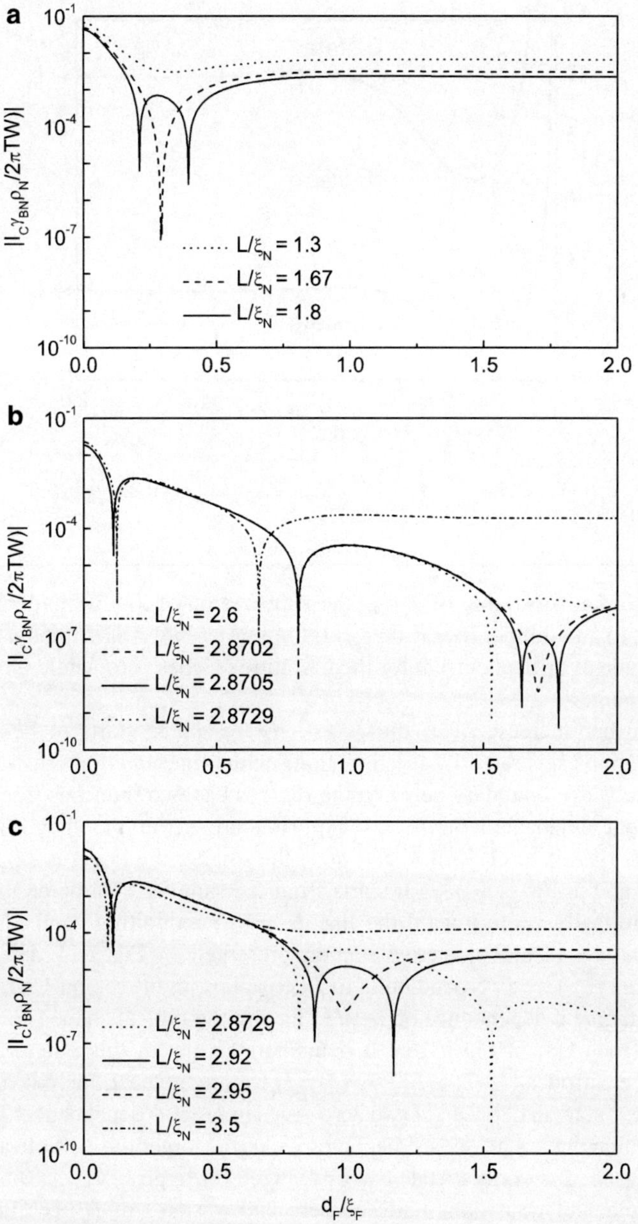

Fig. 7.16 Normalized absolute value of critical current versus the thickness of F film d_F/ξ_F for $\gamma = 0.03, \gamma_B = 0.01, \xi_N/\xi_F = 10, d_N/\xi_N = 0.05, h = 10, \gamma_{BN}/\gamma_{BF} = 1, T = 0.5T_C$ for several distances between the superconducting electrodes $L/\xi_N = 1.3, 1.67, 1.8, 2.6, 2.8702, 2.8705, 2.8729, 2.95, 2.95, 3.5$

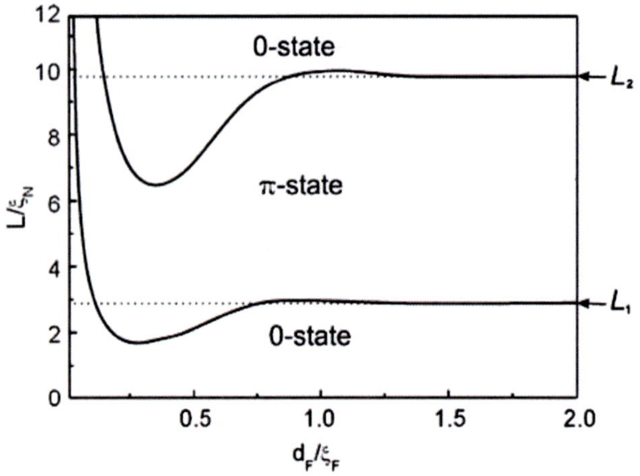

Fig. 7.17 (L, d_F) phase diagram for S-FN-S junction at $\gamma = 0.03, \gamma_B = 0.01, \xi_N/\xi_F = 10, d_N/\xi_N = 0.05, h = 10, \gamma_{\mathrm{BN}}/\gamma_{\mathrm{BF}} = 1, T = 0.5T_C$

It is seen that at this value of L/ξ_N the suppression of I_C is smaller for thicker ($d_F > 0.3\xi_F$) and thinner ($d_F < 0.3\xi_F$) F films. At $L/\xi_N \approx 1.67$ and $d_F \approx 0.3\xi_F$, the magnitude of critical current for the first time reaches zero, while the sign of I_C does not change.

With further increase of L the L/ξ_N, d_F/ξ_F plane starts to be subdivided into two regions separated by the line along which the junction critical current is equal to zero. The boundary between the regions has two branches (see Fig. 7.17). The first one is located at $d_F < 0.3\xi_F$. It starts from the first critical point $(L_{c1}/\xi_N, d_{F,c1}/\xi_F) \approx (1.67, 0.3)$ and $d_{F,c1}$ the smaller the larger is L. The second branch located at $d_F > 0.3\xi_F$. It starts from the same critical point and for large d_F asymptotically verge toward the line $L = L_1$ exhibiting damped oscillation around it. As a result, any cross-section presented in Fig. 7.14 dependence of $I_C(L/\xi_N, d_F/\xi_F)$ by a perpendicular to L axis plane in the region $1.67 < L/\xi_N < 2.87$ should give a dependence of $I_C(d_F)$ having the typical shape shown by solid line ($L/\xi_N = 1.8$) in Fig. 7.16a. It demonstrates that in this range of distances between S electrodes ($1.67 < L/\xi_N < 2.87$) for any given L there is a nucleation of only one π state in between of two zero states in $I_C(d_F)$ dependence. The interval of d_F in which the π state exists, becomes wider the larger the L. Note also that for $1.67 < L/\xi_N < 2.87$ only 0-state can be realized for large $d_F/\xi_F \gtrsim 0.8$.

The number of transitions between zero and π states in $I_C(d_F)$ increases by asymptotically approaching the line $L = L_1$. This is illustrated in Fig. 7.16b. At ($L/\xi_N = 2.6$), there are still only two transitions, namely, from 0-state to π-state and from π to zero state. At ($L/\xi_N \approx 2.8702$) in the zero state domain, there is nucleation of the next critical point at $d_F/\xi_F \approx 1.7$. In it $I_C = 0$, while the sign of I_C is kept positive for all $d_F/\xi_F \gtrsim 0.8$. Further increase of L leads to generation of additional π state in the vicinity of $d_F/\xi_F \approx 1.7$ as it is shown in Fig. 7.16b by solid line. The closer L to L_1 the larger is the amount of zero to π transitions.

As it was already pointed above, this behavior of critical current at $L/\xi_N \approx L_1$ is a result of small oscillations of $\mathrm{Im}(q)$, which occur at large d_F. Note that in the region $L = L_1 - 0$ S-FN-S junction always is in the zero state at $d_F \to \infty$.

Contrary to that, for $L = L_1 + 0$ it is π-state that is finally established in the limit of large d_F (see the dotted line for $L/\xi_N = 2.8729$ in Fig. 7.16c). Further increase of L leads to the reduction of thickness intervals in which the zero states exist. They collapse one by one with L. The last stage of this process is shown in Fig. 7.16c. It is seen that transition from ($L/\xi_N = 2.8729$) to ($L/\xi_N = 2.92$) leads to reduction of the zero state located in vicinity of $d_F/\xi_F = 1$. At ($L/\xi_N \approx 2.95$) it completely shrinks, so that I_C becomes always negative at $d_F/\xi_F \gtrsim 0.2$. As a result in the distance interval $3.5 \lesssim L/\xi_N \lesssim 6.5$ the typical shape of $I_C(d_F)$ dependence for a fixed L has the form of the curve presented in Fig. 7.16c by the line calculated for $L/\xi_N \approx 3.5$. There is only one 0-π transition, which occurs at $d_F/\xi_F \approx 0.2$. It is the first branch of the locus of point at which $I_C = 0$ on $L/\xi_N, d_F/\xi_F$ plane.

It is seen from Figs. 7.14, 7.17 that at $L/\xi_N \approx 6.5$ and $d_F/\xi_F \approx 0.32$ there is a nucleation of the next critical point. Again, the two branches start to propagate from it. They produce the next boundary on $L/\xi_N, d_F/\xi_F$ plane, thus subdividing this plane into three regions.

The first branch is located at $d_F/\xi_F \lesssim 0.32$. It propagates along L nearly parallel to the already existing in this domain branch generated at critical point $L_{c1}/\xi_N, d_{F,c1}/\xi_F) \approx (1.67, 0.3)$. In the narrow zone between these branches, the junction is in the π state. The second branch is located at $d_F > 0.32\xi_F$. Starting from the second critical point for large d_F it asymptotically verges toward the line $L = L_2 = 10.089$ exhibiting damped oscillations around it. Quantitatively, the behavior of $I_C(L, d_F)$ in the vicinity of $L = L_2$ and at slightly larger L is the same as we discuss above. There is increasing number of zero to π transitions as soon as $L \to L_2 - 0$ and the collapses of π states in the $L = L_2 + 0$ region with L increase. Finally in the interval $10.32 \lesssim L/\xi_N \lesssim 11.5$, there are only two transitions of $I_C(d_F)$ and at $d_F/\xi_F \gtrsim 0.012$ there is only zero state of the critical current.

Appendix

7.3.3 *Solution of Linearized Usadel Equations*

It is convenient to write the general solution of the boundary value problem in the form

$$\Phi_N(x, y) = \Phi_N(y) + \sum_{n=-\infty}^{\infty} A_n(x) \cos \frac{\pi n(y - L)}{L}, \tag{7.52}$$

$$\Phi_F(x, y) = \Phi_F(y) + \sum_{n=-\infty}^{\infty} B_n(x) \cos \frac{\pi n(y - L)}{L}, \tag{7.53}$$

where $\Phi_N(y)$ and $\Phi_F(y)$ are asymptotic solutions of (7.1), (7.2) at the distance far from FN interface

$$\Phi_N(y) = \frac{G_S\Delta}{\sqrt{\Omega}\gamma_{\mathrm{BN}}}\left(\frac{\cos\frac{\varphi}{2}\cosh\frac{L-2y}{2\xi_{N\Omega}}}{\sinh\frac{L}{2\xi_{N\Omega}}} - \frac{\mathrm{i}\sin\frac{\varphi}{2}\sinh\frac{L-2y}{2\xi_{N\Omega}}}{\cosh\frac{L}{2\xi_{N\Omega}}}\right), \tag{7.54}$$

$$\Phi_F(y) = \frac{\sqrt{\widetilde{\Omega}}G_S\Delta}{\Omega\gamma_{\mathrm{BF}}}\left(\frac{\cos\frac{\varphi}{2}\cosh\frac{L-2y}{2\xi_{F\Omega}}}{\sinh\frac{L}{2\xi_{F\Omega}}} - \frac{\mathrm{i}\sin\frac{\varphi}{2}\sinh\frac{L-2y}{2\xi_{F\Omega}}}{\cosh\frac{L}{2\xi_{F\Omega}}}\right), \tag{7.55}$$

where $\xi_{N\Omega} = \xi_N/\sqrt{\Omega}$, $\xi_{F\Omega} = \xi_F/\sqrt{\widetilde{\Omega}}$, while functions $A_n(x)$ and $B_n(x)$ satisfy the following boundary problem

$$\xi_N^2\frac{\partial^2}{\partial x^2}A_n(x) - u_n^2A_n(x) = 0, \tag{7.56}$$

$$\xi_F^2\frac{\partial^2}{\partial x^2}B_n(x) - v_n^2B_n(x) = 0, \tag{7.57}$$

$$\frac{\gamma_B}{\gamma}\xi_N\frac{\widetilde{\Omega}}{\Omega}\frac{\partial}{\partial x}A_n(0) - \frac{\widetilde{\Omega}}{\Omega}A_n(0) + B_n(0) = R_n, \tag{7.58}$$

$$\gamma_B\xi_F\frac{\partial}{\partial x}B_n(0) + B_n(0) - \frac{\widetilde{\Omega}}{\Omega}A_n(0) = R_n, \tag{7.59}$$

$$R_n = \frac{\widetilde{\Omega}}{\Omega}\frac{G_S\Delta}{L}\kappa_n\left(\mathrm{e}^{\frac{\mathrm{i}\varphi}{2}} + (-1)^n\mathrm{e}^{\frac{-\mathrm{i}\varphi}{2}}\right), \tag{7.60}$$

$$\frac{\partial}{\partial x}A_n(d_N) = 0, \qquad \frac{\partial}{\partial x}B_n(-d_F) = 0. \tag{7.61}$$

Here $\Omega = |\omega|/\pi T_C$, $\widetilde{\Omega} = \widetilde{\omega}/\pi T_C$ and

$$\kappa_n = \frac{1}{\gamma_{\mathrm{BN}}}\frac{\xi_N}{u_n^2} - \frac{1}{\gamma_{\mathrm{BF}}}\frac{\xi_F}{v_n^2}, \tag{7.62}$$

$$u_n = \sqrt{\left(\frac{\pi n\xi_N}{L}\right)^2 + \Omega}, \quad v_n = \sqrt{\left(\frac{\pi n\xi_F}{L}\right)^2 + \widetilde{\Omega}}. \tag{7.63}$$

Solution of (7.56)–(7.61) has the form

$$A_n(x) = -\frac{R_n\gamma v_n\Omega\cosh\left\{\frac{x-d_N}{\xi_N}u_n\right\}}{\widetilde{\Omega}\delta_n\sinh\frac{u_nd_N}{\xi_N}}, \tag{7.64}$$

$$B_n(x) = \frac{R_nu_n\cosh\left\{\frac{x+d_F}{\xi_F}v_n\right\}}{\delta_n\sinh\frac{v_nd_F}{\xi_F}}, \tag{7.65}$$

where δ_n is defined as:

$$\delta_n = \gamma_B v_n u_n + u_n \coth \frac{v_n d_F}{\xi_F} + \gamma v_n \coth \frac{u_n d_N}{\xi_N} \tag{7.66}$$

7.3.4 Calculation of Critical Current

To calculate the supercurrent across the S-FN-S junction we substitute of the expressions (7.52)–(7.53), (7.64)–(7.66) into formula for the general formula for supercurrent

$$I_S(x, y) = \frac{-i\pi T W}{e\rho_F} \sum_{\omega=-\infty}^{\infty} \frac{1}{\widetilde{\omega}^2} \int_{-d_f}^{0} \left[\Phi^*_{-\omega,F} \frac{\partial}{\partial y} \Phi_{\omega,F} \right]$$
$$- \frac{i\pi T W}{e\rho_N} \sum_{\omega=-\infty}^{\infty} \frac{1}{\omega^2} \int_{0}^{d_n} \left[\Phi^*_{-\omega,N} \frac{\partial}{\partial y} \Phi_{\omega,N} \right].$$

The calculations gives $I_S = I_C \sin \varphi$, where

$$I_C = \frac{\Delta^2 \pi T W \gamma}{e\rho_N} \sum_{\omega=-\infty}^{\infty} \frac{G_S^2}{\omega^2} \left[\sum_{j=1}^{6} k_j S_j \right. \tag{7.67}$$
$$\left. + \frac{d_F}{\sqrt{\widetilde{\Omega}} \xi_N \gamma_{\text{BF}}^2 \sinh \frac{L}{\xi_{F\Omega}}} + \frac{d_N/\gamma}{\sqrt{\Omega} \xi_N \gamma_{\text{BN}}^2 \sinh \frac{L}{\xi_{N\Omega}}} \right]$$

and W is a width of junction in the direction perpendicular to axes $0y$ and $0x$. The last two items in (7.67) determine the critical current of the structures with ether ferromagnetic (SFS) or normal (SNS) interlayers. By S_j we define the ordinary and double sums:

$$S_1 = \sum_{n=-\infty}^{\infty} \frac{n\kappa_n u_n \sin \frac{\pi n}{2}}{v_n \delta_n}, S_2 = \sum_{n=-\infty}^{\infty} \frac{n\kappa_n v_n \sin \frac{\pi n}{2}}{u_n \delta_n},$$
$$S_3 = \sum_{n=-\infty}^{\infty} \frac{\kappa_n u_n \cos \frac{\pi n}{2}}{v_n \delta_n}, S_4 = \sum_{n=-\infty}^{\infty} \frac{\kappa_n v_n \cos \frac{\pi n}{2}}{u_n \delta_n},$$
$$S_5 = \sum_{n,m=-\infty}^{\infty} \frac{C_{\text{nm}} u_n u_m}{\sinh \frac{v_n d_F}{\xi_F} \sinh \frac{v_m d_F}{\xi_F}} I_v,$$
$$S_6 = \sum_{n,m=-\infty}^{\infty} \frac{C_{\text{nm}} v_n v_m}{\sinh \frac{u_m d_N}{\xi_N} \sinh \frac{u_n d_N}{\xi_N}} I_u, \tag{7.68}$$

$$I_v = \frac{\sinh\frac{(v_n+v_m)d_F}{\xi_F}}{v_n+v_m} + \frac{\sinh\frac{(v_n-v_m)d_F}{\xi_F}}{v_n-v_m},$$

$$I_u = \frac{\sinh\frac{(u_n+u_m)d_N}{\xi_N}}{u_n+u_m} + \frac{\sinh\frac{(u_n-u_m)d_N}{\xi_N}}{u_n-u_m},$$

$$C_{\mathrm{nm}} = \frac{n\sin\frac{\pi n}{2}\cos\frac{\pi m}{2}\kappa_m\kappa_n}{\delta_m\delta_n},$$

and coefficients k_j are

$$k_1 = \frac{\sqrt{\widetilde{\Omega}}\xi_{F\Omega}^2\pi}{\gamma_{\mathrm{BF}}L^2\sinh\frac{L}{2\xi_{F\Omega}}},\; k_2 = \frac{-\sqrt{\Omega}\xi_{N\Omega}^2\pi}{\gamma_{\mathrm{BN}}L^2\sinh\frac{L}{2\xi_{N\Omega}}},$$

$$k_3 = \frac{\sqrt{\Omega}\xi_{N\Omega}^2\pi}{\gamma_{\mathrm{BN}}L^2\sinh\frac{L}{2\xi_{N\Omega}}},\; k_4 = \frac{-\xi_N}{\gamma_{\mathrm{BN}}L\cosh\frac{L}{2\xi_N\Omega}}, \tag{7.69}$$

$$k_5 = \frac{\pi\xi_N\xi_F^2}{L^3},\; k_6 = \gamma\frac{\pi\xi_N^3}{L^3}.$$

To calculate the sums (7.68) we may use the procedure known from the theory of functions of complex variables

$$\sum_{n=-\infty}^{\infty} f(n)\sin(\pi n/2) = \frac{\pi}{2}\sum_k \frac{res(f(z_k))}{\cos(\pi/2z_k)}, \tag{7.70}$$

$$\sum_{n=-\infty}^{\infty} f(n)\cos(\pi n/2) = -\frac{\pi}{2}\sum_k \frac{res(f(z_k))}{\sin(\pi/2z_k)}, \tag{7.71}$$

where $res(f(z_k))$ is residue of function $f(z)$ at the critical point z_k. From (7.68) it follows that there are critical points $z_u = \pm\,\mathrm{i}L\sqrt{\Omega}/(\xi_N\pi)$ and $z_v = \pm\,\mathrm{i}L\sqrt{\widetilde{\Omega}}/(\xi_F\pi)$, which are the roots of equations $u(z)=0$, and $v(z)=0$, respectively. In addition there is also an infinite number of z_k, which are the roots of equation:

$$\delta(z) = 0. \tag{7.72}$$

Applying the procedure (7.70)–(7.71) to calculation of (7.68) it is possible to show that the last two terms in expression for I_C (7.67) are exactly compensated by the parts of these sums, which are calculated from the residue at critical points $z = z_u$ and $z = z_v$. Therefore critical current (7.67) can be expressed as the sum of terms resulting from the application of the rule (7.70)–(7.71) to (7.68) at $z = z_k$.

Our analysis have shown that the value part of z_k consists of a root having the lowest real part, $z = z_{\min}$, and the two systems of roots. In the first system, $z_{k,N}$, there is an item in the real part of $z_{k,N}$, which at large k increase with the number k of the root as $k\xi_N/d_N$, while in the second, $z_{k,F}$, this increase is proportional to $k\xi_F/d_F$. Below we will restrict ourselves to the consideration of the limit at which

$z_{\min}$ makes the major contribution to the junction critical current, i.e. $|z_{\min}| << |z_k|$. It can be shown that the lowest value among the roots of $z_{k,F}$ group is achieved at the limit of large d_F and is bounded by $\sqrt{\widetilde{\Omega}}L/(\pi\xi_F)$. The lowest value of second group of the roots, $z_{k,N}$, is bounded by $\sqrt{\xi_N^2/d_N^2 - \Omega(L/(\pi\xi_N))}$, the value at which $z_{k,N}$ are approached to in the limit of small γ. Thus under the condition

$$|z_{\min}| << |\sqrt{\widetilde{\Omega}}L/(\pi\xi_F)|, \tag{7.73}$$

$$|z_{\min}| << |\sqrt{\xi_N^2/d_N^2 - \Omega(L/(\pi\xi_N))}| \tag{7.74}$$

we can rewrite (7.72) in the form

$$u^2 = -\frac{\xi_N}{d_N}\frac{\gamma\sqrt{\widetilde{\Omega}}}{\gamma_B\sqrt{\widetilde{\Omega}} + \coth\left\{\frac{d_F}{\xi_F}\sqrt{\widetilde{\Omega}}\right\}} \tag{7.75}$$

and for $z_{\min}$ finally get

$$z_{\min} = \mathrm{i}\frac{L}{\pi\xi_N}\sqrt{\frac{\gamma\sqrt{\widetilde{\Omega}}}{\gamma_B\sqrt{\widetilde{\Omega}} + \coth\left\{\frac{d_F}{\xi_F}\sqrt{\widetilde{\Omega}}\right\}}\frac{\xi_N}{d_N} + \Omega}. \tag{7.76}$$

Note, that the imaginary parts of the roots of both groups ($z_{k,F}$, $z_{k,N}$) do not exceed their real parts. It means that inequality (7.74) guarantees the smallness of $\mathrm{Re}\, z_{\min}$ compared to $\mathrm{Re}(z_{k,F}, z_{k,N})$.

Assuming further that the total contribution to I_C from the all the residues at critical points $z_{k\geq 1}$ is small compared to that at $z = z_{\min}$

$$\left|\mathrm{Re}\,\frac{1}{z_{\min}\sinh \pi z_{\min}}\right| >> \Sigma_F + \Sigma_N, \tag{7.77}$$

$$\Sigma_{F(N)} = \left|\mathrm{Re}\sum_k \frac{1}{z_{k,F(N)}\sinh \pi z_{k,F(N)}}\right| \tag{7.78}$$

we arrive at the following expression for I_C:

$$I_C = \frac{4\pi T}{e}\frac{W}{\gamma_{\mathrm{BN}}^2\rho_N}\gamma\,\mathrm{Re}\sum_{\omega>0}^{\infty}\frac{G_S^2\Delta^2k^2\,(S_F + S_N)}{D^2\omega^2 q\xi_N\sinh(qL)}, \tag{7.79}$$

$$S_F = \frac{\xi_F^2}{\xi_N^2}\left[1 + \frac{2vd_F}{\xi_F}\sinh^{-1}\frac{2vd_F}{\xi_F}\right]\frac{u^2}{v}\coth\frac{vd_F}{\xi_F},$$

$$S_N = \gamma\left[1 + \frac{2ud_N}{\xi_N}\sinh^{-1}\frac{2ud_N}{\xi_N}\right]\frac{v^2}{u}\coth\frac{ud_N}{\xi_N},$$

where $u = \sqrt{\Omega - q^2\xi_N^2}$, $v = \sqrt{\widetilde{\Omega} - q^2\xi_F^2}$, and inverse coherence lengths q is given by

$$q = \frac{1}{\xi_N}\sqrt{\frac{\xi_N}{d_N}\frac{\gamma\sqrt{\widetilde{\Omega}}}{\gamma_B\sqrt{\widetilde{\Omega}} + \coth\left\{\frac{d_F}{\xi_F}\sqrt{\widetilde{\Omega}}\right\}} + \Omega}. \tag{7.80}$$

The coefficients k and D in (7.79) have the form

$$k = \frac{1}{u^2} - \frac{\gamma_{\mathrm{BN}}}{\gamma_{\mathrm{BF}}}\frac{\xi_F}{\xi_N}\frac{1}{v^2},$$

$$\begin{aligned} D = {} & \left(\frac{d_N}{\xi_N} + \gamma\frac{\xi_F^2}{\xi_N^2}\frac{d_F}{\xi_F}\right)\coth\frac{ud_N}{\xi_N}\coth\frac{vd_F}{\xi_F} \\ & + \left(\gamma_B v\frac{d_N}{\xi_N} + \frac{\xi_F^2}{\xi_N^2}\frac{\gamma}{v}\right)\coth\frac{ud_N}{\xi_N} \\ & + \left(\gamma_B u\frac{d_F}{\xi_F}\frac{\xi_F^2}{\xi_N^2} + \frac{1}{u}\right)\coth\frac{vd_F}{\xi_F} \\ & + \frac{v}{u}\left(\gamma_B + \gamma\frac{d_N}{\xi_N}\right) + \frac{\xi_F^2}{\xi_N^2}\frac{u}{v}\left(\gamma_B + \frac{d_F}{\xi_F}\right). \end{aligned} \tag{7.81}$$

From (7.74) and (7.76) it follows that the approximation (7.80) for q is valid if

$$\frac{\xi_F^2}{\eta h \xi_N^2} << \frac{d_N}{\xi_N} << \eta, \tag{7.82}$$

where

$$\eta = \begin{cases} \frac{1}{\gamma}\sqrt{\gamma_B^2 + \gamma_B\sqrt{2h^{-1}} + h^{-1}}, & \frac{d_F}{\xi_F} >> 1/\sqrt{h}, \\ \frac{1}{\gamma}\sqrt{\gamma_B^2 + 2\gamma_B\frac{\xi_F}{d_F}\frac{\Omega}{h^2} + \frac{\xi_F^2}{d_F^2 h^2}}, & \frac{d_F}{\xi_F} << 1/\sqrt{h}. \end{cases} \tag{7.83}$$

where $h = H/\pi T_C \operatorname{sgn}\omega$. To get (7.82) we additionally restricted ourselves by considering the most interesting from practical point of view situation when

$$h >> T/T_C, \quad \xi_F << \xi_N. \tag{7.84}$$

It follows from inequalities (7.82) and (7.83) that the range of validity of expression (7.76) is the larger, the smaller is the parameter γ and thickness of F film d_F or the larger is γ_B. At $\gamma = 0$ or $\gamma_B \to \infty$ rigid boundary conditions take place at NF interface and expression (7.76) is valid for arbitrary thickness of the normal film.

From (7.77) to (7.78) it follows that in the limit of thin F and N films, $d_F/\xi_F << 1/\sqrt{h}$ and $d_N/\xi_N << 1/\sqrt{\Omega}$, the result (7.79)–(7.81) is valid if the conditions (7.82) are fulfilled.

The inequality (7.77) provides also the restriction on the thickness d_F and d_N of F and N films. Physically, it comes from the fact that with d_F (d_N) increase the full supercurrent flowing across F (N) film is enlarged proportionally to d_F (d_N). Therefore the smallness of this current component compared to contribution to I_C, which is accumulated in vicinity of FN interface, results in

$$d_F \operatorname{Re}(q) << \exp\left\{\frac{L}{\xi_F}\sqrt{h}\right\}. \tag{7.85}$$

$$d_N \operatorname{Re}(q) << \exp\left\{L\left(\frac{\sqrt{\Omega}}{\xi_N} - q\right)\right\}. \tag{7.86}$$

Finally we should take into account that the form of the boundary conditions is valid for relatively large γ_{BN} thus providing additional restriction for application of (7.79), which sets the limit on the distance between superconducting electrodes

$$L >> \operatorname{Re}\left(\frac{1}{q}\operatorname{arctanh}\frac{1}{\gamma_{\mathrm{BN}} q \xi_N}\right). \tag{7.87}$$

From (7.80), (7.84), (7.87) it follows that inequality (7.85) is always fulfilled for experimentally reasonable thickness of F layer and does not apply a serious restriction on the use of (7.79).

Taking into account the inequality (7.82) we can further simplify the expression for the critical current (7.79) and transform it into the formula

$$I_C = \frac{2\pi T}{e}\frac{d_N}{\xi_N}\frac{W}{\gamma_{\mathrm{BN}}^2 \rho_N}\operatorname{Re}\sum_{\omega>0}^{\infty}\frac{G_S^2\Delta^2}{\omega^2 q \xi_N \sinh(qL)}, \tag{7.88}$$

in which the dependence of $I_C(d_F)$ enters only via functional dependence $q(d_F)$ determined by (7.80). It is important to note that to use expression (7.88) it is enough to be in the range of parameters, which guarantees the implementation of (7.82).

7.4 New Geometry of SFNS Junctions

The choice of junction geometry considered in previous sections was based on existing concept [46,47], that in such ramp-type configuration the critical current I_C is larger than in overlap geometry, when S electrodes are located on top of the weak link multilayer. In this section, this statement is reconsidered and it is demonstrated that its valid only for the fully transparent interfaces between S electrodes and a

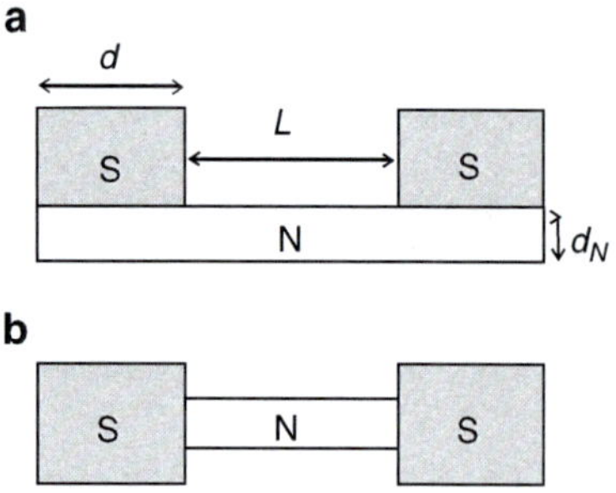

Fig. 7.18 (**a**) $SN - N - NS$ junction, (**b**) the SNS junction

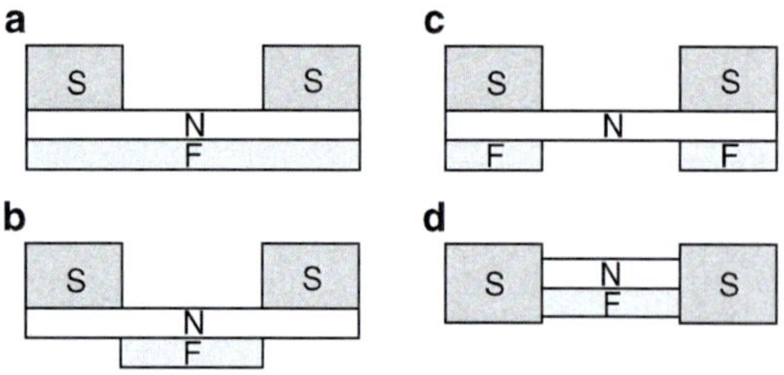

Fig. 7.19 (**a**) $SNF - NF - FNS$ junction, (**b**) the $SN - NF - NS$ junction, (**c**) the $SNF - N - FNS$ junction, (**d**) the $S - NF - S$ junction

weak link region. Three different geometries of Josephson junctions: (1) SN-NF-NS devices, which consist of two SN complex electrodes connected by NF weak region; (2) SNF-N-FNS structures, in which N film connects two SNF complex electrodes and (3) SNF-NF-FNS junctions with S electrodes located on top of the FN bilayer are studied. Critical currents of these Josephson structures in the framework of linearized Usadel equations for arbitrary length of complex electrodes will be discussed and then it will be compared with the results for the above three cases and those obtained in previous sections to determine the geometry, which provides the largest magnitude of the critical current. Also in this section, we will demonstrate that 0-π transition in the considered structures can be driven not only by variation of distance L between S electrodes, as predicted by known models, but also by changing the length d of the SNF overlap region.

We consider multilayered structures presented in Figs. 7.18 and 7.19. They consist of superconducting electrodes with the length d deposited on the top either a single N film or on NF bilayer. The bilayer consists of ferromagnetic (F) and normal metal (N) films having a thickness d_F, and d_N, respectively (see Fig. 7.19). The junctions shown in Figs. 7.18b and 7.19d are the structures having ramp-type geometry intensively studied previously (see [46, 48–54]).

The x- and y-axes have been chosen in the directions perpendicular and parallel to the plane of N film and put the origin in the middle of structure at FN interface (Fig. 7.19a,b,d) or at the lower free interface of N film (Figs. 7.18, 7.19c).

We will solve linearized Usadel equations (7.1), (7.2) for these cases in the limit of small thicknesses of N and F films

$$d_N \ll \xi_N, \quad d_F \ll \xi_F \tag{7.89}$$

as in Sect. 7.2. The details of calculations are summarized in appendices.

7.4.1 Critical Current of SN-N-NS Josephson Junction

The expressions for the critical currents, I_C^{SNS}, $I_C^{\mathrm{SN-N-NS}}$, of SNS junction shown in Fig. 7.18 are well known in the considered model [52]. They have the form

$$I_C^{\mathrm{SNS}} = K\frac{d_N}{\xi_N}\sum_{n=0}^{\infty}\frac{\Gamma}{q\sinh(qL)}, \tag{7.90}$$

$$I_C^{\mathrm{SN-N-NS}} = \frac{K}{\xi_N d_N}\sum_{n=0}^{\infty}\frac{\Gamma\sinh^2(qd)}{q^3\sinh(q(L+2d))}, \tag{7.91}$$

where coefficient $K = (2\pi T W)/(R_{\mathrm{BN}}\mathcal{A}_{\mathrm{BN}}\gamma_{\mathrm{BN}}e)$, $q = \xi_N^{-1}\sqrt{\Omega}$ is inverse decay length and $\Gamma = \Delta^2/(\omega^2+\Delta^2)$.

As it is shown in Appendices expression, (7.91) is also followed from the more general formula for critical current of SNF-NF-SNF devices shown in Fig. 7.19a in the limit of small thickness of F film ($d_F \to 0$).

The ratio of these two critical currents, $I_C^{\mathrm{SN-N-NS}}/I_C^{\mathrm{SNS}}$, is visualized in Fig. 7.20 as a function of thickness of normal layer, d_N, for several lengths of complex electrode $d/\xi_N = 0.5, 1, 10$. It is clearly seen that there are intervals of parameters

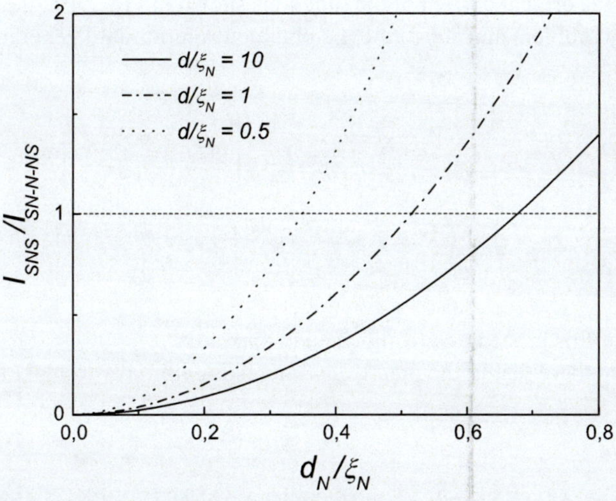

Fig. 7.20 $I_C^{\mathrm{SN-N-NS}}/I_C^{\mathrm{SNS}}$ versus thickness of N film d_N/ξ_N for $d/\xi_N = 0.5, 1, 10, L/\xi_N = 2$, $T/T_C = 0.5$

under which critical current of SN-N-NS junction can essentially exceed of I_C^{SNS}. The physics of this effect is evident.

In the considered limit of small SN interfaces transparency for ramp-type geometry (Fig. 7.18a) under condition $L >> 1/q$ the magnitude of induced into N metal Φ_N functions at SN interfaces is close to

$$\Phi_N(d_N) = \frac{G_S\Delta}{\gamma_{\mathrm{BN}}\xi_N q}, \quad \gamma_{\mathrm{BN}}\xi_N q >> 1, \tag{7.92}$$

while in the case of overlap geometry (Fig. 7.18b) for $d_N << 1/q$ magnitude of Φ_N functions induced into N metal is in the first approximation on d_N/ξ_N independent on coordinate x and is equal to

$$\Phi_N(d_N) = \frac{G_S\Delta}{\gamma_{\mathrm{BN}}\xi_N q^2 d_N}, \quad \gamma_{\mathrm{BN}}\xi_N q^2 d_N >> 1. \tag{7.93}$$

From (7.92), (7.93), it immediately follows that the large factor γ_{BN} in (7.93) can be renormalized by a small ratio of d_N/ξ_N, thus leading to effective increase of superconductivity at the interface between the N film and SN composite electrode compare to the strength of superconducting correlations at SN boundary of SNS ramp-type devices.

7.4.2 Critical Current of Devices with F Film in Weal Link Region

To calculate the critical current of the junctions shown in Fig. 7.19, one has to solve the boundary problem and substitute the obtained solution into general formula for supercurrent:

$$I_S = \frac{-i\pi T W}{e\rho_F}\sum_{\omega=-\infty}^{\infty}\frac{1}{\widetilde{\omega}^2}\int_{-d_F}^{0}\left[\Phi^*_{-\omega,F}\frac{\partial}{\partial y}\Phi_{\omega,F}\right] - \frac{i\pi T W}{e\rho_N}\sum_{\omega=-\infty}^{\infty}\frac{1}{\omega^2}\int_{0}^{d_N}\left[\Phi^*_{-\omega,N}\frac{\partial}{\partial y}\Phi_{\omega,N}\right]. \tag{7.94}$$

The details of this procedure are allocated in appendix.

It is shown there that in the practically interesting limit of strong N film

$$\zeta_N >> \zeta_F, \quad \xi_N >> \xi_F \tag{7.95}$$

the critical current of SNF-NF-FNS (Fig. 7.19a), SN-NF-NS (Fig. 7.19b), and SNF-N-FNS (Fig. 7.19c) structures

$$I_C^{\text{SNF-NF-FNS}} = \frac{K}{\xi_N d_N} \operatorname{Re} \sum_{n=0}^{\infty} \frac{\Gamma U q_1 \sinh^2(q_1 d)}{\sinh(q_1(L+2d))}, \tag{7.96}$$

$$I_C^{\text{SN-NF-NS}} = \frac{K}{\xi_N d_N} \operatorname{Re} \sum_{n=0}^{\infty} \frac{\Gamma q_1}{q^4(Q_{q,q_1} + P_{q,q_1})}, \tag{7.97}$$

$$I_C^{\text{SNF-N-FNS}} = \frac{K}{\xi_N d_N} \operatorname{Re} \sum_{n=0}^{\infty} \frac{\Gamma U q}{Q_{q_1,q} + P_{q_1,q}} \tag{7.98}$$

can be expressed by formulas (7.96), (7.97), and (7.98), respectively. Here, functions $Q(\alpha, \beta)$, $P(\alpha, \beta)$, and U are defined as

$$Q_{a,b} = \frac{2\coth(ad)\cosh(bL)\,b}{a}, \tag{7.99}$$

$$P_{a,b} = \sinh(bL)\left(1 + \frac{b^2\coth^2(ad)}{a^2}\right), \tag{7.100}$$

$$U = \left(\frac{v^2\zeta_F^2\zeta_N^2}{1 - v^2u^2\zeta_F^2\zeta_N^2}\right)^2, \tag{7.101}$$

where q_1 is fundamental inverse coherence length of the problem:

$$q_1^2 = \frac{1}{2}\left[u^2 + v^2 - \sqrt{(u^2 - v^2)^2 + 4\zeta_F^{-2}\zeta_N^{-2}}\right], \tag{7.102}$$

while u and v

$$u^2 = \left(\frac{1}{\zeta_N^2} + \frac{\Omega}{\xi_N^2}\right), \quad v^2 = \left(\frac{1}{\zeta_F^2} + \frac{\Omega}{\xi_F^2} + \mathrm{i}\frac{h}{\xi_F^2}\right), \tag{7.103}$$

are partial inverse coherence lengths. The parameters ζ_F and ζ_N are the coupling constants $\zeta_F^2 = \gamma_B d_F \xi_F$, $\zeta_N^2 = \gamma_B d_N \xi_N / \gamma$, which describe the mutual influence of N and F films on superconducting correlations in the junction.

Strictly speaking, the formulas (7.96), (7.97), and (7.98) are valid in the limit of thin N and F films (7.89). However, making use of the formalism developed in [51] it is possible to prove that all of them can be also valid for arbitrary thickness of F film if one simply use in (7.96), (7.97), and (7.98) the more general expression for inverse coherence length q_1, namely 7.46.

Expressions (7.96), (7.97), and (7.98) can be simplified in several practically interesting cases.

In the limit of large d ($d >> 1/q, 1/q_1$) both $\coth(qd) \to 1$ and $\coth(q_1 d) \to 1$. As a result for SN-NF-NS and SNF-N-NFS junctions, one may use the same formulas (7.97), (7.98) with more simple forms of functions $Q(\alpha, \beta)$ and $P(\alpha, \beta)$

$$Q_{a,b} = \frac{2\cosh(bL)\, b}{a}, \tag{7.104}$$

$$P_{a,b} = \sinh(bL)\left(1 + \frac{b^2}{a^2}\right), \tag{7.105}$$

while for SNF-NF-FNS junction

$$I_C^{\text{SNF-NF-FNS}} = \frac{1}{2}\frac{K}{\xi_N d_N}\,\mathrm{Re}\sum_{n=0}^{\infty} \Gamma U q_1 \exp(-q_1 L). \tag{7.106}$$

In the limit of large distance L between S electrodes, $L >> 1/q, 1/q_1$ the main contribution to the sums in (7.96), (7.97), and (7.98) comes from the first item and for the critical current of SNF-NF-FNS, SN-NF-NS, and SNF-N-NFS one can get, respectively:

$$I_C^{\text{SNF-NF-FNS}} = \frac{2K}{\xi_N d_N}\,\mathrm{Re}\sum_{n=0}^{\infty} \frac{\Gamma U q_1 \exp(-q_1 L)}{(1 + \coth(qd))^2} \tag{7.107}$$

$$I_C^{\text{SN-NF-NS}} = \frac{2K}{\xi_N d_N}\,\mathrm{Re}\sum_{n=0}^{\infty} \frac{\Gamma q_1 \exp(-q_1 L)}{q^2 (q + q_1 \coth(qd))^2}, \tag{7.108}$$

$$I_C^{\text{SNF-N-FNS}} = \frac{2K}{\xi_N d_N}\,\mathrm{Re}\sum_{n=0}^{\infty} \frac{\Gamma U q q_1^2 \exp(-qL)}{(q_1 + q\coth(q_1 d))^2}. \tag{7.109}$$

Below the obtained results (7.96), (7.97), and (7.98) will be compared with the value of the critical current calculated in [48] for ramp-type SFNS junction

$$I_C^{\text{SFNS}} = K\frac{d_N}{\xi_N}\,\mathrm{Re}\sum_{n=0}^{\infty} \Gamma \frac{\left(1 - \frac{1}{q_1^2 - \nu^2}\frac{\gamma_{\text{BN}}}{\gamma_{\text{BF}}}\frac{\xi_N}{\xi_F \zeta_N^2}\right)^2}{q_1 \sinh(q_1 L)}. \tag{7.110}$$

It is necessary to mention that in the limit of decoupled F and N films ($\gamma_B \to \infty$) expressions for the critical currents (7.96)–(7.98) reduce to the formula for SN-N-NS devices (7.91), while the critical current of SFNS ramp-type structure (7.110) transforms to that (7.90) valid for SNS junctions.

Figures 7.21–7.26 show the phase diagrams for critical current, which in (L/ξ_N,d/ξ_N) plane gives the information about the sign of I_c. In the areas marked in Figs. 7.21–7.26 by 0 and π, the critical current is positive (0-state) and negative (π-state), correspondingly, while the lines give the point curves at which $I_c = 0$. The position of these curves in (L/ξ_N,d/ξ_N) plane also depends on relative thickness (d_F/ξ_F and d_N/ξ_N) of both F and N films.

The phase diagrams for SNF-N-FNS structures are given in Figs. 7.21 and 7.22. In this geometry, there is the only N film in the region between SNF multilayers.

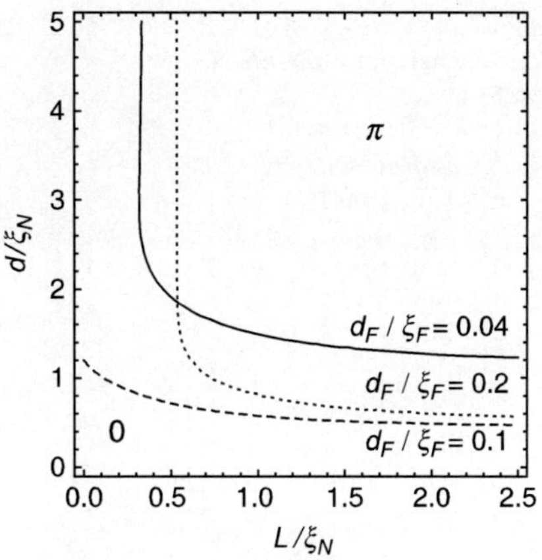

Fig. 7.21 $(L/\xi_N, d/\xi_N)$ phase diagram for SNF-N-FNS structure for $d_F/\xi_F = 0.04, 0.1, 0.2$ (*solid, dashed, dotted lines*) at $\gamma = 0.1$, $\gamma_B = 0.1$, $d_N/\xi_N = 0.1$, $\xi_N/\xi_F = 10$, $T/T_C = 0.5$, $H/\pi T_C = 30$

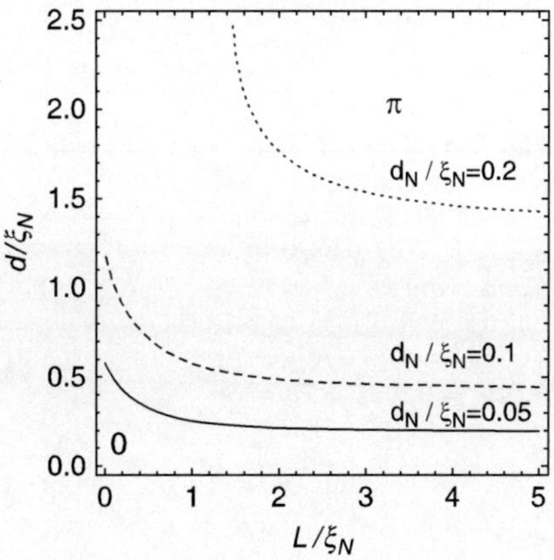

Fig. 7.22 $(L/\xi_N, d/\xi_N)$ phase diagram for SNF-N-FNS structure for $d_N/\xi_N = 0.05, 0.1, 0.2$ (*solid, dashed, dotted lines*) at $\gamma = 0.1$, $\gamma_B = 0.1$, $d_F/\xi_F = 0.1$, $\xi_N/\xi_F = 10$, $T/T_C = 0.5$, $H/\pi T_C = 30$

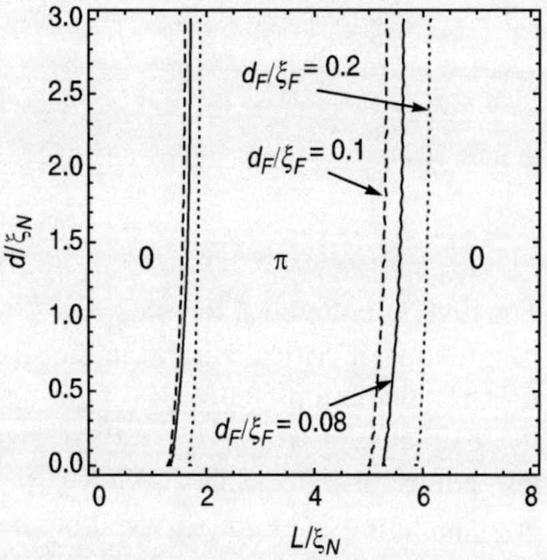

Fig. 7.23 $(L/\xi_N, d/\xi_N)$ phase diagram for SN-FN-NS structure for $d_F/\xi_F = 0.08, 0.1, 0.2$ (*solid, dashed, dotted lines*) at $\gamma = 0.1$, $\gamma_B = 0.1$, $d_N/\xi_N = 0.1$, $\xi_N/\xi_F = 10$, $T/T_C = 0.5$, $H/\pi T_C = 30$

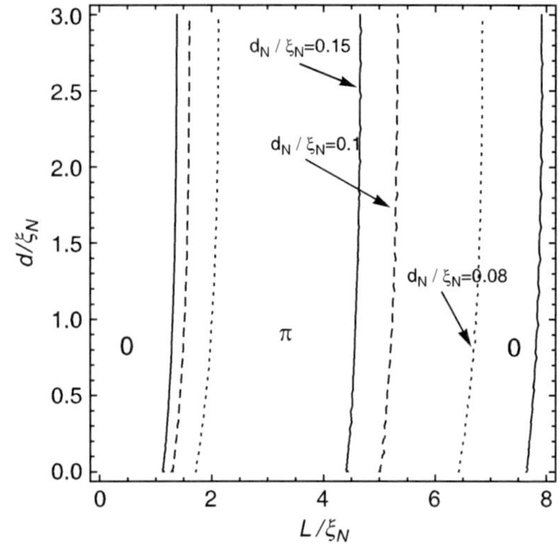

Fig. 7.24 $(L/\xi_N, d/\xi_N)$ phase diagram for SN-FN-NS structure for $d_N/\xi_N = 0.08, 0.1, 0.15$ (*solid, dashed, dotted lines*) at $\gamma = 0.1$, $\gamma_B = 0.1$, $d_F/\xi_F = 0.1$, $\xi_N/\xi_F = 10$, $T/T_C = 0.5$, $H/\pi T_C = 30$

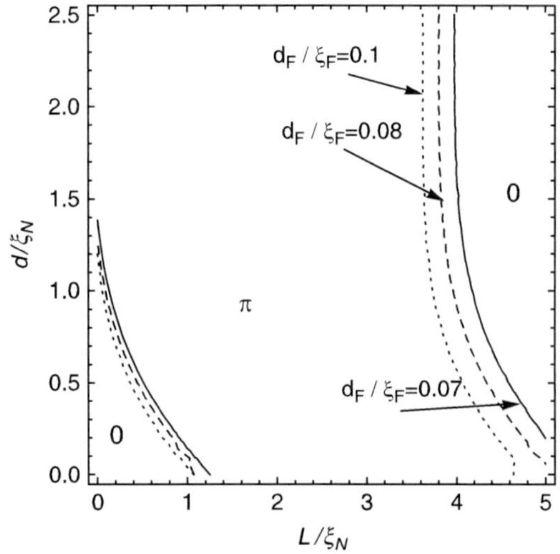

Fig. 7.25 $(L/\xi_N, d/\xi_N)$ phase diagram for SNF-FN-FNS structure for $d_F/\xi_F = 0.07, 0.08, 0.1$ (*solid, dashed, and dotted lines*) at $\gamma = 0.1$, $\gamma_B = 0.1$, $d_N/\xi_N = 0.1$, $\xi_N/\xi_F = 10$, $T/T_C = 0.5$, $H/\pi T_C = 30$

The inverse coherence length $q = \xi_N^{-1}\sqrt{\Omega}$ in N film is real; therefore, there are no oscillations of critical current in the structure.

The calculations show that in this case there can be only one curve on the $(L/\xi_N, d/\xi_N)$ plane, at which $I_c = 0$ for fixed other parameters. The existence of only one point curve for SNF-N-FNS structure can be understood from the following argumentations. Contrary to well-studied SFS junctions, the coherence length in the part of weak link region of SNF-N-FNS devices located between SNF electrodes is real, thus preventing the oscillations of function Φ_N in that region of the N film. The oscillations of condensate function exist only in the NF part of weak link located

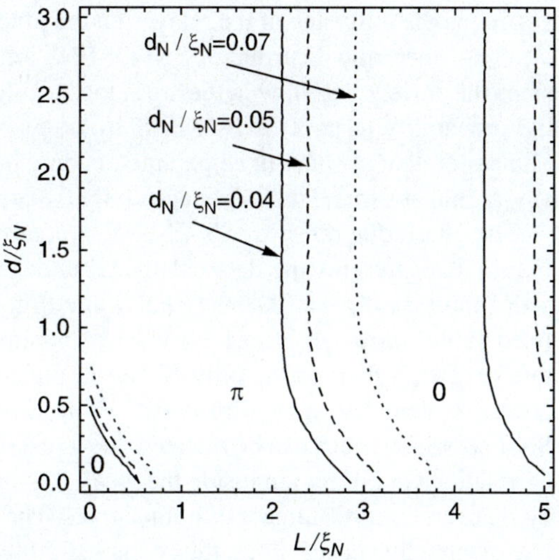

Fig. 7.26 $(L/\xi_N, d/\xi_N)$ phase diagram for SNF-FN-FNS structure for $d_N/\xi_N = 0.04, 0.05, 0.07$ (*solid, dashed, and dotted lines*) at $\gamma = 0.1$, $\gamma_B = 0.1$, $d_F/\xi_F = 0.1$, $\xi_N/\xi_F = 10$, $T/T_C = 0.5$, $H/\pi T_C = 30$

under the S electrodes. Obviously, under these circumstances the sign of I_c must be only controlled by value of condensate function at the boundary between the SNF electrodes and the N film connecting them. This value of condensate function determines two complex coefficients, A_1, and, A_2, (see (7.137) and (7.138)). In combination with nonoscillatory decay of Φ_N function into the N film from the SNF electrodes, these coefficients provide only two choices for the sign of I_c and only one curve at which $I_c = 0$. This is in contrast to SFS devices with F film in between of S-electrodes. In the latter case, the sign of I_c depends also on relation between the geometrical size of a junction and the imaginary part of the coherence length (the period of oscillations of the order parameter). It is combination of these two factors that provides the opportunity to have multiple changes of I_c sign and infinite number of curves at $(L/\xi_N, d/\xi_N)$ plane at which $I_c = 0$.

Therefore in the considered SNF-N-FNS structures there is only one of these two factors and only one opportunity for I_c to change its sign, which can be realized or not depending on the parameters of the structure.

The position of the transition curve calculated for fixed ratio $d_N/\xi_N = 0.1$ and several values of $d_F/\xi_F = 0.04, 0.1, 0.2$ is shown in Fig. 7.21. The location of the curve depends on d_F by nonmonotonic way. At $d_F = 0$, there is only 0-state in the structure. With the increase of d_F, the curve first shifts to the left bottom corner of the phase diagram, then it turns back and at some critical value of d_F it tends to infinity, thus providing only 0-state in the structure with further d_F increase. Such nonmonotonic behavior is due to nonmonotonic behavior of q_1 from (7.46).

Figure 7.22 shows $(L/\xi_N, d/\xi_N)$ phase diagram calculated for fixed ratio $d_F/\xi_F = 0.1$ and several values of $d_N/\xi_N = 0.05, 0.1, 0.2$. It is seen that with d_N increase the point curves at which $I_c = 0$ shifts in the direction to the right corner of diagram providing the increase of area for 0 state. This fact can be understood if one takes into account that under fixed d_F the larger is N layer thickness the smaller is the influence of the F layer on the junction properties. It is obvious that at $d_N \gtrsim \xi_N$ the critical current of SNF-N-FNS junction will tend to that of SN-N-NS since the current will flow in the areas located closer to S electrodes thus decreasing the probability to have SNF-N-FNS structure in the π-state. Also, it is important to mention that at some fixed parameters only 0-state or π-state can be realized for any L, and at some fixed parameters only 0-state can be realized for any d.

The phase diagrams for SN-FN-NS structures are given in Figs. 7.23 and 7.24. Figure 7.23 presents the data calculated under fixed value of $d_N/\xi_N = 0.1$ for a set of ratio $d_F/\xi_F = 0.08, 0.1, 0.2$, while Fig. 7.24 gives diagram obtained under fixed value of $d_F/\xi_F = 0.1$ for a set of parameters $d_N/\xi_N = 0.08, 0.1, 0.15$. In this geometry, there is the only N film in the complex SN electrodes. The inverse coherence length q in N film is real value. Consequently, both 0 and π-states in SN-FN-NS junctions can be realized due to oscillatory behavior of superconducting correlations in NF region inside the weak link area, which connects SN electrodes. So there are infinite number of point curves. The point curves at which $I_c = 0$ looks like practically vertical lines, thus demonstrating weak influence of overlap distance d on alternation of 0 and π states in the junction.

Finally, Figs. 7.25, 7.26 give $(L/\xi_N, d/\xi_N)$ phase diagrams for SNF-FN-FNS junctions. In these structures, coherence lengths are complex both under superconductor in complex SNF electrodes and in NF part of weak link region. The appearance of 0 or π state in this case depends also on matching these oscillations at SNF/NF boundary. As a result, the point curves at which $I_c = 0$ are not as vertical as them one can see in Figs. 7.23, 7.24, thus demonstrating their strong dependence on both lengths L/ξ_N and d, and the $0 - \pi$ transition with d increase is not so sensitive to L variations as for SN-FN-NS structure.

Figure 7.27 shows dependence of absolute value of normalized critical currents of SNF-NF-FNS, SNF-N-FNS, SN-NF-NS, and SFNS junctions as a function of L/ξ_N for infinite length of SN interface d. It is seen that at given magnitude of L/ξ_N critical current of SN-N-NS junction, $I_C^{\text{SN-N-NS}}$, has the maximum value among all others. This fact is obvious since in this structure there is no additional suppression of superconductivity provided by the F film. If we compare the value of I_C far from the 0 - π transition points for all other considered structures, then we may have $I_C^{\text{SN-FN-NS}} > I_C^{\text{SNF-N-FNS}} > I_C^{\text{SNF-FN-FNS}} > I_C^{S-FN-S}$. This sequence of values are due to consecutive increase of suppression of superconductivity provided by F film.

In SN-FN-NS junctions in the considered region of parameters, the superconducting correlations are suppressed by F film only in weak link region, thus providing the large value of I_C. In SNF-N-FNS junctions, the critical current is smaller than in SN-FN-NS devices due to suppression of superconductivity in SNF part of the structure. Due to it, the decay of superconducting correlations into N part of weak link starts from the values, which are smaller than in SN-FN-NS devices. In SNF-FN-FNS devices, there is suppression of superconductivity in all parts of structure by F film. Finally in SNFS ramp-type structures, the critical current has the smallest value. In SN-N-NS and SNF-N-FNS junctions, I_C decays with L without oscillations. However as it follows from the phase diagram presented in Fig. 7.21,

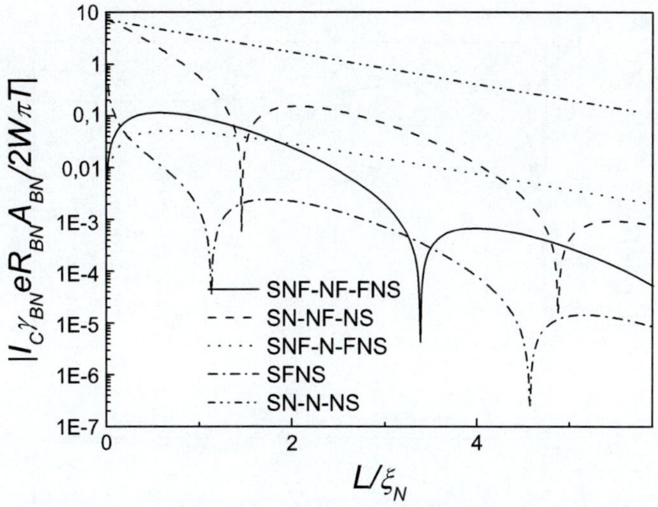

Fig. 7.27 I_C for SN-N-NS, SFNS, SNF-N-FNS, SN-FN-NS, SNF-FN-FNS structures versus L/ξ_N calculated for $\xi_N/\xi_F = 10$, $T/T_C = 0.5$, $H/\pi T_C = 30$, $d_N/\xi_N = 0.1$, $d_F/\xi_F = 0.1$

under the chosen set of parameters the SNF-N-FNS structure is the π state, so that its critical current is negative, while SN-N-NS is always in 0-state.

The decay length in NF part of SN-FN-NS, SNF-FN-FNS, and SNFS devices is complex providing damping oscillations of I_C as a function of L. The period of these oscillations and their decay length are the same for all the junctions and controlled by bulk properties of NF part of weak link. The initial conditions for these oscillations at SN/NF, SNF/FN, and S/NF interfaces are different resulting in shift of the oscillations along L axis.

Figure 7.28 shows the amplitude of critical current of SNF-FN-FNS, SN-FN-NS, SNF-N-FNS structures versus the length of SN interface in complex electrodes calculated under fixed ratio of $L/\xi_N = 2$.

From the presented curves, it follows that critical currents have a tendency to increase with d as $\tanh^2(d/\xi_N)$, while at large d they arrive at independent on d values.

The continuous support of superconductivity from the S electrodes along all the SN interfaces results in considerable difference between $I_C(L)$ and $I_C(d)$ dependencies. The last may have only one change sign of I_C as a function of d.

It is also necessary to note that maximum of I_C in $I_C(d)$ dependence maybe not necessarily achieved in the limit of $d \to \infty$. For instance, at $d_N/\xi_N = 0.2$ and $d/\xi_N = 0.5$ (see Fig. 7.28c) the magnitude of I_C fifty times larger compare to the value, which is reached at $d \to \infty$. This strong enhancement may be important for some practical applications of these structures.

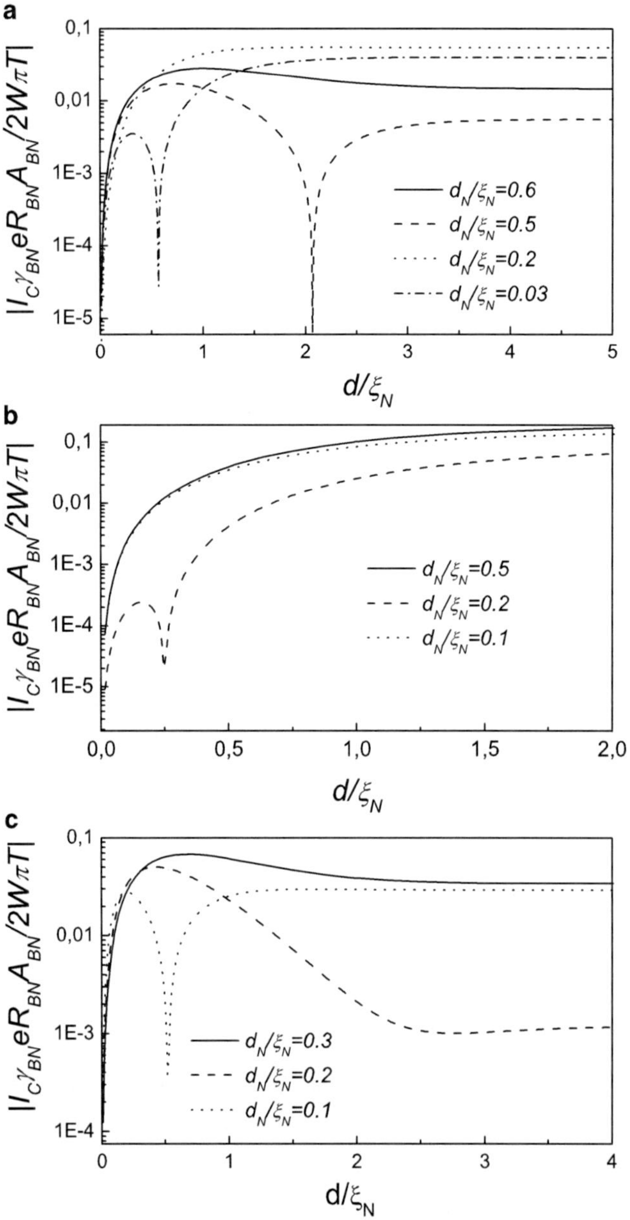

Fig. 7.28 I_C for (**a**) SNF-FN-FNS, (**b**) SN-FN-NS, (**c**) SNF-N-FNS structures versus d/ξ_N calculated at $\xi_N/\xi_F = 10$, $T/T_C = 0.5$, $H/\pi T_C = 30$, $d_F/\xi_F = 0.1$, $L/\xi_N = 2$

7.4.3 Calculation of Supercurrent for SNF-NF-FNS Junction

To calculate critical current of SNF-NF-FNS Josephson junction it is enough to solve linearized Usedel equations for condensate functions of normal (Φ_N) and ferromagnetic (Φ_F) films in weak link region, as well as for condensate functions in N films under left (Φ_{N1}) and right (Φ_{N2}) superconducting electrodes. These equations have the form

$$\xi_N^2 \left\{ \frac{\partial^2}{\partial x^2} + \frac{\partial^2}{\partial y^2} \right\} \Phi_{N,N1,N2} - \Omega \Phi_{N,N1,N2} = 0,$$

$$\xi_F^2 \left\{ \frac{\partial^2}{\partial x^2} + \frac{\partial^2}{\partial y^2} \right\} \Phi_{F,F1,F2} - \tilde{\Omega} \Phi_{F,F1,F2} = 0. \tag{7.111}$$

They should be supplemented by the boundary conditions on SN interfaces at $x = d_N$

$$\gamma_{\mathrm{BN}} \xi_N \frac{\partial}{\partial x} \Phi_{N1,N2} = G_S \Delta \exp^{\mp i\varphi/2}, \tag{7.112}$$

where sign minus (plus) should be chosen for left (right) S electrode. At FN interface located at $x = 0$ the boundary conditions have the form:

$$\frac{\xi_N}{\Omega} \frac{\partial}{\partial x} \Phi_{N,N1,N2} = \gamma \frac{\xi_F}{\widetilde{\Omega}} \frac{\partial}{\partial x} \Phi_{F,F1,F2}, \tag{7.113}$$

$$\gamma_B \xi_F \frac{\partial}{\partial x} \Phi_{F,F1,F2} + \Phi_{F1} = \frac{\widetilde{\Omega}}{\Omega} \Phi_{N,N1,N2}, \tag{7.114}$$

while at free interfaces

$$\frac{\partial}{\partial y} \Phi_{N1,N2} = 0, \quad y = \mp(L/2 + d) \tag{7.115}$$

$$\frac{\partial}{\partial y} \Phi_{F1,F2} = 0, \quad y = \mp(L/2 + d)$$

$$\frac{\partial}{\partial x} \Phi_{F,F1,F2} = 0, \quad x = -d_F \tag{7.116}$$

$$\frac{\partial}{\partial x} \Phi_N = 0, \quad x = d_N \tag{7.117}$$

they are followed from the demand of preventing a current flow across them. Finally at the interfaces between complex electrodes and weak link region (at $y = \mp L/2$) all the functions and their first derivatives should be uninterrupted:

$$\frac{\partial}{\partial y} \Phi_{N1,N2} = \frac{\partial}{\partial y} \Phi_N, \tag{7.118}$$

$$\Phi_{N1,N2} = \Phi_N, \tag{7.119}$$

$$\frac{\partial}{\partial y}\Phi_{F1,F2} = \frac{\partial}{\partial y}\Phi_F, \tag{7.120}$$

$$\Phi_{F1,F2} = \Phi_F. \tag{7.121}$$

In the considered limit of thin F and N films

$$d_N \ll \xi_N, \quad d_F \ll \xi_F,$$

the two-dimensional boundary problem (7.111)–(7.121) can be reduced to a one-dimensional. To do this we can suppose that in the main approximation condensate functions do not depend on coordinate x,

$$\Phi_{N,N1,N2} = A_{N,N1,N2}(y), \ \Phi_{F,F1,F2} = B_{F,F1,F2}(y), \tag{7.122}$$

and that their derivatives with respect to x can be expressed as follows:

$$\frac{\partial \Phi_{N,N1,N2}}{\partial x} = \left\{ \frac{\Omega}{\xi_N^2} A_{N,N1,N2} - \frac{\partial^2 A_{N,N1,N2}}{\partial y^2} \right\} (x - d_N),$$

$$\frac{\partial \Phi_{F,F1,F2}}{\partial x} = \left\{ \frac{\widetilde{\Omega}}{\xi_F^2} B_{F,F1,F2} - \frac{\partial^2 B_{F,F1,F2}}{\partial y^2} \right\} (x + d_F). \tag{7.123}$$

After substitution of (7.122) and (7.123) into the boundary conditions (7.113), (7.114) we arrive at one-dimensional differential equations in respect to functions $A_{N,N1,N2}(y)$ and $B_{F,F1,F2}(y)$. Solution of thus obtained one-dimensional boundary problem for the weak link region can be expressed in the form:

$$\begin{aligned} A_N = {} & A_1 \cosh(q_1 y) + A_2 \sinh(q_1 y) \\ & + \frac{\beta}{\zeta_N^2}\frac{\Omega}{\widetilde{\Omega}} (B_1 \cosh(q_2 y) + B_2 \sinh(q_2 y)) \end{aligned} \tag{7.124}$$

$$\begin{aligned} {}_F = {} & B_1 \cosh(q_2 y) + B_2 \sinh(q_2 y) \\ & - \frac{\beta}{\zeta_F^2}\frac{\widetilde{\Omega}}{\Omega} (A_1 \cosh(q_1 y) + A_2 \sinh(q_1 y)) \end{aligned} \tag{7.125}$$

where inverse coherence lengths of the problem:

$$q_{1,2}^2 = \frac{1}{2}\left[u^2 + v^2 \mp \sqrt{(u^2 - v^2)^2 + 4\zeta_F^{-2}\zeta_N^{-2}} \right], \tag{7.126}$$

$$u^2 = \left(\frac{1}{\zeta_N^2} + \frac{\Omega}{\xi_N^2} \right), \quad v^2 = \left(\frac{1}{\zeta_F^2} + \frac{\widetilde{\Omega}}{\xi_F^2} \right), \tag{7.127}$$

and $\zeta_F^2 = \gamma_B d_F \xi_F, \quad \zeta_N^2 = \gamma_B d_N \xi_N/\gamma, \beta = (q_1^2 - v^2)^{-1}$.

The appropriate solutions for F and N films located under S electrodes are

$$A_{N1,N2} = A_{11,12}\cosh(q_1 y) + A_{12,22}\sinh(q_1 y) + \frac{\beta}{\zeta_N^2}\frac{\Omega}{\widetilde{\Omega}}(B_{11,12}\cosh(q_2 y) + B_{12,22}\sinh(q_2 y)) - N\mathrm{e}^{\mp \mathrm{i}\varphi/2} \tag{7.128}$$

$$B_{F1,F2} = B_{11,12}\cosh(q_2 y) + B_{12,22}\sinh(q_2 y) - \frac{\beta}{\zeta_F^2}\frac{\widetilde{\Omega}}{\Omega}(A_{11,12}\cosh(q_1 y) + A_{12,22}\sinh(q_1 y)) - F\mathrm{e}^{\mp \mathrm{i}\varphi/2} \tag{7.129}$$

The integration coefficients in (7.124), (7.125), (7.128), and (7.129) can be found by substituting these expressions into the boundary conditions. This procedure leads to

$$A_1 = \frac{\cos(\varphi/2)\sinh(q_1 d)}{\sinh(q_1(L/2+d))}\frac{F\beta\zeta_N^{-2} - N}{\nu+1}, \tag{7.130}$$

$$A_2 = \frac{\mathrm{i}\sin(\varphi/2)\sinh(q_1 d)}{\cosh(q_1(L/2+d))}\frac{F\beta\zeta_N^{-2} - N}{\nu+1},$$

$$B_1 = -\frac{\cos(\varphi/2)\sinh(q_2 d)}{\sinh(q_2(L/2+d))}\frac{N\beta\zeta_F^{-2} + F}{\nu+1},$$

$$B_2 = -\frac{\mathrm{i}\sin(\varphi/2)\sinh(q_2 d)}{\cosh(q_2(L/2+d))}\frac{N\beta\zeta_F^{-2} + F}{\nu+1},$$

where

$$N = \frac{1}{\xi_N d_N}\frac{2\nu^2\zeta_N^2\zeta_F^2}{1 - 4\nu^2 u^2\zeta_N^2\zeta_F^2}\frac{G_S\Delta}{\gamma_{\mathrm{BN}}},$$

$$F = \frac{\widetilde{\Omega}}{\Omega}\frac{1}{\xi_N d_N}\frac{\zeta_N^2}{1 - 4\nu^2 u^2\zeta_N^2\zeta_F^2}\frac{G_S\Delta,}{\gamma_{\mathrm{BN}}},$$

and $\nu = \beta^2\zeta_N^{-2}\zeta_F^{-2}$.

By substituting of the solution (7.124), (7.125), (7.130) into general formula for supercurrent (7.94) we obtain the expression for supercurrent in the SNF-NF-FNS structure:

$$I_S = (I_{C1} + I_{C2})\sin(\varphi), \tag{7.131}$$

where

$$I_{C2} = \frac{K\zeta_F^2\zeta_N^2}{\xi_N d_N}\,\mathrm{Re}\sum_{\omega=0}^{\infty}\frac{q_2\Gamma\left(1+\nu^2\beta\right)^2(\nu+1)^{-1}\sinh(q_2 d)^2}{(1-\nu^2u^2\zeta_F^2\zeta_N^2)^2\sinh(q_2(L+2d))},$$

$$I_{C1} = \frac{K}{\xi_N d_N}\,\mathrm{Re}\sum_{\omega=0}^{\infty}\frac{q_1\Gamma\left(\nu^2\zeta_F^2\zeta_N^2-\beta\right)^2(\nu+1)^{-1}\sinh(q_1 d)^2}{(1-\nu^2u^2\zeta_F^2\zeta_N^2)^2\sinh(q_1(L+2d))},$$

and $\Gamma = \Delta^2/(\Omega^2+\Delta^2)$.

In the limit $\zeta_N >> \zeta_F, \xi_N >> \xi_F$, the part of the full critical current, I_{C2}, is small, so that the magnetidute of I_C of SNF-NF-FNS structure is reduced to

$$I_C^{\text{SNF-NF-FNS}} = \frac{K}{\xi_N d_N} \sum_{\omega=0}^{\infty} \operatorname{Re} \frac{\Gamma U q_1 \sinh(q_1 d)^2}{\sinh(q_1(L+2d))}.$$

7.4.4 Calculation of Supercurrent for SNF-N-FNS Junction

To calculate supercurrent across SNF-N-FNS junction we should slightly change the procedure described in Appedix A taking into account the appearance of additional three interfaces in the structure. Since the current can not flow across them instead of (7.121) we should use

$$\frac{\partial}{\partial y}\Phi_{F1,F2} = 0,\ y = \mp L/2,\ -d_F \leq z \leq 0, \tag{7.132}$$

$$\frac{\partial}{\partial x}\Phi_N = 0,\ x = 0,\ -\frac{L}{2} \leq y \leq \frac{L}{2}. \tag{7.133}$$

In the limit of thin F and N films $d_N \ll \xi_N, d_F \ll \xi_F$, the of solution of Usadel equations in the N film of weak link has more simple form compare to (7.124):

$$A_N = A_1 \cosh(qy) + A_2 \sinh(qy), \tag{7.134}$$

while in FN bilayer under S electrodes it closes to that of (7.128), and (7.129)

$$\begin{aligned} A_{N1,N2} = {} & A_{11,12} \cosh(q_1 y) + A_{12,22} \sinh(q_1 y) \\ & + \frac{\beta}{\zeta_N^2} \frac{\Omega}{\widetilde{\Omega}} \left(B_{11,12} \cosh(q_2 y) + B_{12,22} \sinh(q_2 y)\right) - N e^{\mp i\varphi/2}, \end{aligned} \tag{7.135}$$

$$\begin{aligned} B_{F1,F2} = {} & B_{11,12} \cosh(q_2 y) + B_{12,22} \sinh(q_2 y) \\ & - \frac{\beta}{\zeta_F^2} \frac{\widetilde{\Omega}}{\Omega} \left(A_{11,12} \cosh(q_1 y) + A_{12,22} \sinh(q_1 y)\right) - F e^{\mp i\varphi/2}. \end{aligned} \tag{7.136}$$

The integration constants in (7.134), (7.135), (7.136) can be found from the boundary conditions. In particular for A_1 and A_2 one can get

$$A_1 = \frac{-N \cos\{\varphi/2\}}{\left(Q_{q_1,\frac{q}{2}} + \nu Q_{q_2,\frac{q}{2}}\right) / (\nu+1) \tanh\frac{qL}{2} + \cosh\frac{qL}{2}}, \tag{7.137}$$

$$A_2 = \frac{-iN \sin\{\varphi/2\}}{\left(Q_{q_1,\frac{q}{2}} + \nu Q_{q_2,\frac{q}{2}}\right) / (\nu+1) + \sinh\frac{qL}{2}}. \tag{7.138}$$

Substitution of (7.134), (7.137), (7.138) into general formula for supercurrent:

$$I_S = -\frac{i\pi T W}{e\rho_N} \sum_{n=-\infty}^{\infty} \frac{1}{\omega^2} \int_0^{d_N} \left[\Phi^*_{-\omega,N} \frac{\partial}{\partial y} \Phi_{\omega,N} \right]$$

in the limit $\zeta_N >> \zeta_F, \xi_N >> \xi_F$ leads to

$$I_C^{\text{SNF-N-FNS}} = \frac{K}{\xi_N d_N} \operatorname{Re} \sum_{n=0}^{\infty} \frac{\Gamma U q}{Q_{q_1,q} + P_{q_1,q}}, \tag{7.139}$$

where functions $Q_{q_1,q}$ and $P_{q_1,q}$ are determined by (7.99) and (7.100), respectively.

7.4.5 *Calculation of Supercurrent for SN-NF-NS Junction*

To calculate of supercurrent across SN-FN-NS junction we should change the procedure described in Appedix A by taking into account the absence of F film in complex SN electrode. To do this the boundary conditions (7.121) in appropriate regions should be replaced by

$$\frac{\partial}{\partial y} \Phi_F = 0, \; y = \mp L/2, . - d_F \leq x \leq 0, \tag{7.140}$$

$$\frac{\partial}{\partial x} \Phi_{N1,N2} = 0, \; x = 0, . \frac{L}{2} \leq |y| \leq \frac{L}{2} + d \tag{7.141}$$

In the limit of thin F and N films $d_N \ll \xi_N, d_F \ll \xi_F$, solution of the boundary problem in the weal link can be found in the form

$$\begin{aligned} A_N = {} & A_1 \cosh(q_1 y) + A_2 \sinh(q_1 y) \\ & + \frac{\beta}{\zeta_N^2} \frac{\Omega}{\widetilde{\Omega}} \left(B_1 \cosh(q_2 y) + B_2 \sinh(q_2 y) \right), \end{aligned} \tag{7.142}$$

$$\begin{aligned} B_F = {} & B_1 \cosh(q_2 y) + B_2 \sinh(q_2 y) \\ & - \frac{\beta}{\zeta_F^2} \frac{\widetilde{\Omega}}{\Omega} \left(A_1 \cosh(q_1 y) + A_2 \sinh(q_1 y) \right), \end{aligned} \tag{7.143}$$

while for it in the N films located under S electrodes they are

$$A_{N1,N2} = A_{11,12} \cosh(q_1 y) + A_{12,22} \sinh(q_1 y) - N e^{\mp i\varphi/2}. \tag{7.144}$$

Integration constants A_1, A_2, B_1, B_2 in (7.142)–(7.144) can be found from the boundary conditions resulting in

$$A_1 = \frac{\cos\{\varphi/2\} q^{-2} G_S \Delta/(\xi_N d_N \gamma_{\text{BN}})}{\cosh\frac{q_1 L}{2} + \left(\nu Q_{\frac{q_2}{2},\frac{q_1}{2}}/2 + Q_{q,\frac{q_1}{2}}(\nu+1)\right)\tanh\frac{q_1 L}{2}},$$

$$A_2 = \frac{\mathrm{i}\sin\{\varphi/2\} q^{-2} G_S \Delta/(\xi_N d_N \gamma_{\text{BN}})}{\sinh\frac{q_1 L}{2} + \nu\tanh^2\left(\frac{q_2 L}{2}\right) Q_{\frac{q_2}{2},\frac{q_1}{2}}/2 + Q_{q,\frac{q_1}{2}}(\nu+1)}, \tag{7.145}$$

$$B_1 = A_1 \frac{1}{\zeta_F^2} \beta \frac{\widetilde{\omega}}{|\omega|} \frac{q_1 \sinh(q_1 L/2)}{q_2 \sinh(q_2 L/2)}, \tag{7.146}$$

$$B_2 = A_2 \frac{1}{\zeta_F^2} \beta \frac{\widetilde{\omega}}{|\omega|} \frac{q_1 \cosh(q_1 L/2)}{q_2 \cosh(q_2 L/2)}. \tag{7.147}$$

Substituting this result into general formula for supercurrent (7.94) in the limit $\zeta_N >> \zeta_F, \xi_N >> \xi_F$ we arrived at the following formula for critical current of the SN-FN-NS junction

$$I_C^{\text{SN-NF-NS}} = \frac{K}{\xi_N d_N} \operatorname{Re} \sum_{n=0}^{\infty} \frac{\Gamma q_1}{q^4 (Q_{q,q_1} + P_{q,q_1})}.$$

7.5 Conclusion

In this chapter, Josephson effect in S-FN-S structures under condition of relatively large suppression parameters γ_{BN} and γ_{BF} at SN and SF interfaces was discussed. Thus, it has been shown that the use of a bilayer thin-film FN structure as a weak-link material can lead to the effective decrease in H and to a significant increase in both the damping length and oscillation period of the dependence $I_C(L)$ of the S-(FN)-S junctions as compared to the respective values for similar structures containing only the ferromagnetic film.

Also the effect of finite thickness of the ferromagnetic films and normal S-FN-S transition in the critical current was analyzed. It was shown that for arbitrary film thickness qualitatively preserved oscillatory behavior of the critical current with changing the distance between superconducting electrodes, but the characteristic scale of the damping and period of oscillations depend strongly on the thickness of the films. Thus, for arbitrary thickness of the films all the earlier results for the limit of thin films are qualitatively correct. At the thickness of the ferromagnet, which is comparable with the coherence length, the characteristic scale of the damping and period of oscillations cease to depend on the thickness of the F film. Also very important is the fact that near the critical distance between superconducting electrodes, that is those in which the critical current is zero for an infinitely thick ferromagnetic film, there is rapid change as a sign, and value of the critical current with small changes in distance between the superconducting electrodes. Off from such narrow critical areas as a sign, and the value of critical current do not depend on the thickness of the F film. These results are very important from a practical point of view, because one can create junctions with parameters that do not depend on the spread of both geometric and transport parameters of material structures that are inherent in any technological process.

A second major technological constraints on the practical production of proposed S-FN-S and S-FNF-S structures was that they are ramp-type junctions. From a technological point of view, it is much more convenient to work with structures that have only parallel boundaries between layers. Therefore the geometry in which the S electrodes are placed on top FN structure was studied. Three different geometries were considered. The advantages given by for the practical realization of the geometry, in which the S electrodes are located on the FN structure, to the geometry of the ramp-type SFNS junctions were analyzed. It was proved that amplitude of the critical current for SN-FN-NS-type structure reaches the highest values compared with other geometries because of the absence of additional suppression of superconductivity under electrodes from ferromagnetic film. The effect of finite area of SN interface of SFNS junctions in which the electrodes are located on the top of FN structure on realization of states with negative and positive sign of the critical current was studied. It was proved that besides multiple $0-\pi$ transition for ramp-type structures with increase of distance between the electrodes in the structures with the new geometry a single $0-\pi$ transition with increase of length of the SN boundary is realized. This $0-\pi$ transition can exist even in the SFN-N-FNS structures with nonferromagnetic region of weak link.

Acknowledgements This work was supported by RFBR grant 09-02-12176-ofi_m and partially supported by Moldavian State Program grant 11.836.05.01A.

References

1. S.M. Frolov, D.J. Van Harlingen, V.A. Oboznov, V.V. Bolginov, V.V. Ryazanov, Phys. Rev. B **70**, 144505 (2004)
2. S.M. Frolov, D.J. Van Harlingen, V.V. Bolginov, V.A. Oboznov, V.V. Ryazanov, Phys. Rev. B **74**, 020503 (R) (2006)
3. H. Sellier, C. Baraduc, F. Lefloch, R. Calemczuck, Phys. Rev. B **68**, 054531 (2003)
4. Y. Blum, A. Tsukernik, M. Karpovski, A. Palevski, Phys. Rev. B **70**, 214501 (2004)
5. C. Surgers, T. Hoss, C. Schonenberger, et al., J. Magn. Magn. Mater. **240**, 598 (2002)
6. C. Bell, R. Loloee, G. Burnell, M.G. Blamire, Phys. Rev. B **71**, 180501 (R) (2005)
7. V. Shelukhin, A. Tsukernik, M. Karpovski, Y. Blum, K. B. Efetov, A.F. Volkov, T. Champel, M. Eschrig, T. Lofwander, G. Schon, A. Palevski, Phys. Rev. B 73 174506 (2006)
8. V.A. Oboznov, V.V. Bol'ginov, A.K. Feofanov, V.V. Ryazanov, A. Buzdin, Phys. Rev. Lett. **96**, 197003 (2006)
9. M. Weides, K. Tillmann, H. Kohlstedt, Phys. C **437-438**, 349–352 (2006)
10. M. Weides, M. Kemmler, H. Kohlstedt, A. Buzdin, E. Goldobin, D. Koelle, R. Kleiner, Appl. Phys. Lett. **89**, 122511 (2006)
11. M. Weides, M. Kemmler, E. Goldobin, H. Kohlstedt, R. Waser, D. Koelle, R. Kleiner, Phys. Rev. Lett. v. **97**, 247001 (2006)

12. H. Sellier, C. Baraduc, F. Lefloch, R. Calemczuck, Phys. Rev. Lett. **92**(25), 257005 (2004)
13. F. Born, M. Siegel, E.K. Hollmann, H. Braak, A.A. Golubov, D.Yu. Gusakova, M.Yu. Kupriyanov, Phys. Rev. B. **74**, 140501 (2006)
14. J.W.A. Robinson, S. Piano, G. Burnell, C. Bell, M.G. Blamire, Phys. Rev. Lett. **97**, 177003 (2006)
15. J.W.A. Robinson, S. Piano, G. Burnell, C. Bell, M.G. Blamire, Phys. Rev. B. **76**, 094522 (2007)
16. M.Yu. Kupriyanov, A.A. Golubov, M. Siegel, Proc. SPIE **6260**, 227–238 (2006)
17. L.B. Ioffe, V.B. Geshkenbein, M.V. Feigelman, A.L. Fauchere, G. Blatter, Nature (London) **398**, 679 (1999)
18. A.V. Ustinov, V.K. Kaplunenko, J. Appl. Phys. **94** 5405 (2003)
19. G. Blatter, V.B. Geshkenbein, L.B. Ioffe, Phys. Rev. B. **63**, 174511 (2001)
20. E. Terzioglu, M.R. Beasley. IEEE Trans. Appl. Superc. Phys. Rev. B. **8**, 48 (1998)
21. Hans Hilgenkamp, M.R. Supercond. Sci. Technol. **21**, 034011 (5pp) (2008)
22. F.S. Bergeret, A.F. Volkov, K.B. Efetov, Rev. Mod. Phys. **77**, 1321 (2005)
23. A. Kadigrobov, R.I. Shekhter, M. Jonson, Eurorhys. Lett. **54**(3), 394 (2001)
24. A.F. Volkov, F.S. Bergeret, K.B. Efetov, Phys. Rev. Lett. **90**(11), 117006 (2003)
25. F.S. Bergeret, A.F. Volkov, K.B. Efetov, Phys. Rev. B **64**, 134506 (2001)
26. F.S. Bergeret, A.F. Volkov, K.B. Efetov, Phys. Rev. B **68**, 064513 (2003)
27. R.S. Keizer, S.T.B. Goennenwein, T.M. Klapwijk, G. Miao, G. Xiao, A. Gupta, Nature **439**, 825 (2006)
28. I. Sosnin, H. Cho, V.T. Petrashov, A.F. Volkov, Phys. Rev. Lett. **96**, 157002 (2006)
29. N.M. Chtchelkatchev, I.S. Burmistrov, Phys. Rev. B **68**, 140501(R) (2003)
30. I.S. Burmistrov, N.M. Chtchelkatchev, Phys. Rev. B **72**, 144520 (2005)
31. T. Champel, M. Eschrig, PRB **72**, 054523 (2005)
32. M. Houzet, A.I. Buzdin, Phys. Rev. B **74**, 214507 (2006)
33. M.A. Maleki, M. Zareyan, Phys. Rev. B **74**, 144512 (2006)
34. Y.V. Fominov, A.F. Volkov, K.B. Efetov, Phys. Rev. B **75**, 104509 (2007)
35. A.F. Volkov, A. Anishchanka, Phys. Rev. B **71**, 024501 (2005)
36. A.F. Volkov, K.B. Efetov, Phys. Rev. B **78**, 024519 (2008)
37. B. Crouzy, S. Tollis, D.A. Ivanov, Phys. Rev. B **76**, 134502 (2007)
38. L. Usadel, Phys. Rev. Lett. textbf25, 507 (1970)
39. M.Yu. Kuprianov, V.F. Lukichev, Zh. Eksp. Teor. Fiz. **94**, 139 (1988) [Sov. Phys. JETP **67**, 1163 (1988)]
40. E.A. Koshina, V.N. Krivoruchko, Fiz. Nizk. Temp. **157**, 26 (2000) [Low Temp. Phys. **26**, 115 (2000)]
41. A.A. Golubov, M.Yu. Kupriyanov, Pis'ma Zh. Eksp.Teor.Fiz. **75**, 709 (2002) [JETP Lett. **75**, 588 (2002)]
42. A.I. Buzdin, Pis'ma Zh. Eksp.Teor.Fiz. **78**, 1073 (2003) [JETP Lett. **78**, 583 (2003)]
43. A.F. Volkov, Ya.V. Fominov, K.B. Efetov, Phys. Rev. B **72**, 184504 (2005);
44. Ya.V. Fominov, A.F. Volkov, K.B. Efetov, Phys. Rev. B **75**, 104509 (2005)
45. T. Champel, T. Lofwander, M. Eschrig, Phys. Rev. Lett. **100**, 077003 (2008)
46. K.K. Likharev, Rev. Mod. Phys. **51**, 101 (1979)
47. A.A. Golubov, M.Yu. Kupriyanov, V.F. Lukichev, Mikroelektronika **12**, 342 (1983) [Soviet-Microelectronics, **12**, 180 (1983)]
48. T.Yu. Karminskaya, M.Yu. Kupriyanov, Pis'ma Zh. Eksp. Teor. Fiz. **85**, 343 (2007) [JETP Lett. **85**, 286 (2007)]
49. T.Yu. Karminskaya, M.Yu. Kupriyanov, Pis'ma Zh. Eksp. Teor. Fiz. **85**, 343 (2007) [JETP Lett. **86**, 61 (2007)]
50. T.Yu. Karminskaya, M.Yu. Kupriyanov, A.A. Golubov, Pis'ma Zh.Eksp. Teor. Fiz. **87**, 657 (2008) [JETP Lett. **87**, 570 (2008)]
51. T.Yu. Karminskaya, M.Yu. Kupriyanov, A.A. Golubov, Phys. Rev. B **79**, 214509 (2009)
52. A.A. Golubov, M.Yu. Kupriyanov, E. Il'ichev, Rev. Mod. Phys. **76**, 411 (2004)
53. F.S. Bergeret, A.F. Volkov, K.B. Efetov, Rev. Mod. Phys. **77**, 1321 (2005)
54. A.I. Buzdin, Rev. Mod. Phys. **77**, 935 (2005)

Chapter 8
Physics and Applications of Superconducting Phase Inverters Based on Superconductor–Ferromagnet–Superconductor Josephson Junctions

V.V. Ryazanov

Abstract Coexistence of superconductivity and magnetism is one of the most actively progressing fields in the condensed matter physics. Undoubtedly the most impressive phenomena observed in this field recently are related to spatial nonuniform superconductivity in a ferromagnet close to superconductor/ferromagnet interface. One of them is phase difference inversion in Josephson junctions with ferromagnetic weak links, SFS π-junctions. This review is devoted to physics and applications of these novel Josephson structures.

8.1 Introduction

Josephson "π-junctions" [1] are weakly coupled superconducting structures with the ground state phase difference of the macroscopic superconducting wave function $\varphi = \pi$. They are characterized by the inverted Josephson current-phase relation (*CPR*): $I_s = I_c \sin(\varphi + \pi) = -I_c \sin\varphi$, negative critical current $-I_c$, and negative Josephson coupling energy $E_\pi = -E_J \cos\varphi$, where $E_J = I_c\Phi_0/(2\pi)$, $\Phi_0 = h/2e$ is the quantum of magnetic flux. The inverse *CPR* and a total π-junction coupling energy $E = E_J(1 + \cos\varphi)$ are shown on the right side in Fig. 8.1.

The origin of the π-state in a superconductor/ferromagnet/superconductor (*SFS*) junction is an oscillating and sign-reversing superconducting order parameter in the ferromagnet close to a superconductor/ferromagnet interface [2, 3] (see [4] as a review). Owing to these oscillations, different signs of the order parameter can occur at the two banks of the *SFS* sandwich when the ferromagnetic layer thickness is of the order of half an oscillation period (see Fig. 8.2), which corresponds to a sign change of the supercurrent and a negative Josephson coupling energy. Spatial

V.V. Ryazanov (✉)
Institute of Solid State Physics, Russian Academy of Sciences, Chernogolovka, Moscow District, 142432, Russia
e-mail: valery.ryazanov@gmail.com

A. Sidorenko (ed.), *Fundamentals of Superconducting Nanoelectronics*, NanoScience and Technology, DOI 10.1007/978-3-642-20158-5_8,

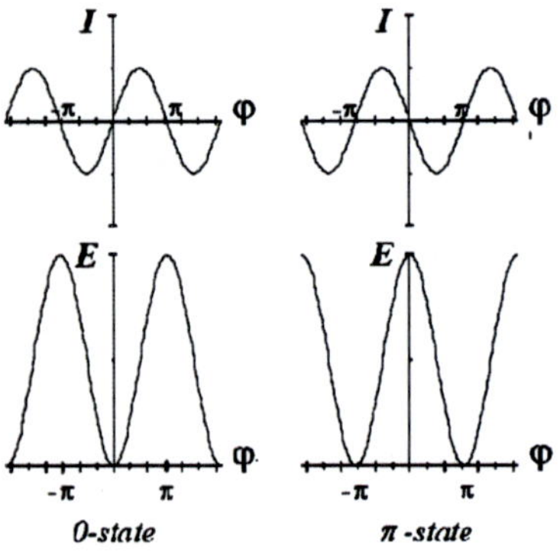

Fig. 8.1 The supercurrent and the coupling energy vs. junction phase difference for Josephson junctions in conventional (0-) and π-states

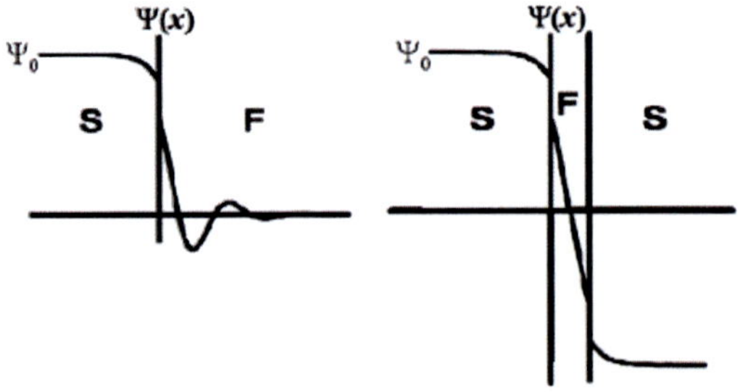

Fig. 8.2 Sign-reversing superconducting wave function Ψ (superconducting order parameter) in a SF-bilayer close to SF-interface (*left panel*) and in SFS π-junction (*right panel*)

oscillations of the superconducting order parameter in a ferromagnet close to an *SF* interface were predicted in [2, 3]. The physical origin of the oscillations is the exchange splitting of spin-up and spin-down electron subbands in ferromagnets [5]. Conventional superconductors used in the SFS π-junctions contain Cooper pairs, two electrons with opposite spin and momentum $(+\mathbf{k}\uparrow, -\mathbf{k}\downarrow)$. Such a system is described by an isotropic excitation gap or a superconducting order parameter. It was predicted long ago by Larkin and Ovchinnikov [6] and by Fulde and Ferrel [7] that pairing still can occur when the electron energies and momenta at the Fermi energy are different for the two spin directions, for instance as the result of an exchange field in magnetic superconductors. The "LOFF" state is qualitatively different from the zero-momentum state: it is spatially inhomogeneous and the order parameter contains nodes, where the phase changes by π. It was not observed reliably in bulk materials still, but it can be induced in a weak ferromagnet sandwiched between two superconductors. The spatial variation of the superconducting order parameter in the ferromagnet arises as a response of the Cooper pair to the energy difference between the two spin directions [8]. The electron with the energetically favorable spin increases its momentum by $Q \sim E_{\text{ex}}/v_{\text{F}}$, where E_{ex} is the exchange energy and

v_F is the Fermi velocity, while the other electron decreases its momentum by Q. Since the original momentum of each electron can be positive or negative, the total pair momentum inside the ferromagnet is $2Q$ or $-2Q$. Combination of the two possibilities leads to the oscillating order parameter $\Psi(x)$ in the junction along the direction normal to the SF interfaces: $\Psi(x) \sim \cos(2Qx)$ [8, 9]. The same picture applies in the diffusive limit [3]. Now, the oscillations are superimposed on the decay of the order parameter due to pair breaking by impurities in the presence of the exchange field. In [2, 3], it was predicted that an *SFS* junction can yield a phase shift of π between the superconducting banks. The π-state offers new ways for studying the coexistence of superconductivity and magnetism and may also be important for superconducting electronics.

A review of experimental works concerned with investigations of Josephson structures and carried out with the author participation is presented below.

8.2 SFS Junctions: Thickness and Temperature Dependences of Josephson Ground States

To observe the manifestations of the transition into the π-state, one should fabricate *SFS* sandwiches with the F-layer thicknesses d_F close to integer numbers of half periods of the order parameter spatial oscillations $\lambda_{ex}/2$. The period is $\lambda_{ex} = 2\pi\xi_{F2}$, where the oscillation (or "imaginary") length ξ_{F2} can be extracted from the complex coherence length ξ_F in a ferromagnet: $1/\xi_F = 1/\xi_{F1} + i/\xi_{F2}$. In the case of large exchange energy and negligible magnetic scattering in a diffusive F-layer, the imaginary length ξ_{F2} and the order parameter decay length ξ_{F1} are equal [3]: $\xi_{F1} = \xi_{F2} = (\hbar D/E_{ex})^{1/2}$, where D is the diffusion coefficient for electrons in a ferromagnet and E_{ex} is the exchange energy responsible for sign-reversal superconductivity in a ferromagnet. However, antagonism of superconductivity and ferromagnetism differing in spin ordering is a cause of the strong suppression of superconductivity in the contact area of the S- and F-materials. The order parameter decay length ξ_{F1} is as small as 1 nm if typical strong ferromagnets such as Fe, Co, and Ni are used like F-interlayers in *SFS* junctions. Our approach is to fabricate Josephson *SFS* junctions with weak ferromagnets. The experimental studies of the π-state were carried out by us on thin-film sandwiches Nb-$Cu_{1-x}Ni_x$-Nb, with x in the range 0.52–0.57 and the Curie temperature, T_{Curie}, of the copper-nickel layers in the range 30–150 K. The onset of ferromagnetism is around $x = 0.44$; above this concentration, the *Ni* magnetic moment increases with about 0.01 μ_B/at%Ni [10], which allows precise tuning of the magnetism. The weak ferromagnetism of the CuNi-alloy made possible flowing of supercurrents through the F-layers up to 30 nm in thickness, prepared with the roughness of 1–2 nm. The first observation of the supercurrent through a ferromagnet was carried out in [11] and the first experimental evidence for the π-state in the *SFS* junction was obtained in [12, 13].

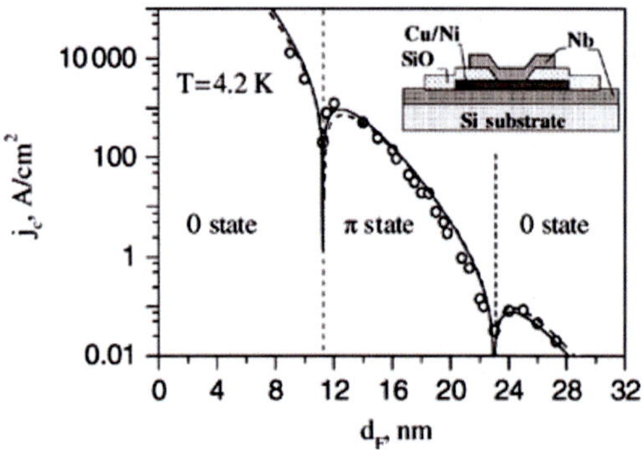

Fig. 8.3 The F-layer thickness dependence of the critical current density for Nb–$Cu_{0.47}Ni_{0.53}$–Nb junctions at temperature 4.2 K [14]. *Open circles* represent experimental results; *solid and dashed lines* show model calculations. The *inset* shows a schematic cross-section of our *SFS* junctions

An insulating SiO layer is used between the top S-electrode of SFS junction and the bottom SF sandwich. The window in this layer determines the junction area (of $10 \times 10\,\mu m^2$, typically). A schematic sandwich cross section is given in Fig. 8.3 (inset). First, the bottom Nb electrode (110 nm) of the SFS junction is sputtered by dc-magnetron and structured by lift-off process. Then $Cu_{1-x}Ni_x/Cu$ bilayer with 50 nm Cu is sputtered on top of the Nb electrode by RF-diode sputtering system. Topology of the ferromagnet bilayer is formed by physical argon plasma etching process. The next technological step is a thermal evaporation of isolating SiO (150 nm) film followed by lift-off process. Final technological step consists of two in-situ operations: the physical argon plasma etching process helping to clean the ferromagnet bilayer surface (30 nm of Cu film is removed) and the dc-magnetron sputtering of the Nb (240 nm) film. Pattering of the Nb film is done by lift-off process. The junction normal resistance R did not exceed $5\times10^{-4}\,\Omega$, so the transport characteristics of the junctions were measured by the picovoltmeter based on SQUID with sensitivity better than 10^{-11} V. I–V and $I_c(H)$ characteristics for Nb-$Cu_{0.47}Ni_{0.53}$-Nb junctions have shown in Fig. 8.4 (insets). The I–V characteristics are well described by the expression $V = R\left(I^2 - I_c^2\right)^{1/2}$, with R values presented in the main panel in Fig. 8.4. The linear approximation of the $R(d_F)$ dependence yields the interface resistance $r_B = 30\,\mu\Omega$ for junctions with the area of $10 \times 10\,\mu m^2$ and F-layer resistivity $\rho_F = 62\,\mu\Omega$ cm. It allows us to estimate the following parameters in our ferromagnet: the electron elastic mean free path $l = 1$ nm, the diffusion coefficient $D = 5\,cm^2/s$.

Figure 8.4 (lower inset) shows that the magnetic field dependence $I_c(H)$ for SFS junctions yields the classical "Fraunhofer" pattern. The oscillation period is in reasonable agreement with the cross-section of the junction. Note that the central peak is at zero field, even though the alloy is ferromagnetic. This signifies that on average there is no change in the phase difference over the junction along the different directions in the plane of the junction, presumably due to a small-scale magnetic domain structure of the magnetic layer with zero net magnetization. The central peaks were found shifted when SFS junctions were heated above T_c (but below the ferromagnetic transition temperature, T_{Curie}) and a small field briefly applied, leading to a finite magnetization [11]. Sometimes the peak was found

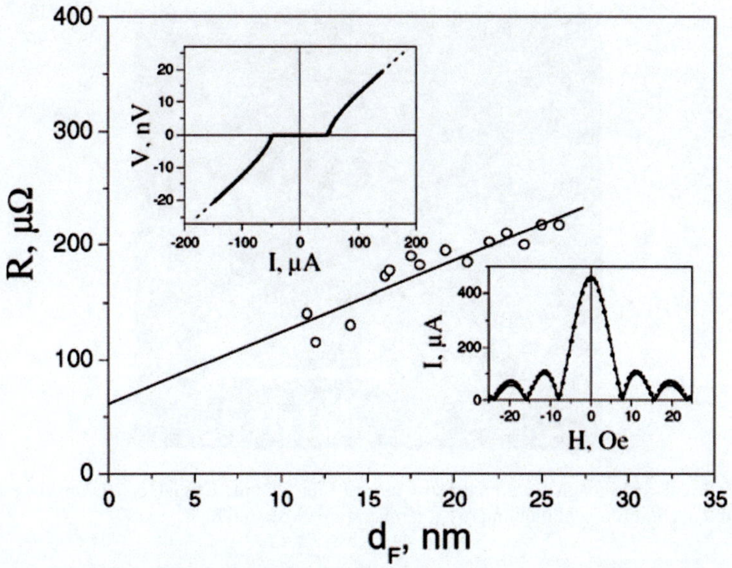

Fig. 8.4 Resistance of Nb–$Cu_{0.47}Ni_{0.53}$–Nb sandwiches normalized to the junction area of $10 \times 10\ \mu m^2$ vs. the F-layer thickness [14]. *Insets* show typical I–V and Fraunhofer ($I_c(H)$) dependences of *SFS* junctions fitted to the conventional Josephson expressions

shifted in zero applied field, probably due to flux trapping in the superconducting banks during cooling down. This could be remedied by reheating and recooling. The starting point for all experiments was a central peak at zero field. The small-scale magnetic domain structure of thin (20 nm) magnetic $Cu_{0.47}Ni_{0.53}$-layer was visualized by means of a decoration technique in Ref. [15] at temperature close to 6 K. The results of the work show that the magnetic domains have perpendicular magnetic anisotropy and the domain structure period is about 0.1 μm (see Fig. 8.5).

In [14], we have investigated the thickness dependence of the *SFS* junction critical current density in a wide F-layer thickness range for more than twenty sandwiches. All junctions were prepared with lateral sizes smaller than the Josephson penetration depth to ensure uniform supercurrent distribution. Weakly ferromagnetic $Cu_{0.47}Ni_{0.53}$ interlayers had the Curie temperature of about 60 K. In the barrier thickness interval of 8–28 nm, the critical current density varied by 6 orders of magnitude and had nodes at two d_F values as presented in Fig. 8.3.

One can see that the curve in Fig. 8.3 demonstrates both direct $0 - \pi$ transition and reverse transition from π- to conventional (0-) state. At transition points, $d_{c1,c2}$, the critical current is equal to zero and then should formally change its sign. Since in our transport experiments, we could measure only the magnitude of the critical current, the negative region of $I_c(d_F)$ between the two sharp cusps (that corresponds to the π-state) is reflected into the positive region.

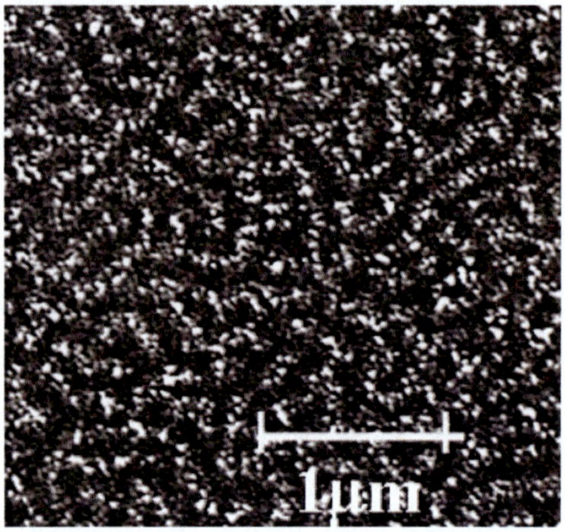

Fig. 8.5 Small-scale magnetic domain structure of thin (20 nm) $Cu_{0.47}Ni_{0.53}$-layer visualized by means of a decoration technique in [15] at temperature close to 6 K

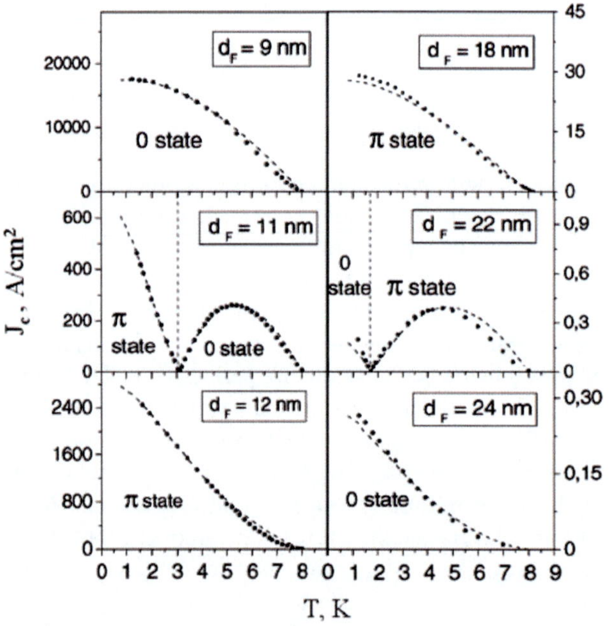

Fig. 8.6 Temperature dependences of the *SFS* junction critical current density at several F-layer thicknesses close to the critical ones [14]. The *dashed lines* show calculation results

Because of a slight temperature dependence of the order parameter oscillation period and other temperature dependent processes discussed below, we could pass through the transition points using samples with critical F-layer thicknesses $d_{c1} = 11$ nm and $d_{c2} = 22$ nm by changing temperature. Temperature-driven $0 - \pi$ and $\pi - 0$ transitions are presented in the middle panels of Fig. 8.6. The upper and lower panels show the temperature dependences of the critical current for samples with F-layer thicknesses around $d_{c1,c2}$. For barrier thicknesses over 1–2 nm from $d_{c1,c2}$ $0-\pi$ transitions are not observed in the experimental temperature range. This implies that the temperature decrease from 9 K down to 1 K is accompanied by the decrease of 1–2 nm in the spatial oscillation period and by the decrease of about 0.3 nm in the imaginary length. In this temperature range, the change of ξ_{F1} is about 0.2 nm as it has been estimated from $I_c(d_F)$ curves at different temperatures. The possibility of manipulating the coherence length by temperature was demonstrated in [13], in which the temperature-driven $0-\pi$ transition was observed for the first time. We supposed that the condition for having the temperature as a parameter is $k_B T \approx E_{ex}$ in this first our work. The exchange field and the temperature then are equally important, and the complex coherence length ξ_F in a ferromagnet is described by the following expression:

$$\xi_F = \sqrt{\frac{\hbar D}{2(\pi k_B T + iE_{ex})}}, \tag{8.1}$$

which yields for ξ_{F1} and ξ_{F2}:

$$\xi_{F1,2} = \sqrt{\frac{\hbar D}{\left[E_{ex}^2 + (\pi k_B T)^2\right]^{1/2} \pm \pi k_B T}}. \tag{8.2}$$

However already for $Cu_{0.47}Ni_{0.53}$ exchange energy $E_{ex}/k_B T$ estimated from data presented in Fig. 8.3 is about 850 K. The lowest reentrant curve in Fig. 8.7 demonstrates temperature-driven $\pi-0$ transition for even larger Ni content in CuNi-alloy [16]. For the case $E_{ex} >> k_B T$, the expressions (8.2) for $\xi_{F1}(T)$ and $\xi_{F2}(T)$ are given by:

$$\xi_{F1,2} = \sqrt{\frac{\hbar D}{E_{ex}}}\left(1 \mp \frac{\pi k_B T}{2E_{ex}}\right). \tag{8.3}$$

A correction $k_B T/(2E_{ex})$ in (8.3) is smaller than 10^{-2} for $E_{ex}/k_B T = 850$ K. However, values of $\xi_{F1} = 1.3$ nm and $\xi_{F2} = 3.5$ nm obtained from the slope of the $I_c(d_F)$ data (Fig. 8.3) and from the interval between the two minima of $I_c(d_F)$ correspondingly differ by almost a factor of 3, which cannot be explained simply by thermal terms in (8.3).

We have described the data shown in Figs. 8.3 and 8.6 by A. Buzdin's theoretical model [4, 14] that includes additional depairing processes, which increase ξ_{F2} and decrease ξ_{F1}. Because our F layer is an alloy, the role of magnetic scattering in the junction barrier is important [16]. Magnetic inhomogeneities due to Ni-rich clusters [17, 18] are known to exist in $Cu_{1-x}Ni_x$ films for x close to 0.5. For such Ni concentrations, the Curie temperature is small, and we may expect that the inverse spin-flip scattering time $\hbar\tau_s^{-1}$ could be of the order of the average exchange field E_{ex} or even larger. Spin-flip scattering due to ferromagnetic domain walls may contribute too [19]. This circumstance strongly modifies the proximity effect in

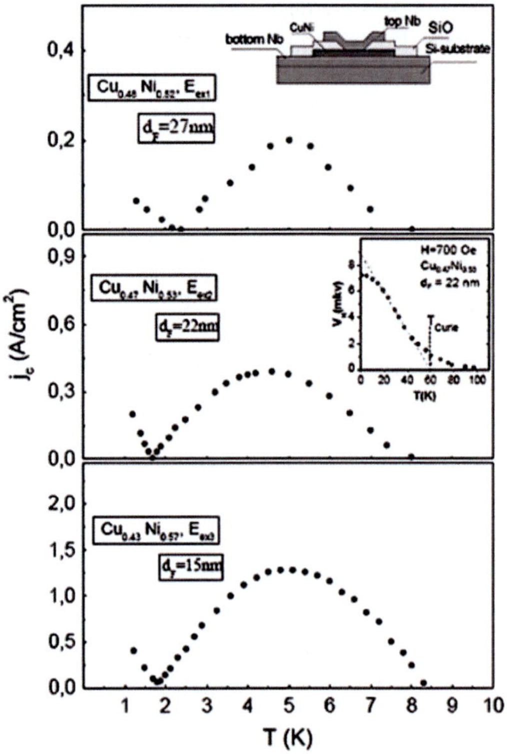

Fig. 8.7 Anomalous temperature dependences of the SFS junction critical current density at various Ni content ($x = 0.52$, 0.53 and 0.57) in $Cu_{1-x}Ni_x$-interlayers of Nb–$Cu_{1-x}Ni_x$–Nb sandwiches with thickness d_F close to value corresponding to the reverse transition from the π-to 0-state. The *inset* in the *upper panel* shows a schematic picture of the cross-section of the SFS junction. The *inset* in the *middle panel* is an example of Curie temperature detection by means of the anomalous Hall effect measurements of saturation magnetization

SF systems. The role of spin-orbit scattering should be neglected for the $Cu_{1-x}Ni_x$ alloys since it is only substantial in ferromagnets with large atomic numbers Z. To take into account the magnetic scattering in the framework of the Usadel equations, it is necessary to add the term $G\hbar/\tau_s$ together with temperature (the Matsubara frequencies ω) and exchange energy [20], where G is the normal Green's function. Then the Usadel equation for sufficiently high interface transparency is written as

$$\left(\omega + \mathrm{i}E_{\mathrm{ex}} + \frac{\hbar\cos\Theta}{\tau_s}\right)\sin\Theta - \frac{\hbar D}{2}\frac{\partial^2\Theta}{\partial x^2} = 0, \quad (8.4)$$

where $\cos\Theta$ and $\sin\Theta$ appear due to usual parameterization of the Green's functions: $G = \cos\Theta(x)$ and $F = \sin\Theta(x)$. The normal Green's function is temperature dependent and so the spin-flip term $G\hbar/\tau_s$ has also temperature dependence. An exact expression for the thickness and temperature dependence of the critical current in *SFS* junctions can be found in [14]. Calculated curves based on this exact expression shown by dashed lines in Fig. 8.6 and by solid line in Fig. 8.3. The dashed line in Fig. 8.3 shows a fitting by the following simple expression obtained from the linearized Usadel equation:

$$j_c = \exp\left(-\frac{d_F}{\xi_{F1}}\right)\left[\cos\left(\frac{d_F}{\xi_{F2}}\right) + \left(\frac{\xi_{F1}}{\xi_{F2}}\right)\sin\left(\frac{d_F}{\xi_{F2}}\right)\right]. \tag{8.5}$$

One can see that (8.5) satisfactorily describes the thickness dependence of the SFS junction critical current. However, it cannot describe significant temperature dependence close to $d_{c1,c2}$ and temperature-driven $0-\pi$ transitions. Really, expressions for coherence lengths obtained from the linearized Usadel equation have not significant temperature dependence:

$$\frac{1}{\xi_{F1,2}} = \sqrt{\frac{E_{ex}}{\hbar D}}\sqrt{\sqrt{1+\left(\frac{\omega}{E_{ex}}+\frac{\hbar}{E_{ex}\tau_s}\right)^2} \pm \left(\frac{\omega}{E_{ex}}+\frac{\hbar}{E_{ex}\tau_s}\right)}. \tag{8.6}$$

For the case $E_{ex} >> k_B T$, the expressions (9.6) can be written as

$$\frac{1}{\xi_{F1,2}} = \sqrt{\frac{E_{ex}}{\hbar D}}\sqrt{\sqrt{1+\left(\frac{\hbar}{E_{ex}\tau_s}\right)^2} \pm \frac{\hbar}{E_{ex}\tau_s}}. \tag{8.7}$$

The values of $\xi_{F1} = 1.3$ nm and $\xi_{F2} = 3.5$ nm estimated above can be obtained for $\hbar/\tau_s = 1.33\,E_{ex}$ and $E_{ex}/k_B = 850$ K. These parameters also yield good agreement in the case of fitting of the experimental data on base of the exact expression (shown by dashed lines in Fig. 8.6 and by solid line in Fig. 8.3).

8.3 Phase-Sensitive Experiments: Phase Inversion and Spontaneous Magnetic Flux

We have suggested and applied several experimental phase-sensitive methods for direct observations of the transition to the π-state. The most direct and simple way to detect the phase difference shift introduced by the SFS π-junctions is to study the magnetic field dependences of π-junction interferometer characteristics. Figure 8.8 shows real and schematic pictures of the network of SFS sandwiches (Nb-$Cu_{0.46}Ni_{0.54}$-Nb) that was used in the first our phase-sensitive experiment [21]. We have proposed this multiply-connected structure consisting of five identical SFS junctions arranged into two loops like a simplest symmetric intrinsically frustrated interferometer. The total phase change along the each interferometer loop should be a multiple of 2π.

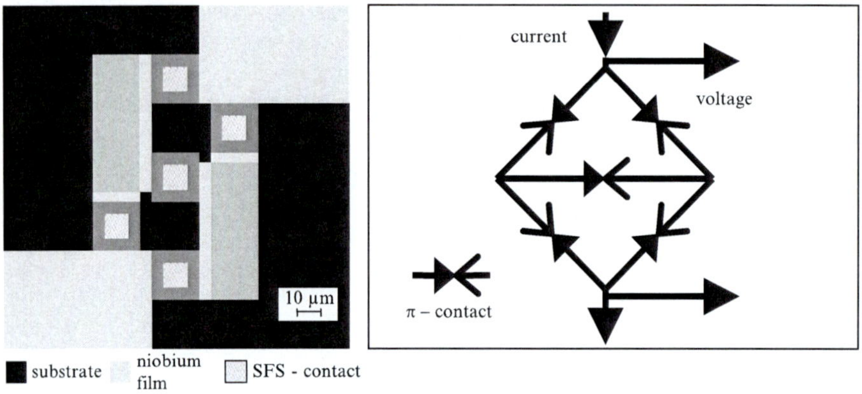

Fig. 8.8 Real (*left panel*) and schematic (*right panel*) pictures of the network of five SFS sandwiches, which was used in the phase-sensitive experiment [21]

Therefore, additional (to an intrinsic 3π-shift provided by the π-junctions) phase change should appear across the junctions, resulting in the circulating supercurrent of two possible directions. The SFS junctions in the fabricated structure demonstrate transition to the π-state at $T_\pi = 2.2\,\text{K}$. Above this temperature (0-state), the magnetic field dependence of the interferometer critical current, $I_m(H)$, was practically the same like one for dc-SQUID (see Fig. 8.9a). Periodical maximal peaks were observed at external fields corresponding to an integer number of the magnetic flux quanta Φ_0 per cell, i.e., integer frustration parameters $f = \Phi/\Phi_0$. Also small peaks can be seen at half-integer values of the frustration parameter f. These secondary peaks result from the phase interference over the outer loop of the net structure, which is twice the unit cell [22, 23]. At temperatures below T_π, i.e. in the π-state, the $I_m(H)$ pattern was found to be shifted by exactly half a period (Fig. 8.9b).

A physical picture of the effect observed in the π-state is the following. At $f = 0$ and other integer values of f, there is a circulating spontaneous current close to the SFS junctions critical current, I_c, which flows in the outer double loop of the interferometer. The current induces the extra phase shift $2 \times \pi/2$ in each interferometer cell and compensates the odd number of π-shifts in them. Because initially the interferometer is in the spontaneous fully frustrated state, the maximal interferometer transport supercurrent, I_m, is close to zero at $H = 0$. The external magnetic flux equal to half-integer quanta Φ_0 per cell produces the necessary phase shift of π in each cell without any circulating currents in the structure, so I_m reaches maxima at half-integer frustration parameters. The $I_m(T)$ dependence at $H = 0$, shown in Fig. 8.10a, mimics the $I_c(T)$ dependence for single junctions (see Fig. 8.6) and demonstrates a sharp cusp at the temperature T_π of the transition into the π-state. However, one should keep in mind that the left and right branches of the $I_m(H = 0, T)$ dependence are intrinsically asymmetric: while the high-temperature branch corresponds to the temperature dependence of

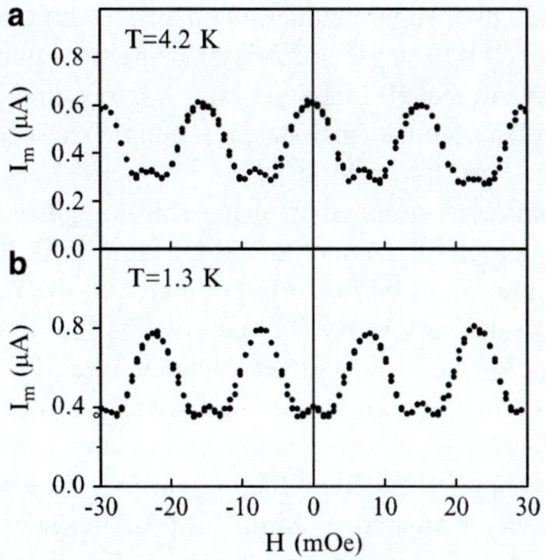

Fig. 8.9 Magnetic field dependences of the critical transport current for the structure depicted in Fig. 8.8 at temperature above (**a**) and below (**b**) T_π [21]

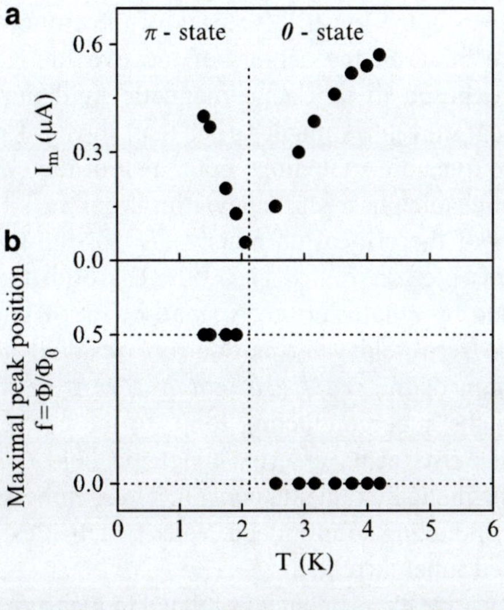

Fig. 8.10 (**a**) Temperature dependence of the interferometer critical transport current in the absence of applied magnetic field; (**b**) temperature dependence and jump of the maximal peak position on the $I_m(\Phi)$-dependences [21]

the doubled junction critical current, the low-temperature branch corresponds to the temperature dependence of the small peak amplitude only. Figure 8.10b shows the maximal peaks positions before and after the $0 - \pi$ transition and demonstrates the sharpness of this transition.

Structures including exclusively SFS-junctions hardly could find practical applications due to their low normal-state resistance. In [24], a Josephson junction inverter was suggested as a superconducting analog of the complementary metal-oxide-semiconductor (CMOS) logic. It relies on using two-junction interferometers (dc-SQUIDs) of conventional (0-junctions) and π-types and requires that 0- and π-junctions have similar I_c and normal-state resistance. These technologically stringent requirements can be softened by using an alternative "asymmetric" approach, which employs π-junctions as passive phase shifters (phase inverters) in basic cells of the modified single-flux-quantum (SFQ) logic [25]. Here, the π-junction critical current I_c is chosen to be much larger than that of conventional tunnel 0-junctions employed in the very same SFQ cell, so the phase difference across the π-junction is always close to π even at zero magnetic field. As the total change of the order parameter's phase over the closed path must become a multiple of 2π, the "missing" phase difference of π or $-\pi$ is induced on the remaining part of the cell by a spontaneously generated superconducting current. To verify the operation of π-junction phase shifters in an analog regime [26], we fabricated two geometrically identical superconducting loops (see schematic in Fig. 8.11a, b) on a single Si substrate (see Fig. 8.11c). The circuit Fig. 8.11b is a two-junction interferometer (dc-SQUID). The configuration of the circuit in Fig. 8.11a is nominally identical to that in Fig. 8.11b, except that an SFS π-junction has been inserted in the left branch of the loop, seen in the lower left corner of the circuit image in Fig. 8.11c. The on-chip distance between the centers of the two loops is 140 μm, so both interferometers are exposed to the same magnetic field during the experiment. The π-junction critical current is much larger than those of the tunnel junctions. Therefore, during the dynamic switchings in the rest of the circuit, π-junctions do not introduce any noticeable phase shifts deviating from π.

The dependencies of the critical currents $I_c(H)$ of the two devices shown in Fig. 8.11a, b are presented in Fig. 8.11e. Whereas both curves have the same shape, they are shifted by a half-period. A small offset of the symmetry axes for both curves from the zero-field value is due to a residual magnetic field in the cryostat. The minimum of the red $I_c(H)$ curve at zero field is due to inclusion of the π-junction in the superconducting loop. In the conventional SQUID, the same frustrated state exists at an external magnetic field corresponding to half-integer numbers of magnetic flux quanta per cell. Thus, embedding an SFS π-phase shifter into a superconducting loop indeed leads to self-biasing of the loop by a spontaneously induced supercurrent.

The next phase-sensitive experiment was related to measurements of the current-phase relation (CPR) of an SFS Josephson junction [27] that demonstrated directly the sign change in the critical current when the junction undergoes a transition into the π-state below a temperature T_π at which the critical current vanishes. The CPR was measured in the rf-SQUID configuration [28] shown in Fig. 8.12a. A dc-SQUID galvanometer was used to measure the current I_L that flows through the superconducting loop as a function of the current I applied across the junction. The CPR function $I_J(\varphi)$ is related to I and I_L by

$$I = I_J(\varphi) + I_L = I_J\left(\frac{2\pi\Phi}{\Phi_0}\right) + \frac{\Phi}{L}, \tag{8.8}$$

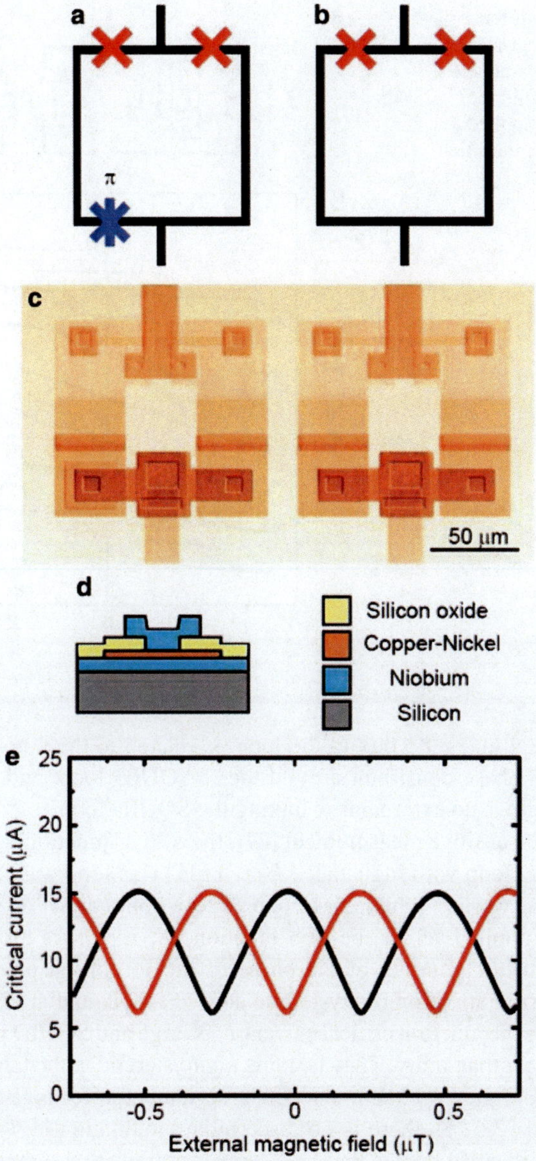

Fig. 8.11 *Complementary dc-SQUIDs*. (**a**) Schematic of a complementary dc-SQUID employing two conventional Josephson junctions (*red crosses*) and a π-junction (*orange star*). (**b**) Schematic of a conventional dc-SQUID used as a reference device. (**c**) An SEM micrograph of the fabricated dc-SQUIDs. The ferromagnetic layer is shown in *orange*. (**d**) Schematic cross-section through an SFS π-junction. (**e**) Dependencies of the critical currents of the devices shown in (**c**) vs. the applied magnetic flux. The *red curve* related to the π-SQUID is shifted by half a period. Modulation amplitude is limited, as the factor $2LI_c \approx 0.85\ \Phi_0$. [26]

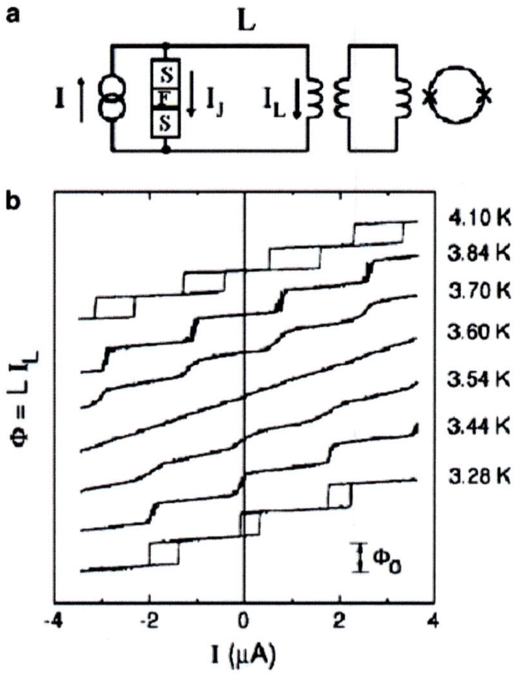

Fig. 8.12 (**a**) Circuit for measuring the current-phase relations of an SFS junction. (**b**) Magnetic flux Φ in the rf-SQUID loop vs. applied current I showing a transition from hysteretic to nonhysteretic curves as $|I_c|$ drops. Curves offset for clarity [27]

where Φ, the total magnetic flux in the loop, is related to the junction phase $\varphi = 2\pi\Phi/\Phi_0$ by the phase constraint around the rf-SQUID loop, and to $I_L = \Phi/L$ provided that there is no external flux linking the SQUID loop.

For this phase-sensitive measurement [27], the SFS π-junction was incorporated into an rf-SQUID loop with inductance $L \approx 1$ nH. This loop was fabricated in the shape of a planar washer, which was coupled to a commercial dc SQUID sensor. As current I is applied across the SFS junction, the magnetic flux in the loop is modulated due to the winding of the phase of the Josephson junction according to (8.8). The curves are strongly hysteretic at $T = 4.2$ K and at low temperatures (Fig. 8.12b) were the junction critical current I_c is large and SQUID parameter $\beta_L = 2\pi LI_c/\Phi_0$ is larger than unity. They become nonhysteretic in the temperature range from 3.7 to 3.5 K. Figure 8.13 shows in detail this temperature range for which $-1 < \beta_L < 1$. At $T = 3.59$ K, there is no discernible modulation in Φ indicating that $I_c = 0$, and we identify this as the $0 - \pi$ junction transition temperature T_π. The most striking feature of the data in Figs. 8.12 and 8.13 is that the relative phase of the modulation abruptly changes by π as the temperature is varied from above to below T_π. Due to the presence of stray residual magnetic fields ($\sim$10 mG) in the cryostat, the phase of the modulation (and hence the junction phase difference) is not in general zero for zero-applied current and varies slightly with temperature.

As can be seen in (8.8), the current-phase relation can be directly extracted from the data in Fig. 8.13 by subtracting the linear flux term and taking account of any phase shifts arising from background fields. The CPR for several temperatures near

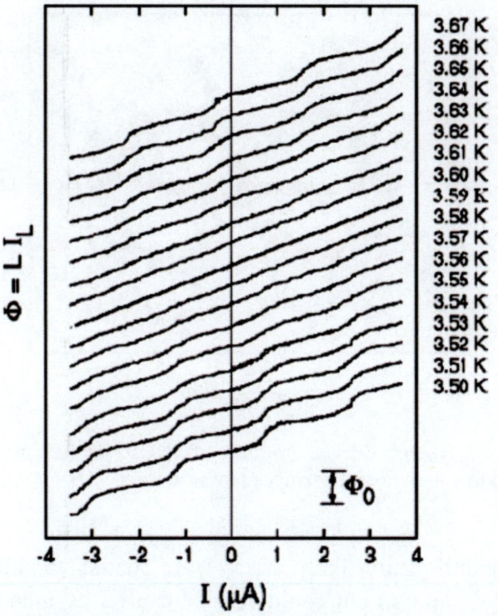

Fig. 8.13 Modulation of the magnetic flux in the rf SQUID loop as a function of current applied across the SFS junction for a series of temperatures. As the temperature is lowered, the critical current vanishes at $T = 3.59\,\mathrm{K}$, below which the modulation shifts phase by π. Curves offset for clarity [27]

T_π is shown in Fig. 8.14. The CPR has a sinusoidal form. At $T_\pi = 3.59\,\mathrm{K}$, only aperiodic fluctuations of the current are observed, which limit the resolution of our critical current measurements to $\sim$10 nA. The CPR curves for temperatures above and below T_π are out of phase by π, verifying that the critical current of the SFS Josephson junction changes sign at T_π.

One more Josephson interferometer in form of distributed $0 - \pi$-junction was realized in [29]. The SFS junction had variations in the effective barrier thickness (as it is shown schematically in Fig. 8.15a), and the ferromagnetic layer thickness was chosen near the first $0-\pi$–transition thickness d_{c1} (see Fig. 8.3). Measurements of the critical current I_c vs. the applied magnetic flux Φ threading the junction barrier reveal that the critical current distribution is not uniform across the junction. Figure 8.15b shows a series of $I_c(\Phi)$ curves in the temperature range 1.4–4.2 K for a $10 \times 10\,\mu\mathrm{m}^2$ junction. At $T = 4.2\,\mathrm{K}$, $I_c(\Phi)$ has a Fraunhofer-like shape but with nonvanishing supercurrents at the side minima. This can occur in a junction with a localized region of high critical current density. In the temperature interval 1.4–1.9 K, a minimum in the critical current is observed at zero field, indicating that regions of opposite-polarity critical current density exist in the junction.

The temperature evolution of the $I_c(\Phi)$ patterns indicates that some fraction of the junction width makes a transition from the 0 state to the π-state as the temperature is lowered, while the remaining part stays in the 0 state. The

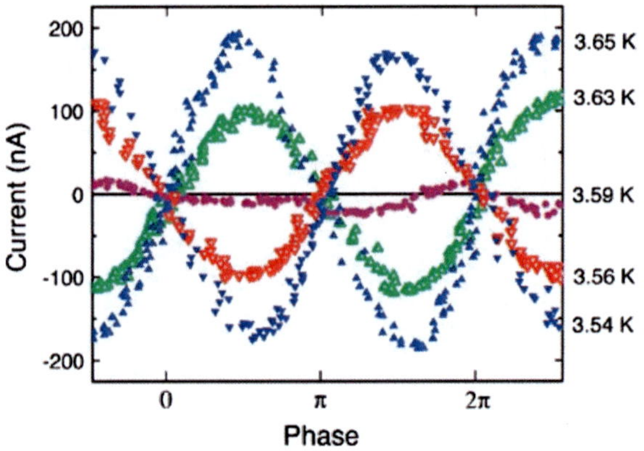

Fig. 8.14 Current-phase relation derived from the rf-SQUID modulation curves of Fig. 8.13 showing the transition to a π-state as the temperature is lowered [27]

critical current nonuniformity likely arises from spatial variations in the barrier thickness across the junction but could also be caused by inhomogeneities in the ferromagnet exchange energy or by variations in the S–F interface transparency [30]. Figure 8.15a shows the step barrier geometry deduced by fitting the measured diffraction patterns in Fig. 8.15b within the short junction approximation in which magnetic fields from the junction current are neglected. We obtain good agreement as demonstrated in Fig. 8.1c. It is not surprising that the small 6 Å step has such a dramatic effect on the diffraction patterns since close to the $0 - \pi$ transition the critical current density in our junctions changes by 1,000 A/cm^2 per 1 nm change in the barrier thickness. For the barrier profile in Fig. 8.15a and the experimental data (Fig. 8.3) for $j_c(d_F)$, we use (8.5) and (8.2) to calculate the temperature dependences of the zero-field critical currents I_{c1} of the thin narrow region, I_{c2} of the wide thick region, and I_c, the total junction critical current. We have obtained that I_{c1} is relatively constant while I_{c2} decreases and changes sign at $T \approx 2.1$ K, causing the I_c to vanish at $T_{\pi 0} \approx 1.55$ K. However, measurements plotted in Fig. 8.15b show that $I_c(\Phi = 0)$ does not go fully to zero at $T_{\pi 0}$, instead reaching a minimum value of $\approx 10\,\mu$A, suggesting that the short-junction approximation is not fully valid. This phenomenon can be explained by self-field effects that must be taken into account in finite-width $0 - \pi$-junctions. At temperatures close to $T_{\pi 0}$, spontaneous currents circulate around interfaces between 0- and π-regions to lower the total energy of the system. These circulating currents generate magnetic flux through the junction that prevents the total critical current from vanishing at any applied magnetic field. For long $0 - \pi$-junctions, this magnetic flux should be equal to $\Phi_0/2$ [31, 32].

All previous experiments described in this chapter show undoubtedly that the π-junction is a source of spontaneous currents and magnetic flux in superconducting networks. In [33], a scanning SQUID-microscope (SSM) was used to image the spontaneous zero-field currents in superconducting networks of temperature-controlled SFS π-junctions.

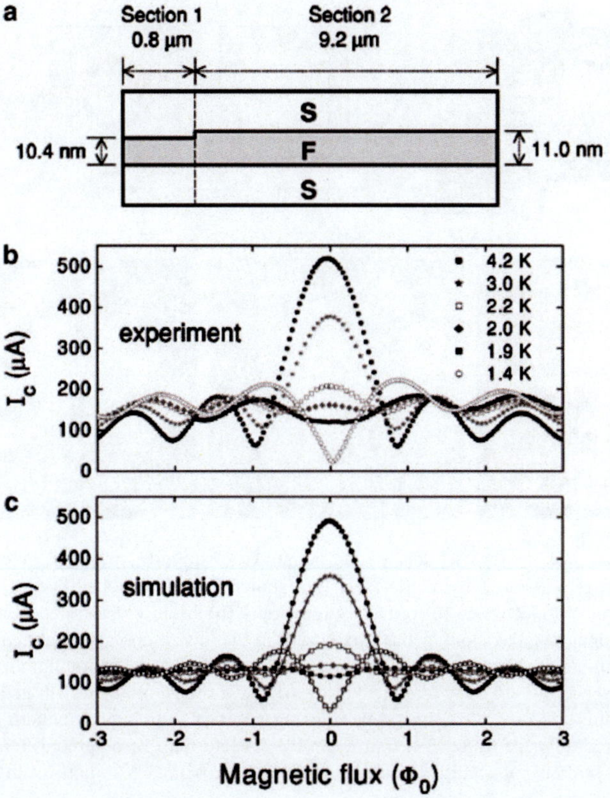

Fig. 8.15 (**a**) Stepped ferromagnetic barrier deduced from critical current vs. applied magnetic flux measurements. (**b**) Diffraction patterns at a series of temperatures showing deviations from Fraunhofer behavior at low temperatures. (**c**) Simulated diffraction patterns using the deduced ferromagnetic barrier profile [29]

As we have shown in [21] (see Figs. 8.8 and 8.9), a remarkable manifestation of the π-state of Josephson junctions is the generation of spontaneous persistent currents in superconducting loops incorporating odd numbers of π-junctions. Such currents are called spontaneous because no applied magnetic fields or power sources are required to create or sustain them. These currents arise to satisfy the fluxoid quantization conditions in the loops in response to the π-shifts across the junctions [1,31].

We fabricated two-dimensional square arrays of SFS Josephson junctions with various geometries. The periodicity of the arrays was 30 µm, with each cell having an open area of $15 \times 15\,\mu\text{m}^2$, corresponding to a geometric inductance of ~ 25 pH. In each array, some of the cells were frustrated with three π-junctions (Fig. 8.16a, right diagram in Fig. 8.16b and red cells in Fig. 8.17), whereas other cells had four π-junctions (left diagram in Fig. 8.16b and white cells in Fig. 8.17) and were therefore unfrustrated (Fig. 8.16b). In the unfrustrated cells, the ground-state

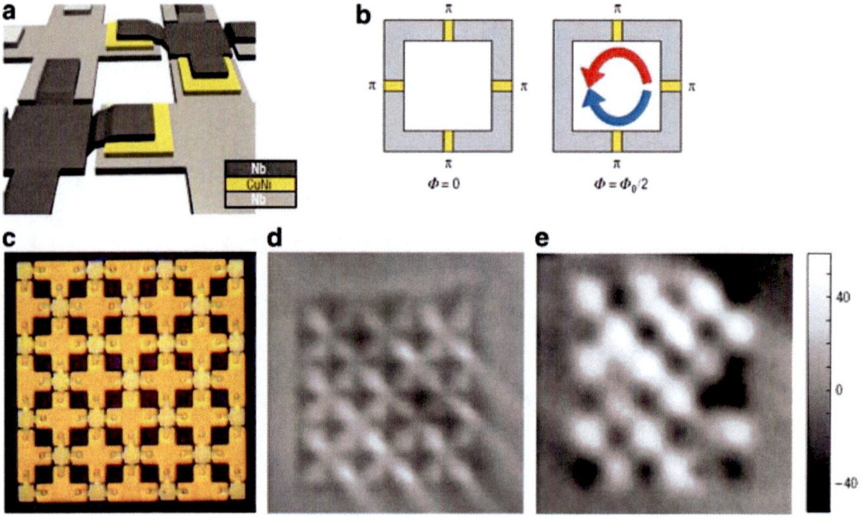

Fig. 8.16 *π-junction arrays*. (**a**) Schematic diagram of a single frustrated array cell. Ferromagnetic CuNi layers windowed by a SiO layer (not shown) are sandwiched between cross-shaped superconducting Nb electrodes to form SFS junctions. (**b**) In an unfrustrated cell with an even number of π-junctions, the spontaneous flux is zero in the lowest-energy state; in a frustrated cell with an odd number of π-junctions, spontaneous currents generate magnetic flux of order $\pm\Phi_0/2$. (**c**) Optical image of a 6×6 fully frustrated array. (**d**) SSM image of an unfrustrated array in small applied magnetic flux ($<< \Phi_0/2$ per cell) showing contrast from superconducting niobium. (**e**) SSM image of a fully frustrated 6×6 array in the π-state with zero-applied field. The vertical magnetic-field scale is not accurately calibrated but is approximately in units of mΦ_0 detected by the SQUID [33]

configuration corresponds to each junction being in its lowest energy state with no circulating current. In contrast, frustrated cells (with sufficient inductance) require a spontaneous current to maintain fluxoid quantization and minimize their energy. Therefore, although screening currents may circulate in all cells in the presence of applied magnetic fields, spontaneous currents appear only in frustrated cells.

The scanning SQUID-microscope (SSM) measured the average vertical magnetic field in a superconductor pickup loop scanned over the surface of a planar sample. The pickup loop, which was coupled to a dc-SQUID detector through a superconducting flux transformer, was fabricated on a Si wafer that was beveled to form a tip. The sensor assembly was hinged so that it rested at a small angle ($\sim 5°$) from the substrate with the tip in contact with the surface, maintaining the pickup coil at a distance of 2–3 μm from the surface. The SSM had a spatial resolution of 5–10 μm, determined by the size of the pickup loop, and a magnetic flux sensitivity of 10^{-5} Φ_0. After the residual magnetic field in the cryostat was compensated with the magnetic field applied from a Helmholtz coil to yield net zero magnetic field, images of unfrustrated arrays showed no contrast at all temperatures. However, in a magnetic field of 1–5 mG, below the threshold field at which magnetic

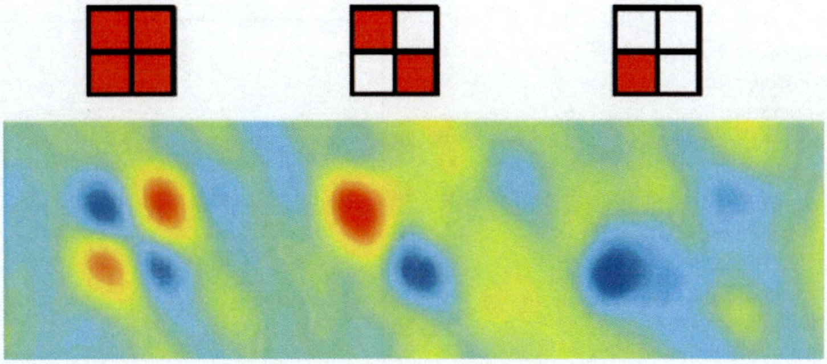

Fig. 8.17 SSM images of various 2×2 arrays in the π-state at $T = 1.5\,\text{K}$ with diagrams indicating the frustration patterns. In zero-applied magnetic field, spontaneous currents are detected only in frustrated (*dark*) cells. Different colors show opposite signs of spontaneous magnetic flux [33]

field-induced vortices enter the array, the array structure became visible because the superconducting islands screen the inductance of the SQUID pickup coil, as shown in Fig. 8.16d for a 6×6 cell array. In contrast, all π-junction arrays with frustrated cells showed spontaneous circulating currents. As an example, Fig. 8.16e shows an SSM image of a uniformly frustrated 6×6 array cooled below the π-state transition temperature in zero magnetic field. Over most of the array, the arrangement of spontaneous currents was antiferromagnetic, with the direction of the spontaneous currents alternating in adjacent array cells to produce a checkerboard magnetic flux pattern, as expected for the ground state configuration (Fig. 8.17). However, deviations from antiferromagnetic patterns were often observed because arrays could cool into metastable excited states in which one or more spontaneous currents are flipped.

We could monitor the onset of spontaneous currents in an array by taking images at a series of different temperatures. We expected such currents to onset only below the π-state transition temperature T_π of the individual junctions in the array. To estimate T_π, we measured the temperature dependence of the critical current of a single isolated SFS junction fabricated on the same chip. The critical current, obtained from the current–voltage characteristics as shown in Fig. 8.18 (upper panel), vanished at $T \approx 2.8 - 2.9\,\text{K}$, which we identify as T_π. The temperature evolution of spontaneous currents in a 6×6 checkerboard-frustrated array is shown in Fig. 8.18 (lower panel). No spontaneous currents were observed at temperatures well above T_π. The nucleation of spontaneous currents in multiply connected circuits incorporating π-junctions depends on the energy balance between the Josephson coupling energies of the junctions $E_\text{J} = I_\text{c}\Phi_0/(2\pi)$ and the magnetic field energy in the loops $E_\text{L} = LI^2/2$ associated with the circulating current I. After we cooled to $T = 2.8\,\text{K}$, just below the expected transition into the π-state, flux from spontaneous currents could be discerned but individual vortices could not be resolved. The reason is that the critical currents of SFS junctions become very small close to T_π so the ratio $E_\text{L}/E_\text{J} << 1$, making the characteristic size of the vortices much larger than a single cell. In this regime, the spontaneous flux generated by each frustrated cell is much smaller than $0.5\Phi_0$ (by roughly a factor of E_J/E_L).

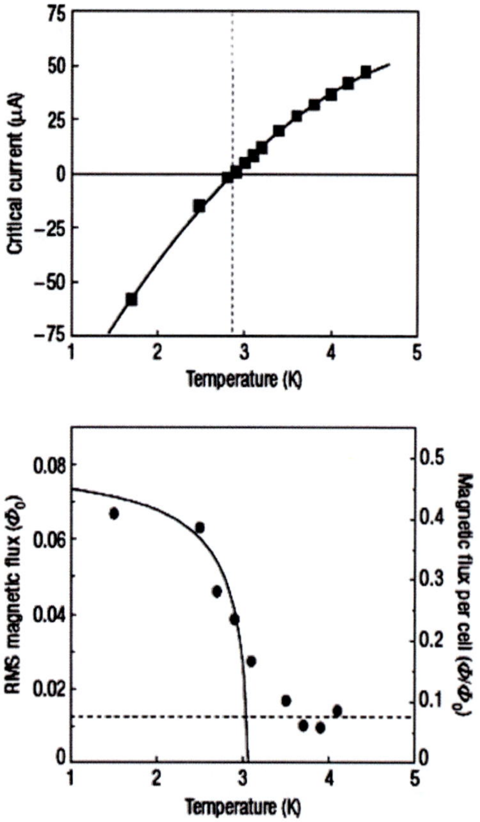

Fig. 8.18 *Temperature-driven onset of spontaneous currents. Upper panel*: temperature dependence of the critical current of a single SFS junction showing a π-junction state below $T_\pi = 2.8$ K. The sign of the critical current is indicated but cannot be determined from current–voltage characteristics alone. *Lower panel*: variance of the magnetic flux generated in the array at different temperatures (*solid dots*) superimposed on the calculated spontaneous-flux-onset curve for a loop of inductance $L = 25$ pH with a single SFS junction (*solid line*). The *dashed line* indicates the baseline set by SQUID detector noise [33]

Because in the vicinity of T_π temperature variations of order 0.1–1.0 K result in two to three orders of magnitude change in E_J, π-junction arrays are suitable for tunable vortex dynamics experiments, which were previously possible only at temperatures close to the superconducting critical temperature T_c. At temperatures well below T_π, the spontaneous currents are bigger, and the checkerboard frustration can be resolved. In Fig. 8.18 (lower panel), we plotted the r.m.s. SQUID voltage obtained by averaging the SSM signal over the array at different temperatures. There is an onset of spontaneous current that occurs at $T \approx 3$ K. For comparison, on the same graph we show the calculated temperature dependence for the onset of spontaneous magnetization in an isolated cell frustrated by a single π-junction [1]. The parameters of the junction were chosen to give an onset of flux near that observed in the array. We found that the onset of flux in the array is substantially broader than that for a single loop owing to variations in the SFS junction barrier thicknesses and corresponding spread in T_π values of individual junctions. The ratio of the Josephson coupling energy E_J to the inductive energy E_L required to generate a magnetic flux of 0.5 Φ_0 is $E_J/E_L = (I_c\Phi_0/2\pi)/\left(\Phi_0{}^2/8\pi^2 L\right) \approx 5$ at our lowest achievable temperature $T = 1.5$ K, from which we estimate that the spontaneous magnetic flux in each frustrated cell reaches 0.4 Φ_0.

8.4 Applications of Superconducting Phase Inverters

High operation speed and low energy consumption may allow the superconducting digital single flux quantum (SFQ-) circuits to outperform traditional semiconductor CMOS logic. The remaining major obstacle to high density of elements on chip is a relatively large cell size necessary to hold a magnetic flux quantum Φ_0. Inserting a Josephson π-junction in the cell is equivalent to applying flux $\Phi_0/2$, and thus makes it possible to solve this problem [25]. Moreover, using π-junctions in superconducting qubits may help to protect them from noise [34, 35]. In this section we demonstrate the operation of both classical (digital) and quantum superconducting circuits, which utilized π-phase inverters realized using SFS junctions.

In superconducting circuits, currents can flow without applying any electric field. The role of the electrostatic potential difference required to drive a current in conventional circuits is played here by a difference φ between the phases of the superconducting order parameters of the electrodes. For conventional superconducting junctions, φ is zero in the absence of current. Thus, the π-junction is a unique element, which can be applied like an intrinsic bias source and a passive phase inverter in superconducting electronics. Several ideas for the realization of a well-defined phase shift have recently been considered [36–38]. The SFQ-circuits with complementary active elements, i.e. π-junctions, were first realized using high-T_c superconductor (HTS) junctions with unconventional d-wave pairing symmetry ensuring the phase shift of π [37]. Operation of a circuit with another (passive) type of the phase shifting element based on a superconducting loop with trapped magnetic flux [36] has been demonstrated in [38]. In contrast to HTS π-junctions, the phase-shifters based on the SFS π-junctions are highly compatible with conventional low-T_c Nb fabrication technology. Moreover, in comparison with the phase-shifting loops with trapped magnetic flux [38], the circuits with the integrated SFS π-junctions do not require setting of the well-defined number of flux quanta in the loop and have more compact and simple design. The main problem of potential integration of the SFQ and quantum (qubit) circuits with the goal of efficient control and readout of the qubit is the reduction of noise, which is generated by SFQ modules [39,40]. This noise may significantly decohere the Josephson qubits, so the problem of the noise reduction requires a profound modification of the schemes and its parameter values, in order to preserve the qubit coherence and reduce the power dissipation. The noise reduction and the safe transfer of rather weak signals between different parts of integrated circuit require the implementation of Josephson junctions with relatively small values of their critical currents I_C. To ensure the Josephson phase drop necessary for the SFQ data storing, the corresponding values of the inductances should be proportionally increased ($L \propto I_C^{-1}$) and, as a consequence, the circuit area is significantly expanded, which make circuits noise sensitive. In this case, the compact circuit design can be realized by replacing the geometrically large storing inductances with relatively compact phase shifting elements. This makes the integrated circuits potentially less sensitive to external electromagnetic noise.

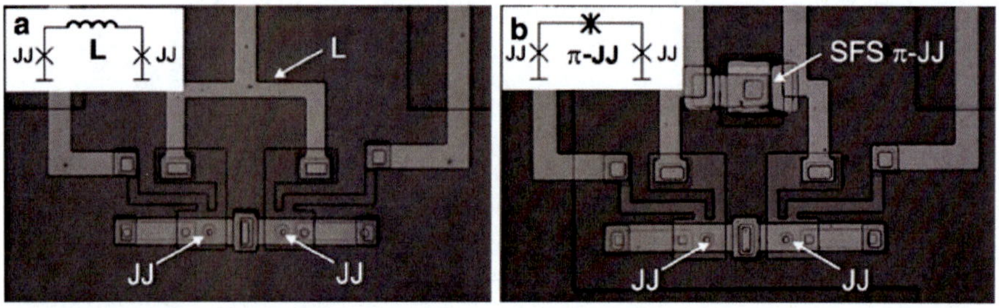

Fig. 8.19 Microphotographs of the dc-interferometers: (**a**) dc-interferometer of conventional type and (**b**) dc-interferometer with integrated SFS π-junction. The *insets* show the electric diagrams of these circuits [41]

The first realization of the SFQ circuits with integrated π-phase shifters based on the SFS Josephson junctions was carried out by us as one can see in [26, 41]. The SFQ logic circuits enable processing of information in the form of single flux quanta Φ_0, which can be stored in elementary superconducting cells including inductors and Josephson junctions. Dynamically, this information is represented by SFQ voltage pulses [42] having a quantized area $\int V(t)dt = \Phi_0$ and corresponding to the transfer of one flux quantum across a Josephson junction. The large storing inductance which is fed by dc current ensuring bistable operation in the conventional SFQ Toggle Flip Flop (TFF) circuit was replaced by the SFS π-junction adding to the TFF loop the well defined phase shift equal to π.

First, to prove the correct functioning of SFS π-junctions themselves, the standalone dc-interferometers were realized on the same chip. Microphotographs of these dc-interferometers with/without integrated π-junction are shown in the Fig. 8.19. Layout and circuit parameters of the dc-interferometer with integrated π-junction were identical to the layout and parameters of the elements forming quantizing interferometer in the TFF circuit. Josephson junctions in the interferometer were designed to be about $I_c = 20\,\mu A$. The critical current of the SFS π-junction exceeded approximately 20 times the critical current of each SIS tunnel junction used in the dc-interferometer.

Small value of a dimensionless SQUID parameter $\beta_L < 1$ provided sufficiently large modulation depth of the interferometer's critical current versus externally applied magnetic field.

Figure 8.20 presents two traces of the measured voltage-flux characteristics of the dc-interferometer without and with integrated π-junction. In this experiment, we fed through the interferometer inductance a ramp current I_{sweep}, which served as an external flux source. The interferometer junctions were biased by additional dc current source. We measured the voltage peaks across the interferometer, which correspond to maximum suppression of the interferometer critical current at $\Phi = (2m + 1)\,\Phi_0/2$, where m is integer number. The shape of the voltage-flux characteristic shown in the Fig. 8.20a is close to the theoretically predicted relation for the two-junctions dc-interferometer. The dc-interferometer with integrated π-junction has the maximum suppression of its critical current at integer-valued

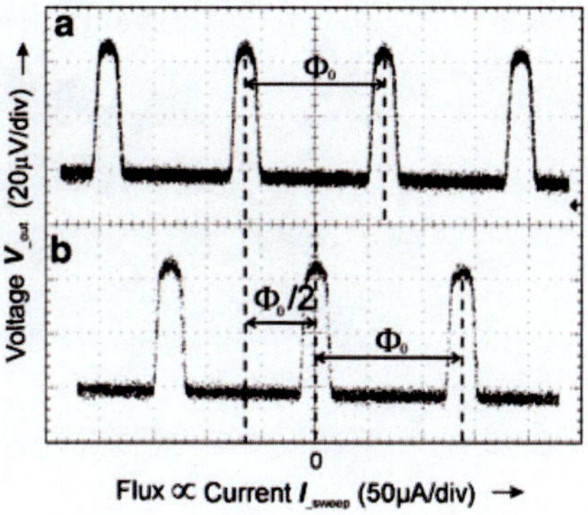

Fig. 8.20 Voltage-flux characteristics of the dc-interferometers: (**a**) of conventional interferometer and (**b**) of the dc-interferometer with SFS π-junction integrated in the interferometer loop [41]

flux, $\Phi = m\Phi_0$. The voltage-flux characteristic in Fig. 8.20b is modified due to the presence of correctly functioning π-phase shifter and the resulting curve is the complementary to the previous one, i.e. $\Phi \rightarrow \Phi + \Phi_0/2$.

Fabrication of the integrated TFF circuits has started with preparation of conventional SFQ-part [43]. The SFS junctions were placed between Nb-wiring nodes and they were co-fabricated directly after the wiring layer definition. The SFS fabrication process includes the standard technological steps described above. The thickness of the ferromagnetic layer was chosen so that the critical current of the SFS π-junction was sufficiently larger than the critical current of the conventional SIS junctions in the circuit. Only upon this condition does the SFS junction operate as a static phase inverter, and the phase difference across it does not change during the phase evolution in the SFQ cell. The critical current density of the Nb–CuNi–Nb π-junction has a sharp maximum $\sim$1,000 A cm^{-2} at $d_F = 12$–13 nm (see Fig. 8.3). This F-layer thickness is very close to the critical thickness $d_{c1} = 11$ nm of the transition from the conventional to the π-state. To realize SFS junctions, which have remained reliably in the π-state in the temperature range $3\,\text{K} < T < 5\,\text{K}$, we have chosen a CuNi layer thickness of 13–15 nm to obtain a critical current density j_c larger than 100 A cm^{-2}. The cross section of an externally shunted Josephson tunnel junction integrated with an SFS junction is shown in Fig. 8.21.

The toggle flip-flop circuit is a one of the key modules of any digital circuit based on the SFQ logic [42, 44]. We have designed and fabricated the TFF circuit, where the large quantizing inductance has been replaced by SFS π-junction (see schematic and image of the TFF core cell in Fig. 8.22a, b).

Our TFF consists of two interferometer loops, one acting as a signal splitter (junctions J_1 and J_2), while the other acting as a memory cell for the TFF states (junctions J_3 and J_4 and π-junction). The TFF has two stable equivalent states, which are differ by direction of currents circulating in the interferometers loops. In the conventional TFF circuit, the equivalence of its states is set up by dc current through the storing inductance providing phase drop equal to π. The TFF with integrated π-junction does not require such additional asymmetric bias and due to

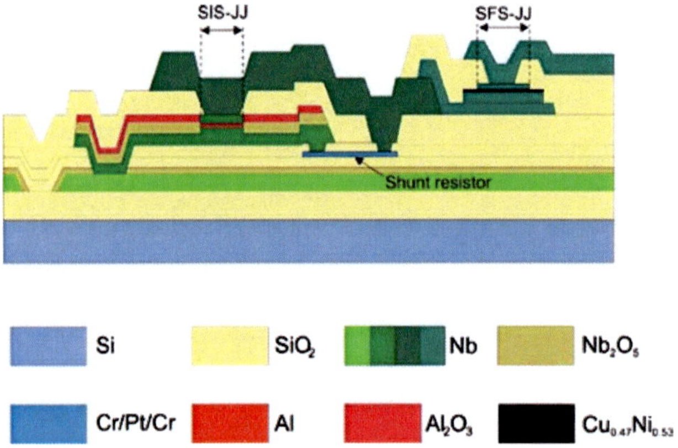

Fig. 8.21 Cross-section (not in scale) of externally shunted Josephson tunnel junction and integrated SFS π-junction [41]

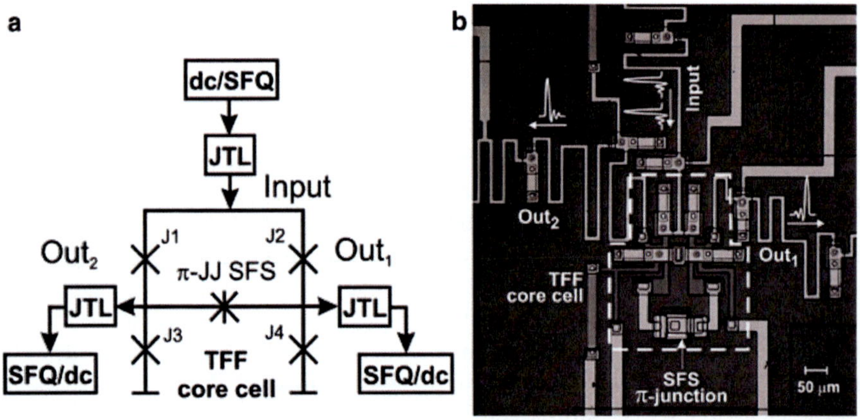

Fig. 8.22 (**a**) Schematic of the TFF circuit with the integrated SFS π-junction junction and (**b**) optical microphotograph of the circuit core [41]

inherent symmetry can stably operate in a wider parameter range. The functionality of the TFF can be considered as a simultaneous switching of the junctions of two decision elements compound by pair of Josephson junctions (junctions J_1, J_3 and J_2, J_4). If the direction of the circulating current in the storing interferometer is anticlockwise and in the interferometer with the junctions J_1 and J_2 and π-junction is clockwise, the junction J_3 in the first (J_1, J_3) decision element and the junction J_2 in the second (J_2, J_4) decision element are forced to switch then SFQ pulse applied. As a result, the directions of the circulating currents in the interferometer loops change and the junctions J_1 and J_4 are forced to switch by the next incoming SFQ pulse. So, in case of the correct circuit functionality, the junctions $J_1 - J_4$ switches at twice lower frequency in comparison with the input frequency, thus the TFF acts as a frequency divider.

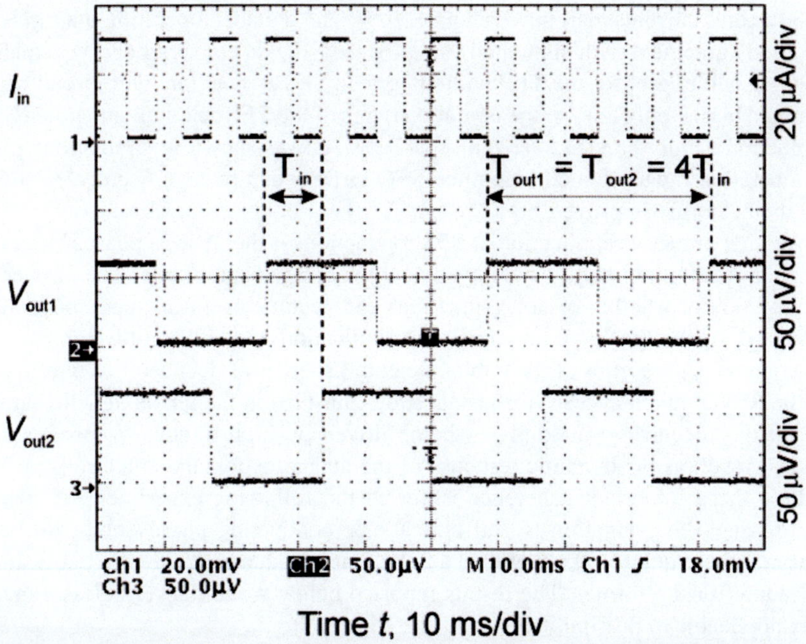

Fig. 8.23 Time traces of the TFF complementary output signals (V_{out1}, V_{out2}). Each of the output traces V_{out1} and V_{out2} has a period four times longer than that of the input signal I_{in}. The mutual time shift of the output signals V_{out1} and V_{out2} corresponds to exactly one period of the input signal I_{in} [41]

The entire integrated circuit consists of 36 Josephson junctions. The input of the TFF circuit is connected to the dc/SFQ converter circuit and the TFF outputs to the SFQ/dc converter circuits [44]. The TFF toggles between its two states under the application of an SFQ pulse. This SFQ pulse is generated by a dc/SFQ converter and is transferred to TFF through a Josephson transmission line (JTL) [42]. Figure 8.23 shows the time traces of the input and output signals confirming correct operation of the TFF circuit. The drive signal applied to dc/SFQ converter (trace I_{in}) creates one SFQ pulse per period of the drive signal, time referenced to its rising ramps. If the first SFQ pulse coming from dc/SFQ converter switches the junctions J_3 and J_2, the second one switches J_4 and J_1. The SFQ pulses generated by J_3 and J_4 are transferred via individual JTLs to SFQ/dc converter circuits and converted into the voltage signals (traces V_{out1}, V_{out2}). Each of the output signal traces V_{out1} and V_{out2} has a period four times longer than that of the input drive signal I_{in}. This behavior originates from the pulse rate division by two sequentially connected TFFs. The first one is the TFF of the novel design with integrated SFS π-junction and the second one is the TFF of conventional design included in the respective SFQ/dc converter. As expected, the mutual time shift of the output signals V_{out1} and V_{out2} corresponds to exactly one period of the input signal I_{in}. The tested circuit including TFF cell and dc/SFQ and SFQ/dc converters demonstrated correct functionality with the bias currents margins of about ±20%. The obtained operation margin is in the good agreement with simulated value for the SFQ/dc converter and yet smaller than one can expect for the TFF with integrated SFS π-junction. Our circuit has a common bias supply line, so the operation range of the TFF cell with integrated SFS π-junction cannot be characterized separately. As was shown in simulations [37], the operation range of such integrated SFQ circuits can be significantly extended and their stability improved.

Another attractive application of SFS π-junctions is their use as phase shifters in coherent quantum circuits realizing superconducting quantum bits. The answer to the question of whether or not π-junctions can become useful in superconducting circuits designed for quantum computing applications depends on their impact on the coherence properties of the qubits. Potential sources of decoherence introduced by the SFS π-junctions can, for instance, be spin-flips in the ferromagnetic barrier [45], either occurring randomly or being driven by high-frequency currents and fields, as well as the dynamic response of the magnetic domain structure [46]. We address these important coherence issues in the following experiment, in which we use an SFS π-junction to self-bias a superconducting phase qubit. We have chosen a phase qubit [47] rather than a flux qubit [48] due to the simpler fabrication procedure for the former. The results reported below would, nevertheless, remain fully applicable to flux qubits.

A phase qubit [47] consists of a single Josephson junction embedded in a superconducting loop. It is magnetically biased close to an integer number of flux quanta in the loop. At such a bias, the potential energy of the qubit exhibits an asymmetric double-well potential, whereas two quantized energy eigenvalues of the phase localized inside the shallow well are used as the logical qubit states $|0\rangle$ and $|1\rangle$. Figure 8.24a shows a circuit schematic and Fig. 8.24b a micrograph of the tested sample. Here, the π-junction is connected in series to the phase qubit's tunnel junction. Coherent qubit operation is demonstrated by the data reported in Fig. 8.25a, showing Rabi oscillation of the excited qubit state population probability

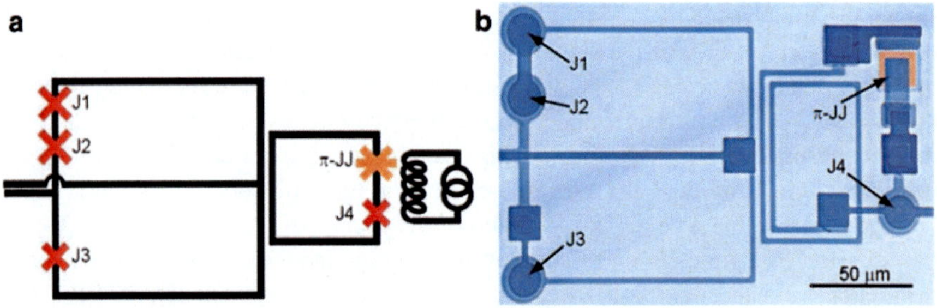

Fig. 8.24 Self-biased phase qubit. (**a**) Schematic of a phase qubit circuit used to test the decoherence properties of the SFS π-junction. The qubit is realized by the central loop with embedded conventional and π-Josephson junctions. The larger loop to its left is a dc-SQUID for qubit readout. To the right of the qubit is a weakly coupled flux bias coil. (**b**) Scanning electron microscope picture of the realized phase qubit employing a π-junction in the qubit loop. The flux bias coil is not shown [26]

in dependence on the duration of a resonant microwave pulse. The oscillations exhibit a decay time of about 4 ns, which is a typical value reachable in samples fabricated using similar fabrication processes [49].

To find out whether the π-junction does introduce additional decoherence, a conventional phase qubit without a π-junction was fabricated on the same wafer. As shown in Fig. 8.25b, this reference qubit shows a nearly identical

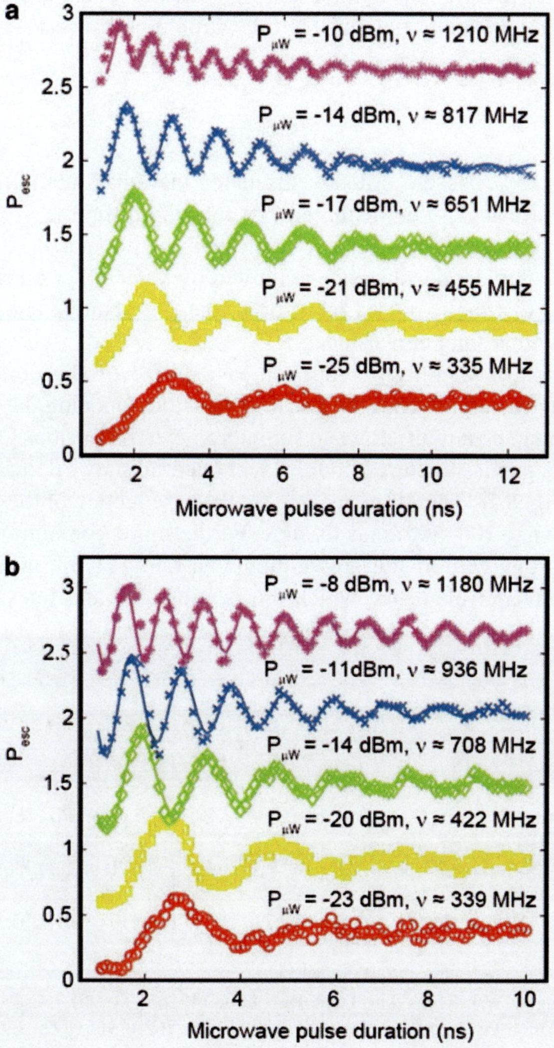

Fig. 8.25 *Rabi oscillations between the ground and the excited qubit states resulted from resonant microwave driving*: (**a**) observed in the phase qubit with embedded π-junction and (**b**) a conventional phase qubit made on the same wafer as a reference. Each data set was taken using the indicated microwave power as delivered by the generator, giving rise to a change in the coherent oscillation frequency as expected for Rabi oscillation [26]

decay time for Rabi oscillations. Thus, we obtained experimentally that the decay time limited by π-junction is significantly larger than 4 ns. We compared the measured decoherence time with the theoretical predictions [50]. We assume here an overdamped SFS π-junction having a normal resistance of $R \approx 500\,\mu\Omega$ and a critical current $I_C \approx 50\,\mu A$. In our case, the qubit level splitting $\Delta >> 2eI_CR$, where $\Delta \approx h^*13.5$ GHz, h is the Plank's constant and e is the elementary charge. Here, the energy $2eI_CR \approx h^*12$ MHz is associated with characteristic Josephson frequency of our SFS π-junction. Simplifying the expression for the relaxation time [50] in this limit, we could theoretically estimate the relaxation time τ_{relax} as

$$\tau_{\text{relax}} = \frac{\Delta}{2I_{\text{C,tun}}R} \approx 2\ \text{ns}. \tag{8.9}$$

Here, $I_{C,tun} \approx 2\,\mu A$ is the critical current of the small tunnel qubit junction. However, the estimated value of the energy relaxation time is of the same order as the measured decoherence time of reference qubit without an SFS π-junction, so we can suppose that the decoherence is limited by some other mechanism in both qubits. We note that the expected relaxation time (8.9) can be enhanced by using SFS junctions with a smaller resistance R.

As an outlook, a significant reduction in the size of the demonstrated SFS π-phase shifters for digital circuits is readily possible, opening the way to scaling superconducting logic circuits down to submicrometer dimensions [25]. The visualization of the magnetic structure of our F-layer material shows domain sizes smaller than 100 nm (Fig. 8.5). Therefore, we believe that a reduction of the junction planar dimensions down to 300–500 nm is feasible. Furthermore, combining the high-j_C π-junction technology with in-situ-grown tunnel barriers [51, 52] may open the way toward active inverter elements, which are in great demand for superconducting electronics.

References

1. L.N. Bulaevskii, V.V. Kuzii, A.A. Sobyanin, Pis'ma Zh. Eksp. Teor. Fiz. **25**, 314 (1977) [JETP Lett. **25**, 290 (1977)]
2. A.I. Buzdin, L.N. Bulaevskii, S.V. Panjukov, Pis'ma Zh. Eksp. Teor. Fiz. **35**, 147 (1982) [JETP Lett. **35**, 178 (1982)]
3. A.I. Buzdin, B. Vujicic, M.Yu. Kupriyanov, Zh. Eksp. Teor. Fiz. **101**, 231 (1992) [JETP **74**, 124 (1992)]
4. A.I. Buzdin, Rev. Mod. Phys. **77**, 935 (2005)
5. M.J.M. de Jong, C.W.J. Beenakker, Phys. Rev. Lett. **74**, 1657 (1995)
6. A.I. Larkin, Yu.N. Ovchinnikov, Sov. Phys. JETP. **20**, 762 (1965) [Zh. Eksp. Teor. Fiz. **47**, 1136 (1964)]
7. P. Fulde, R.A. Ferrel, Phys. Rev. A **135**, 550 (1964)
8. E.A. Demler, G.B. Arnold, M.R. Beasley, Phys. Rev. B **55**, 15 174 (1997)
9. A.V. Andreev, A.I. Buzdin, R.M. Osgood III, Phys. Rev. B **43**, 10124 (1991)
10. A. Rusanov, R. Boogaard, M. Hesselberth, H. Sellier, J. Aarts Phys. C **369**, 300 (2002)

11. V.V. Ryazanov, Usp. Fiz. Nauk **169**, 920 (1999) [Phys.-Usp. **42**, 825 (1999)]
12. A.V. Veretennikov, V.V. Ryazanov, V.A. Oboznov, A.Yu. Rusanov, V.A. Larkin, J. Aarts, Phys. B **284–288**, 495 (2000)
13. V.V. Ryazanov, V.A. Oboznov, A.Yu. Rusanov, A.V. Veretennikov, A.A. Golubov, J. Aarts, Phys. Rev. Lett. **86**, 2427 (2001)
14. V.A. Oboznov, V.V. Bol'ginov, A.K. Feofanov, V.V. Ryazanov, A.I. Buzdin, Phys. Rev. Let. **96**, 197003 (2006)
15. I.S. Veshchunov, V.A. Oboznov, A.N. Rossolenko, A.S. Prokofiev, L.Ya. Vinnikov, A.Yu. Rusanov, D.V. Matveev. Pis'ma Zh. Eksp. Teor. Fiz. **88**, 873 (2008) [JETP Lett. **88**, 758 (2008)]
16. V.V. Ryazanov, V.A. Oboznov, A.S. Prokofiev, V.V. Bolginov, A.K. Feofanov, J. Low Temp. Phys. **136**, 385 (2004)
17. K. Levin, D.L. Mills, Phys. Rev. B **9**, 2354 (1974)
18. R.W. Houghton, M.P. Sarachik, J.S. Kouvel, Phys. Rev. Lett. **25**, 238 (1970)
19. D.A. Ivanov, Ya.V. Fominov, Phys. Rev. B **73**, 214524 (2006)
20. A.I. Buzdin, Pis'ma Zh. Eksp. Teor. Fiz. **42**, 283 (1985) [JETP Lett. **42**, 350 (1985)]
21. V.V. Ryazanov, V.A. Oboznov, A.V. Veretennikov, A.Yu. Rusanov. Phys. Rev. B **65**, 020501 (2001)
22. Tsang Von-Tien, T. Van Duzer, J. Appl. Phys. **46**, 4573 (1975)
23. K.K. Likharev, *"Dynamics of Josephson Junctions and Circuits"*, Chap. 8 (Gordon and Breach Science, New York, 1986)
24. E. Terzioglu, M.R. Beasley, IEEE Trans. Appl. Supercond. **8**, 48 (1998)
25. A.V. Ustinov, V.K. Kaplunenko, J. Appl. Phys. **94**, 5405 (2003)
26. A.K. Feofanov, V.A. Oboznov, V.V. Bol'ginov, J. Lisenfeld, S. Poletto, V.V. Ryazanov, A.N. Rossolenko, M. Khabipov, D. Balashov, A.B. Zorin, P.N. Dmitriev, V.P. Koshelets, A.V. Ustinov, Nature Phys. **6**, 593 (2010)
27. S.M. Frolov, D.J. Van Harlingen, V.A. Oboznov, V.V. Bolginov, V.V. Ryazanov, Phys. Rev B **70**, 144505 (2004)
28. J.R. Waldram, J.M. Lumley, Rev. Phys. Appl. **10**, 7 (1975)
29. S.M. Frolov, D.J. Van Harlingen, V.V. Bolginov, V.A. Oboznov, V.V. Ryazanov, Phys. Rev. B **74**, 020503 (2006)
30. N.G. Pugach, M.Yu. Kupriyanov, A.V. Vedyayev, C. Lacroix, E. Goldobin, D. Koelle, R. Kleiner, A.S. Sidorenko, Phys. Rev. B **80**, 134516 (2009)
31. L.N. Bulaevskii, V.V. Kuzii, A.A. Sobyanin, Solid State Commun. **25**, 1053 (1978)
32. J.R. Kirtley, K.A. Moler, D.J. Scalapino, Phys. Rev. B **56**, 886 (1997)
33. S.M. Frolov, M.J.A. Stoutimore, T.A. Crane, D.J. Van Harligen, V.A. Oboznov, V.V. Ryazanov, A. Ruosi, C. Granata, M. Russo, Nature Physics **4**, 32 (2008)
34. L.B. Ioffe, V.B. Geshkenbein, M.V. Feigelman, A.L. Fauchere, G. Blatter, Nature **398**, 679 (1999)
35. G. Blatter, V.B. Geshkenbein, L.B. Ioffe, Phys. Rev. B **63**, 174511 (2001)
36. J.B. Majer, J.R. Butcher, J.E. Mooij, Appl. Phys. Lett. **80**, 3638 (2002)
37. T. Ortlepp, T. Ariando, O. Mielke, C.J.M. Verwijs, K.F.K. Foo, H. Rogalla, F.H. Uhlmann, H. Hilgenkamp, Science **312**, 1495 (2006)
38. D. Balashov, B. Dimov, M. Khabipov, Th. Ortlepp, D. Hagedorn, A.B. Zorin, F.Im. Buchholz, F.H. Uhlmann, J. Niemeyer, IEEE Trans. Appl. Supercond. **17**, 142 (2007)
39. D.V. Averin, K. Rabenstein, V.K. Semenov, Phys. Rev. B **73**, 094504 (2006)
40. A. Fedorov, A. Shnirman, G. Shön, A. Kidiyarova-Shevchenko, Phys. Rev. B **75**, 224504 (2007)
41. M.I. Khabipov, D.V. Balashov, A.B. Zorin, V.A. Oboznov, V.V. Bolginov, A.N. Rossolenko, V.V. Ryazanov, Supercond. Sci. Technol., **23**, 045032 (2010)
42. K.K. Likharev, V.K. Semenov, IEEE Trans. Appl. Supercond. **1**, 3 (1991)
43. R. Dolata, M.I. Khabipov, F-Im. Buchholz, W. Kessel, J. Niemeyer, Appl.Supercond. 1995 Inst. Phys. Conf. Ser. 148 **2**, 1709–12 (1995)
44. M.I. Khabipov, R. Dolata, F-Im. Buchholz, W. Kessel, J. Niemeyer, Supercond. Sci. Technol. **9**, 822 (1996)

45. R. Melin, Europhys. Lett. **69**, 121 (2005)
46. S. Hikino, M. Mori, S. Takahashi, S. Maekawa, J. Phys. Soc. Jpn. **77**, 053707 (2008)
47. R.W. Simmonds et al, Phys. Rev. Lett. **93**, 077003 (2004)
48. I. Chiorescu, Y. Nakamura, C.J.P.M. Harmans, J.E. Mooij, Science **299**, 1869 (2003)
49. J. Lisenfeld, A. Lukashenko, M. Ansmann, J.M. Martinis, A.V. Ustinov, Phys. Rev. Lett. **99**, 170504 (2007)
50. T. Kato, A.A. Golubov, Y. Nakamura, Phys. Rev. B **76**, 172502 (2007)
51. M. Weides et al. Phys. Rev. Lett. **97**, 247001 (2006)
52. A.A. Bannykh et al. Phys. Rev. B **79**, 054501 (2009)

Chapter 9
Point-Contact Study of the Rare-Earth Nickel-Borocarbide $R\text{Ni}_2\text{B}_2\text{C}$ (R = Y, Dy, Ho, Er, Tm, Lu) Superconductors

Yu.G. Naidyuk and I.K. Yanson

Abstract A brief overview of the achievements in point-contact (PC) study of the superconducting (SC) gap and electron–phonon (boson) interaction (EP(B)I) in the rare-earth nickel-borocarbide compounds $R\text{Ni}_2\text{B}_2\text{C}$ (R = Y, Dy, Ho, Er, Tm, Lu) is presented. Compounds with R = Y, Lu are non-magnetic with the highest critical temperature $T_c \approx 16\,\text{K}$, while the remaining are magnetic. For the latter, the energy scales of antiferromagnetic (AF) and SC order, measured by the Neel temperature T_N, and T_c, vary against R over a wide range between ~ 1 and 10 K. The SC gap, its temperature and magnetic field dependence were studied by PC Andreev reflection (AR) spectroscopy corroborating a multiband structure of the title compounds. For R = Er ($T_N \approx 6\,\text{K} < T_c \approx 11\,\text{K}$) directional PC AR study gives evidence for the two-band (two-gap) nature of SC-ty with a distinct anisotropic decrease of the SC gaps below T_N. The SC gap in R = Ho ($T_N \approx 5.2\,\text{K} < T_c \approx 8.5\,\text{K}$) exhibits below $T^* \approx 5.6\,\text{K}$ a standard single-band BCS-like dependence vanishing above T^*, where a specific magnetic ordering starts to play a role. For R = Tm ($T_N \approx 1.5\,\text{K} < T_c \approx 10.5\,\text{K}$), a decrease in the SC gap is observed by approaching T_N, while for R = Dy ($T_N \approx 10.5\,\text{K} > T_c \approx 6.5\,\text{K}$) the SC gap has a BCS-like dependence in the AF state. PC spectroscopy of EP(B)I spectral function displays both phonon and crystalline-electric-field (CEF) excitation maxima (the latter between 3 and 10 meV) for R = Dy, Ho, Er and Tm, while a dominant phonon maximum around 12 meV and 8.5 meV, correspondingly, is characteristic for R = Y and Lu. Additionally, non-phonon and non-CEF maxima are observed near 3 meV in R = Ho and 6 meV in R = Dy. Specific features of the SC gap behaviour and details of the PC EP(B)I spectra in the mentioned superconductors are discussed.

Y.G. Naidyuk (✉) · I.K. Yanson
B. Verkin Institute for Low Temperature Physics and Engineering, National Academy of Sciences of Ukraine, 47 Lenin ave., 61103, Kharkiv, Ukraine
e-mail: naidyuk@ilt.kharkov.ua

A. Sidorenko (ed.), *Fundamentals of Superconducting Nanoelectronics*, NanoScience and Technology, DOI 10.1007/978-3-642-20158-5_9,

9.1 Introduction

The RNi_2B_2C ($R = Y$, Dy, Ho, Er, Tm, Lu) family of rare-earth nickel-borocarbide compounds is extremely interesting from a perspective of research of fundamental aspects superconductivity and its interplay with magnetic order [1]. These compounds have been attracting much attention since mid 1990s due to their relatively high critical temperature T_c. Superconducting (SC) properties of RNi_2B_2C often exhibit unconventional behaviour. Superconductivity and magnetic order are found to be competing against each other in some of these materials. Thus, these compounds have been described as "a toy box for solid-state physicists" [2]. As it was reported in [3], a multiband scenario is required to properly describe the anomalous temperature dependence of the upper critical field of $R = Y$ and Lu. This concept of strongly coupled two-band superconductivity has been widely used to describe properties of tremendous MgB_2 and nowadays of iron-oxypnictides superconductors. The extend of manifestation of mentioned extraordinary properties of RNi_2B_2C strongly depends on stoichiometric composition or atomic disorder in the samples. Thus, continuous progress in the synthesis of high-quality single crystals was observed and has given a deeper understanding of their fundamental physics during the last decade. This applies also to studying of the nature of Cooper pairing and attractive interaction along with competition of superconducting and magnetic ordered states. In spite of numerous experiments undertaken to study these compounds, however, there is still a room for more detailed investigation.

The study of directional behaviour, temperature and magnetic field dependence of the SC gap can provide an insight into the SC ground state of RNi_2B_2C. This work can be done in the most direct way by point-contact (PC) [4], scanning tunneling or photoemission spectroscopy. The last method has still a low resolution. The tunneling spectroscopy, however, is too sensitive to the surface or even upper layers condition. Therefore, nickel borocarbides are mostly studied using, the remaining one, namely PC spectroscopy. There are more than 20 papers in this field [5–28]. Besides, PC spectroscopy provides a straightforward information as to the PC electron–phonon(boson) interaction (EP(B)I) function $\alpha^2_{PC} F(\varepsilon)$ [4], which can be a test for the phonon-mediated superconductivity. In this work, we provide a brief overview of the efforts in PC studies of the SC gap and EP(B)I spectral function in rare-earth nickel-borocarbide superconductors RNi_2B_2C ($R = Y$, Dy, Ho, Er, Tm, Lu) focusing most attention on the recent papers [17–28].

9.2 Experimental

Mostly, we review in this article data obtained by studying single crystals. For $R =$ Y and Lu c-axis oriented films were also measured [17, 28]. PC preparation and details of the measurement technique are given in the corresponding papers [5–28].

The SC gap was studied by measuring $\mathrm{d}V/\mathrm{d}I(V)$ dependences of PCs between a normal metal and the superconducting borocarbides, showing Andreev reflection (AR) features. The SC gap magnitude was calculated from $\mathrm{d}V/\mathrm{d}I(V)$ using in most cases the standard 1-D BTK theory [29] and a similar theory (in the case of Er and Lu), which includes the pair breaking effect of magnetic impurities [30]. This is important for the mentioned borocarbides due to the presence of magnetic moments in rare-earth ions. Comprehensive analysis of the cases beyond simple 1-D BTK model is given in the very recent review on PC spectroscopy of multiband superconductors [31].

According to the PC spectroscopy theory (see [4], Sect. 3.2), the second derivative of the I–V curve of $R^{-1}\mathrm{d}R/\mathrm{d}V$ type is directly proportional to the EPI spectral function:

$$\frac{1}{R}\frac{\mathrm{d}R}{\mathrm{d}V}(\mathrm{eV}) = \frac{8ed}{3\hbar v_\mathrm{F}}\alpha_\mathrm{PC}^2 F(\varepsilon)\,|_{\varepsilon=\mathrm{eV}}\ , \tag{9.1}$$

where e is the electron charge, d is the PC diameter and v_F is the Fermi velocity. This equation provides an opportunity to study EPI and other bosonic collective excitations simultaneously in the title compounds with the same PCs. It also allows estimate the EPI coupling parameter λ:

$$\lambda = 2\int_0^{\omega_\mathrm{max}} \alpha_\mathrm{PC}^2 F(\varepsilon)\varepsilon^{-1}\mathrm{d}\varepsilon. \tag{9.2}$$

9.3 Point-Contact Andreev-Reflection Spectroscopy of the Superconducting Gap

The SC gap gives rise to the appearance of minima on the $\mathrm{d}V/\mathrm{d}I(V)$ characteristic of a normal metal–superconductor contact at $T << T_\mathrm{c}$ symmetrically placed at $V \approx \pm\Delta/\mathrm{e}$ (see inset in Fig. 9.1). Measured $\mathrm{d}V/\mathrm{d}I$ curves of SC nickel borocarbides exhibit a pair of minima as in the case of a single gap. Therefore, single gap approach is usually used (mainly in the more recent papers) to fit experimental data (see, e.g., [5–11]). As shown in inset in Fig. 9.1, the one-gap fit describes experimental $\mathrm{d}V/\mathrm{d}I(V)$ curve for R = Lu insufficiently, and the correspondence between the curves is much better by applying a two-gap fit. The same was found by fitting of $\mathrm{d}V/\mathrm{d}I(V)$ characteristic measured for R = Lu, Er and Tm single crystals [18, 19, 23, 26, 28]. Moreover, for R = Lu [18, 19] and Er [23, 26] a two-gap fit was applied for $\mathrm{d}V/\mathrm{d}I(V)$ of PCs established along two main crystallographic directions demonstrating applicability of two band(gap) model in both cases (Fig. 9.2).

It turned out that the critical temperature T_c, corresponding to the small gap in $LuNi_2B_2C$ in the BCS extrapolation is 10 K in the ab-plane and 14.5 K in the c-direction (see Fig. 9.2). This unusual temperature dependence of this gap by approaching bulk T_c suggests a weak interband coupling. For the large gap, T_c

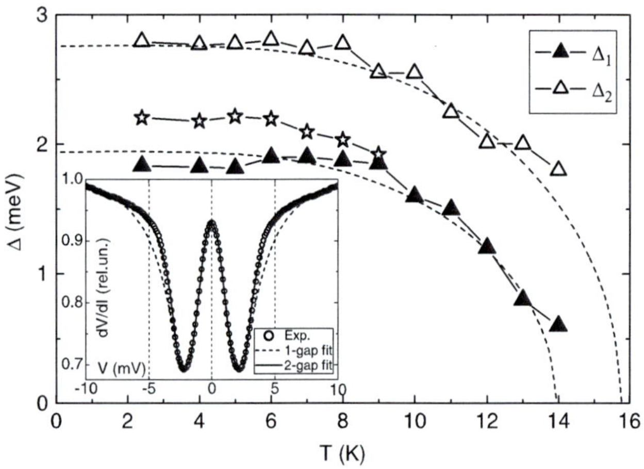

Fig. 9.1 SC gaps behaviour in $LuNi_2B_2C$ established by two gaps BTK fit of PC dV/dI curves for c-axis-oriented film [28]. The fit is done with $Z \approx 0.5$, weight factor $w \approx 0.5$ and $\Gamma = 0$. *Stars* denote minimum position in dV/dI. *Two dashed lines* demonstrate the BCS-like behaviour. *Inset* shows one and two-gap fit for the dV/dI curve at the lowest temperature 2.4 K

coincides with the bulk $T_c = 16.8$ K and absolute gap values extrapolated to $T = 0$ are about 3 meV in both orientations. In the c-direction, the contributions to the conductivity from small and large gaps remain almost identical up to 10–11 K, while the contribution w from the small gap is insignificant and decreases rapidly with increasing of T in the ab-plane [18, 19]. In Figs. 9.1 and 9.2b, it is shown that the gap value and its behaviour are similar for both c-axis oriented film and c-direction in the single crystal, what proves the obtained data.

Noteworthy, at least five parameters (two gaps, "barrier strength" Z, "broadening" Γ and weight factor w showing a relative contribution, e.g., of the small gap to dV/dI) are used by two-gap approach for the fitting procedure, which gives room for its variation. Therefore, to get reasonable data, $dV/dI(V)$ curves should demonstrate clear AR structure with expressed minima (see, e.g., the inset of Fig. 9.1) as well as the absence of non-AR features such as spikes, bumps and other irregularities usually accompanying measured $dV/dI(V)$.

The one-gap fit reasonably describes $dV/dI(V)$ for $R =$ Ho [5, 22] as well as for $R =$ Dy [11]. The obtained $\Delta(T)$ has a BCS-like temperature dependence in both cases, but $\Delta(T)$ vanishes at $T^* \approx 5.6$ K for $R =$ Ho, well below the bulk critical temperature of $T_c \approx 8.5$ K. The AR minimum in $dV/dI(V)$ is drastically weakened between T^* and T_c, what makes impossible any BTK fit to obtain the SC gap value. It was suggested in [22] that superconductivity in the commensurate antiferromagnetic (AF) phase in the $R =$ Ho compound survived at the special nearly isotropic Fermi surface sheet. The SC gap suppression above T^* may be caused by a peculiar magnetic order, which is developed in this compound between T^* and $T_c \approx 8.5$ K. It looks like SC state in this region is very delicate and sensitive to the (local) disturbing.

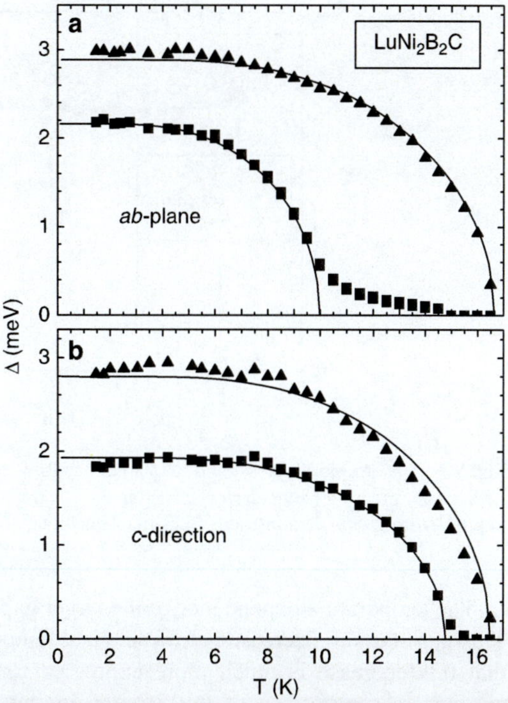

Fig. 9.2 SC gap behaviour in $LuNi_2B_2C$ obtained in [18, 19] by fitting dV/dI curves in the two-gap approximation for two main crystallographic directions. The *solid curves* show the BCS-like behaviour

Recently, an interesting peculiar behaviour of $\Delta(T)$ in $TmNi_2B_2C$ has been found [21, 27]. The SC gap has a maximum around 5 K and decreases below 5 K with the decreasing of temperature. This is in line with the behaviour of the upper critical field along the c-axis. Apparently, AF fluctuations occurring above the magnetically ordered state at $T_N = 1.5\,K$ are responsible for the decrease of the SC gap observed by approaching T_N. Additional interesting behaviour on some AR spectra was observed for this compound in magnetic field [27], which is very similar to the reentrant transition in $\rho(T)$ appearing in magnetic field by lowering the temperature. For this compound, a two-gap fit is also applicable as it is shown in [27], when $\approx$ 10% admixture of the small gap (of about 1 meV) is added to the larger gap (of about 2.6 meV).

As it was shown in [23, 26], the "one-gap" approach to fit the measured high quality $dV/dI(V)$ curves for $ErNi_2B_2C$ results in a visible discrepancy between the fit and $dV/dI(V)$ at the minima position and at zero bias. At the same time, a "two-gap" approach fits better the experimental curves for $ErNi_2B_2C$. As was mentioned already, the upper critical field $H_{c2}(T)$ of non-magnetic borocarbides $R = Y$ and Lu [3] can only be properly described by a two-band model. The finding of two SC gaps in AF state for $ErNi_2B_2C$ testifies to similarities in the electronic band structure in both magnetic and non-magnetic borocarbides, and points to the fact that superconductivity and magnetism develop in $ErNi_2B_2C$ in different bands.

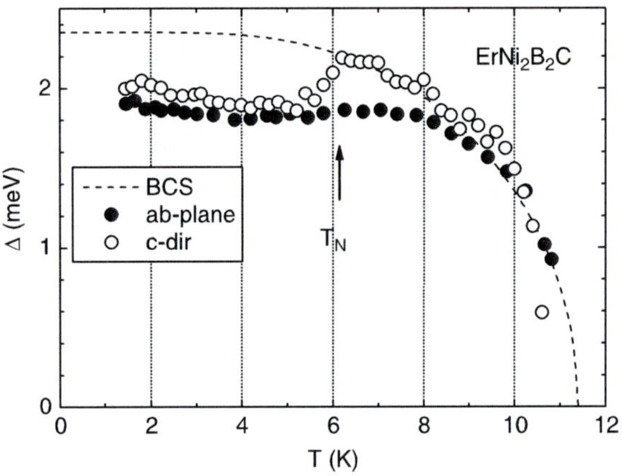

Fig. 9.3 Anisotropic suppression of the larger gap for the c-direction (*open circles*) in comparison with the ab-direction (*solid circles*) for $ErNi_2B_2C$ by transition to the AF state below 6 K according to [26]. *Dashed line* demonstrates BCS-like behaviour

The temperature dependence of the larger gap in $ErNi_2B_2C$ is shown in Fig. 9.3. This gap starts to decrease below the Neel temperature. A detailed analysis shows that this decrease is much more expressed for the c-direction testifying to the anisotropic suppression of the SC gap in this compound. This phenomenon is explained by the pair-breaking exchange field generated by spin-density waves, which reduces the gap for the electrons with wave vectors perpendicular to the propagation spin-density wave vector in the ab-plane. STM measurements in [32] and recent laser-photoemission spectroscopy [33] show also a deviation from the BCS predicted below T_N in $ErNi_2B_2C$. The SC gap decrease in the AF state is also in line with the Machida theory [34] in which a spin-density wave ordering below T_N competes with superconductivity.

It should be noted that the AF structure in R = Er and Tm nickel borocarbides is an incommensurate spin-density wave. As it is mentioned in [21,27], $\Delta(T)$ deviates from the BCS behaviour for R = Tm showing visible decrease below 5 K by approaching $T_N \approx 1.5$ K. On the contrary, R = Ho and Dy compounds with commensurate AF order display a BCS-like gap [11, 22]. Indeed, measurements below $T_N \approx 1.5$ K in case of R = Tm are very desirable to observe the similarity to the Er compound.

Figure 9.4 and Table 9.1 summarize the measurements of the SC gap by PCs in the title compounds. In average, the SC gap values are placed close to the BCS value $\Delta = 1.76\, k_B T_c$ taking into account the vanishing of the SC gap in the Ho borocarbide at $T^* \approx 5.6$ K. A two-gap state is established for R = Lu [18, 19, 28], Er [23, 26] and Tm [27] and a strong anisotropy of the SC gap (probably due to multiband SC state) is observed for R = Y [20].

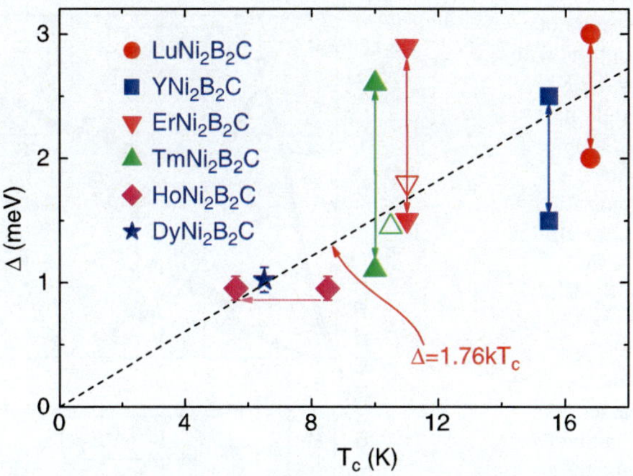

Fig. 9.4 The SC gap (*solid symbols*) established by a PC AR study (see also Table 9.1) as a function of the critical temperature T_c in $R\mathrm{Ni_2B_2C}$. For R = Y, the extremes of the SC gap distribution are presented, while for R = Lu, Tm and Er the small and large gaps within two-gap fit are shown. For R = Ho, T_c is shifted to $T^* = 5.6$ K (see text for explanation). *Open symbols* show the SC gap determined by the tunneling spectroscopy [35, 36]. A *dashed straight line* shows the BCS ratio

Table 9.1 The SC gap(s) value (in meV) extrapolated to $T = 0$ for $R\mathrm{Ni_2B_2C}$ compounds measured by PCAR spectroscopy. Data for films are also given for R = Y and Lu

One-gap approach

$R = Y$	Dy	Ho	Er	Tm	Lu
2.42 ± 0.07 [5]	1.0 [11]	1.04 ± 0.06 [5]	1.7 ± 0.2 [5]	1.1–1.7 [27]	2.6 [110] and 2.4 [001] [24, 25]
0.42 (ab-plane) 1.8 (c-dir) [14]		0.95 [22]	1.95 (ab-plane) 2.36 (c-dir) [26]		
1.5 (ab-plane) 2.3 (c-dir) [20]					
1.5–2.4 (film) [17]					2.6 (film) [28]

Two-gap approach

R = Er	Tm	Lu
2.1 and 1.1 (ab-plane)	2.6 and 1.0 [27]	2.16 and 3.0 (ab-plane)
2.3 and 1.0 (c-axis) [26]		1.94 and 3.0 (c-axis) [18, 19]
		2.14 ± 0.36 and 3 ± 0.3 (film) [28]

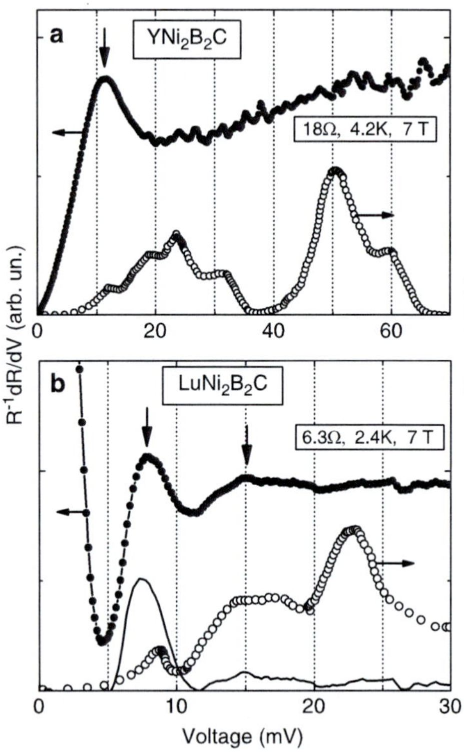

Fig. 9.5 (**a**) PC spectra of YNi_2B_2C homocontacts with $R = 18\,\Omega$ at $T = 4.2\,K$ averaged for two polarities of bias voltage [20]. The superconductivity is suppressed by a magnetic field of 7 T. (**b**) PC spectra of $LuNi_2B_2C$ (film) – Cu contact with $R = 6.3\,\Omega$ at $T = 2.4\,K$ and $B = 7\,T$ [28]. Huge increase in PC spectra intensity below 5 meV is due to not fully suppressed superconductivity in the PC core. In both panels, the phonon DoS (*open circles*) for the corresponding compound is depicted from [37]. *Vertical arrows* show main maxima in the PC spectra corresponding to the first maxima in the phonon DoS

9.4 PC Spectroscopy of Electron–Phonon (BOSON) Interaction

Nonmagnetic compounds. The spectra of the nonmagnetic YNi_2B_2C show a dominant maximum at about 12 mV and a broad shallow hump or a kink positioned close to 50 mV (Fig. 9.5a). These features have a counterpart in the phonon density of states (DoS) of YNi_2B_2C [37]. The PC spectra are similar both for single crystals [20] and films [17]. YNi_2B_2C homocontacts also demonstrate similar spectra (Fig. 9.5a). Very recently [28], the PC spectra of $LuNi_2B_2C$ were measured in the c-axis oriented films (Fig. 9.5b). The most expressed feature is the maximum at about 8.5 meV, which perfectly coincides with the position of the first maximum in the phonon DoS for $LuNi_2B_2C$ [37]. At the same time, a high energy part of PC spectra obtained contains no visible features (except broad maximum around 15 meV). It is unlikely that electron interaction with phonons of higher energy above 20 meV is so weak, so the reason of featureless spectra in this region is probable shortening of inelastic electron mean free path l with increasing energy due to strong EPI, which leads to violation of the spectral regime condition $d << l$ in PC [4].

Compounds with magnetic ordering. PC EP(B)I spectra in magnetic RNi_2B_2C are presented in Fig. 9.6. The most detailed spectra are measured for R = Ho. Here, the maxima (with the exception of maximum at 3 meV) correspond to those

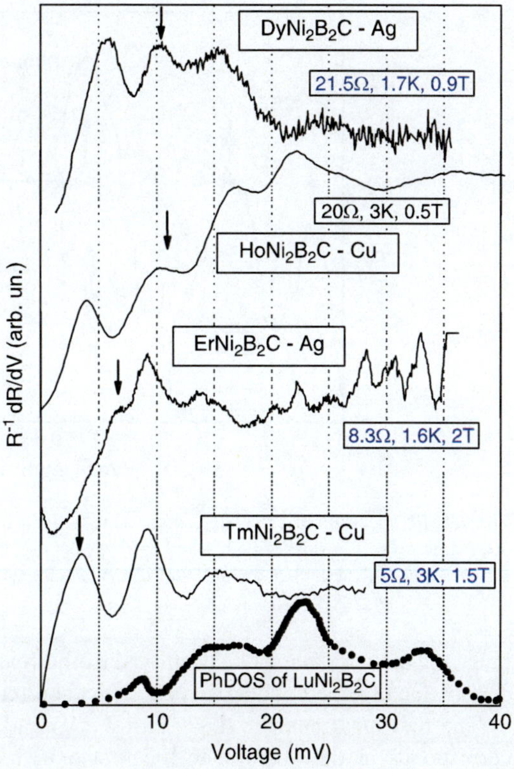

Fig. 9.6 PC spectra of magnetic $R\mathrm{Ni_2B_2C}$ compounds from [12, 13, 22, 27]. The superconductivity is suppressed by a magnetic field. The *bottom curve* shows the phonon DOS for $\mathrm{LuNi_2B_2C}$ [37]. *Vertical arrows* show expected position of the first CEF excitations according to [38]

in the phonon DoS of the isostructural compound $\mathrm{LuNi_2B_2C}$ [37]. The majority of the PC spectra in $\mathrm{HoNi_2B_2C}$ demonstrate a prominent 10-mV peak [22], while the other phonon maxima do not become apparent. The unique spectrum for this compound (see Fig. 9.3 in [22]) displays also pronounced highenergy maxima around 50 and 100 mV, which were not so clearly resolved in the PC spectra of other nickel borocarbides so far. The low energy maximum around 3 mV in R = Ho and 6 mV in R = Dy (Fig. 9.6) has not any respective reference in the phonon DoS nor in the crystal-electric-field (CEF) energy scheme [38]; therefore, it has non-phonon and non-CEF origin. The 3-mV maximum in R = Ho can be suppressed anisotropically by a magnetic field [22] and temperature above 10 K pointing to its "magnetic" origin as discussed in [22]. The 6-mV peak in R = Dy vanishing with increasing of T above 15 K [12] has probably a similar "magnetic" origin. Twice as large 6-mV peak position in R = Dy in comparison with 3-mV peak in R = Ho correlates with two times higher Neel temperature $T_N \approx 10.4$ K in this compound compared to $T_N \approx 5.2$ K in R = Ho. This testifies that 3 mV and 6 mV peculiarities are connected with AF state in some way. It was assumed in [22] that this phenomenon is connected with destruction of the AF order due to "heating" by non-equilibrium phonons in PC. In [39], it was suggested that the long

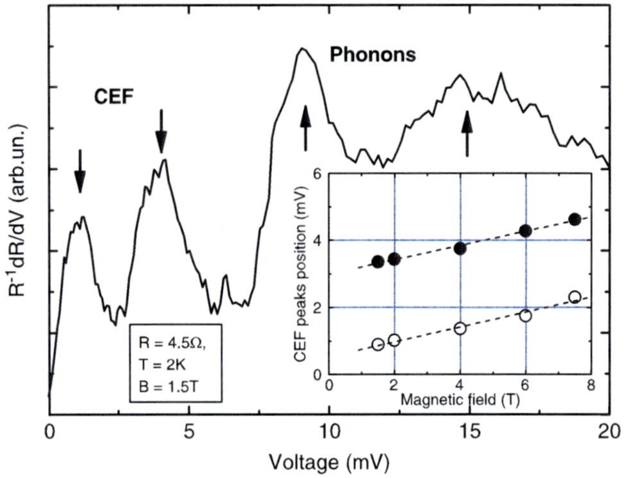

Fig. 9.7 PC spectrum of $TmNi_2B_2C$ compound with an additional low energy peak at about 1 mV. *Inset* shows position shift of both 1 and 3 mV peaks in a magnetic field [27] that testify their CEF nature. Phonon structures at about 9 and 15 mV are robust to a magnetic field

range magnetic order can be destroyed isothermally by current, by analogy with the destruction of superconductivity by critical current.

The first maximum in R = Tm (3 mV) and the shoulder in R = Er (6 mV) corresponds in the position to the first exited CEF level [38]. CEF contributes apparently also to the 10-mV peak in R = Ho and Dy compounds. In the former case, this is seen from the modification of the 10-mV peak in a magnetic field (see Fig. 9.4 in [22]). Interesting phenomena was observed in the PC spectra of R = Tm, where an additional peak at about 1 mV for some PCs is seen (see Fig. 9.7). The authors [27] explained this feature as splitting of the original magnetic Tm-ion ground state doublet into two singlet states due to non-stoichiometry caused by boron/carbon vacancies or as a result of the quadrupolar ordering [1]. Both effects lower CEF symmetry resulting in the splitting of the doublet ground state.

The strength of EP(B)I in the investigated compounds can be estimated by calculation of λ (see (10.2)) from the measured PC spectra. The value of λ for all mentioned compounds lies between 0.1 and 0.25 [8, 12, 20, 22, 27, 28], which is sufficiently smaller than expected for the superconductors. For instance in [40, 41], by comparison of the effective masses obtained by dHvA measurements with calculated bare masses, the coupling strength has been estimated in R = Lu between 0.23 and 2.7 on different Fermi surfaces and for different crystallographic directions. As mentioned by the same authors, specific heat data on polycrystalline samples point to λ = 0.75–1.2 and resistivity measurements allow to estimate the transport coupling constant as λ_{tr} = 0.55–1.1. λ = 0.848 is calculated for R = Y in [42] with numerical accuracy about 10%. The value of λ is almost order of magnitude lower from PC measurements and can be explained in the following manner: (1) using in the PC theory a single band free electron model, (2) short elastic mean free path of electrons $l_\mathrm{i} << d$, what results in decreasing of PC intensity by factor l_i/d, 3) shunting conductivity of multiple PCs containing non-ballistic constrictions. Noteworthy, that discussed parameter λ_PC can be considered to be a kind of the transport EPI constant (see Table 9.1 in [4]) and in general its value can be different from the Eliashberg EPI constant λ, whereas in PC spectra backscattering processes of electrons play the main role [43]. Therefore, it would be very desirable to compute PC EPI function, though it is a difficult task taking into account the complicated band structure of title compounds.

The measurements of PC EP(B)I spectra point to the importance of the CEF excitations in the charge transport as well as to the SC properties of "magnetic" R = Dy, Ho, Er and Tm borocarbides. For example, the relative contribution of the 10-mV peak in the EP(B)I constant λ for R = Ho is evaluated as 20–30% [22], while the contribution to λ of the high frequency modes at 50 and 100 meV amounts up to 10% for each maximum.

9.5 Conclusion

The SC gap has been studied using PC AR spectroscopy for a series of the rare-earth nickel borocarbide superconductors [5–28]. The existence of two SC gaps both in non-magnetic R = Y, Lu [17–20] and in the magnetic compounds R = Tm [27], Er [23, 26] has been shown. A distinct anisotropic decrease of the both gaps in R = Er [26] is observed as the temperature is lowered below T_N. For the R = Ho [22] and Dy [11] compounds with commensurate AF order, the SC gap has a BCS-like behaviour in the AF state, while for R = Tm the gap starts to decrease by approaching a magnetic state with incommensurate AF order. The Er compound has an incommensurate AF order and the gap starts to decrease at lowering temperature slightly above T_N as well [23]. The discrepancy in the magnetically ordered state between R = Ho, Dy (commensurate state) and R = Er, Tm (incommensurate state) results in a different SC gap behaviour. More extensive directional PC measurements for R = Dy and Ho are desirable to check the presence of multi-gap superconductivity in these compounds as well. The measurements of PC AR spectra in R = Tm well below the Neel temperature (1.5 K) are challenging to study interplay of superconductivity and an AF order.

Using PC spectroscopy, a clear coupling of electrons to the low energy phonon and CEF excitations has been shown in the title compounds [12,20,22,27,28]. To our knowledge up to now this is the only experimental technique able to yield energy-resolved information on the EP(B)I in these compounds. In magnetic R = Dy, Ho, Er and Tm compounds, the CEF excitations contribute to the EP(B)I function heavily [22, 27]. For R = Ho, the contribution of two high energy modes (50 and 100 meV) is also notable [22]. The 50-mV mode can be also resolved for other compounds, but it looks to much smeared (usually like a knee) as shown in Fig. 9.5a for R = Y. In this context, we note that the electronic mean free path is shortened with increase of the bias voltage due to the EP(B)I. In this case if the EP(B)I is strong enough, it results in a violation of the ballistic condition (with voltage increase) necessary for energy resolved spectroscopy by PCs. On the contrary, the high energy modes involve vibration of light B and C ions. Since the disorder in the position of B and C light atoms is especially difficult to avoid and control, samples with improved quality are very desirable for further in-depth investigations of SC gap(s) and details of EP(B)I. It should be stressed also that from the theoretical point of view calculation of PC EPI function in the title compounds remains a challenge.

Acknowledgements This work was supported by the National Academy of Sciences of Ukraine. The authors thank D.L. Bashlakov, G. Behr, N.L. Bobrov, P.C. Canfield, S.-L. Drechsler, G. Fuchs, B. Holzapfel, O.E. Kvitnitskaya, D.G. Naugle, K. Nenkov, K.D.D. Rathnayaka, M. Schneider, D. Souptel for the long-term collaboration in the field of the rare-earth nickel borocarbide superconductor investigations and the experimental assistance. Yu. G. N. thanks the Alexander von Humboldt Foundation for generous support.

References

1. K.H. Müller, M. Schneider, G. Fuchs, S.L. Drechsler, in *Handbook on the Physics and Chemistry of Rare Earths*, ed. by K.A. Gschneidner Jr., Jean-Claude Bünzli, Vitalij K. Pecharsky, Vol. 38, Chapter 239, (North-Holland, Amsterdam, 2008)
2. P.C. Canfield, P.L. Gammel, D.J. Bishop, Phys. Today **51**, 40 (1998)
3. S.V. Shulga, S.L. Drechsler, G. Fuchs, K.H. Müller, K. Winzer, M. Heinecke, K. Krug, Phys. Rev. Lett. **80**, 1730 (1998)
4. Yu.G. Naidyuk, I.K. Yanson, *Point-Contact Spectroscopy*, Springer Series in Solid-State Sciences, vol. 145 (Springer Science + Business Media, Inc, New York, 2005)
5. L.F. Rybaltchenko, I.K. Yanson, A.G.M. Jansen, P. Mandal, P. Wyder, C.V. Tomy, D.McK. Paul, Phys. B **218**, 189 (1996)
6. L.F. Rybaltchenko, I.K. Yanson, A.G.M. Jansen, P. Mandal, P. Wyder, C.V. Tomy, D.McK. Paul, Europhys. Lett. **33**, 483 (1996)
7. I.K. Yanson, V.V. Fisun, A.G.M. Jansen, P. Wyder, P.C. Canfield, B.K. Cho, C.V. Tomy, D.McK. Paul, Phys. Rev. Lett. **78**, 935 (1997)
8. I.K. Yanson, V.V. Fisun, A.G.M. Jansen, P. Wyder, P.C. Canfield, B.K. Cho, C.V. Tomy, D.McK. Paul, Fiz. Nizk. Temp. **23**, 951 (1997) [Low Temp. Phys. **23**, 712 (1997)]
9. L.F. Rybaltchenko, A.G.M. Jansen, P. Wyder, L.V. Tiutrina, P.C. Canfield, C.V. Tomy, D.McK. Paul, Phys. C **319**, 189 (1999)
10. I.K. Yanson, in *Symmetry and Pairing in Superconductors*, ed. by M. Ausloos, S. Kruchinin (Kluwer, The Netherlands, 1999), pp. 271–285
11. I.K. Yanson, N.L. Bobrov, C.V. Tomy, D.McK. Paul, Phys. C **334**, 33 (2000)
12. I.K. Yanson, N.L. Bobrov, C.V. Tomy, D.McK. Paul, Phys. C **334**, 152 (2000)
13. I.K. Yanson, in *Rare Earth Transition Metal Borocarbides (Nitrides): Superconducting, Magnetic and Normal State Properties*, ed. by K.H. Müller, V. Narozhnyi, vol. 14, (Kluwer, The Netherlands, 2001), pp. 95–108
14. P. Raychaudhuri, D. Jaiswal-Nagar, G. Sheet, S. Ramakrishnan, H. Takeya, Phys. Rev. Lett. **93**, 156802 (2004)
15. S. Mukhopadhyay, G. Sheet, P. Raychaudhuri, H. Takeya, Phys. Rev. B **72**, 014545 (2005)
16. P. Raychaudhuri, G. Sheet, S. Mukhopadhyay, H. Takeya, Phys. C **460–462**, 95 (2007)

17. D.L. Bashlakov, Yu.G. Naidyuk, I.K. Yanson, S.C. Wimbush, B. Holzapfel, G. Fuchs, S.L. Drechsler, Supercond. Sci.Technol. **18**, 1094 (2005)
18. N.L. Bobrov, S.I. Beloborod'ko, L.V. Tyutrina, I.K. Yanson, D.G. Naugle, K.D.D. Rathnayaka, Phys. Rev. B **71**, 014512 (2005)
19. N.L. Bobrov, S.I. Beloborod'ko, L.V. Tyutrina, V.N. Chernobay, I.K. Yanson, D.G. Naugle, K.D.D. Rathnayaka, Fiz. Nizk. Temp. 32, 641 (2006) [Low Temp. Phys. 32, 489 (2006)]
20. D.L. Bashlakov, Yu.G. Naidyuk, I.K. Yanson, G. Behr, S.L. Drechsler, G. Fuchs, L. Schultz, D. Souptel, J. Low Temp. Phys., **147**, 335 (2007)
21. Y.u.G. Naidyuk, D.L. Bashlakov, N.L. Bobrov, V.N. Chernobay, O.E. Kvitnitskaya, I.K. Yanson, G. Behr, S.L. Drechsler, G. Fuchs, D. Souptel, D.G. Naugle, K.D.D. Rathnayaka, J.H. Ross Jr., Phys. C **460–462**, 107 (2007)
22. Yu.G. Naidyuk, O.E. Kvitnitskaya, I.K. Yanson, G. Fuchs, K. Nenkov, A. Wälte, G. Behr, D. Souptel, S.L. Drechsler, Phys. Rev. B **76**, 014520 (2007)
23. N.L. Bobrov, V.N. Chernobay, Yu.G. Naidyuk, L.V. Tyutrina, D.G. Naugle, K.D.D. Rathnayaka, S.L. Bud'ko, P.C. Canfield, I.K. Yanson, Europhys. Lett. **83**, 7003 (2008)
24. X Lu, W.K. Park, J.D.. Kim, S. Yeo, S. Lee, L.H. Greene, Phys. B **403**, 1098 (2008)
25. X. Lu, W.K. Park, K.Y. Choi, S.I. Lee, S. Yeo, S.G.L. Budko, P.C. Canfield, L.H. Greene, J. Phys. Conf. **150**, 052143 (2009)
26. N.L. Bobrov, V.N. Chernobay, Y.u.G. Naidyuk, L.V. Tyutrina, I.K. Yanson, D.G. Naugle, K.D.D. Rathnayaka, Fiz. Nizk. Temp. **36**, 1228 (2010) [Low Temp. Phys. **36**, 990 (2010)]
27. Yu.G. Naidyuk, O.E. Kvitnitskaya, L.V. Tiutrina, I.K. Yanson, G. Fuchs, K. Nenkov, G. Behr, S.L. Drechsler (to be published)
28. O.E. Kvitnitskaya, Yu.G. Naidyuk, I.K. Yanson, T. Niemeier, G. Fuchs, B. Holzapfel, Supercond. Sci. Technol. **23**, 115001 (2010)
29. G.E. Blonder, M. Tinkham, T.M. Klapwijk, Phys. Rev. B **25**, 4515 (1982)
30. S.I. Beloborodko, Fiz. Nizk. Temp. **29**, 868 (2003) [Low Temp. Phys. **29**, 650 (2003)]
31. D. Daghero, R.S. Gonnelli, Supercond. Sci. Technol. **23**, 043001 (2010)
32. T. Watanabe, K. Kitazawa, T. Hasegawa, Z. Hossain, R. Nagarajan, L. Gupta, J. Phys. Soc. Jpn. **69**, 2708 (2000)
33. T. Baba, T. Yokoya, S. Tsuda, T. Kiss, T. Shimojima, K. Ishizaka, H. Takeya, K. Hirata, T. Watanabe, M. Nohara, H. Takagi, N. Nakai, K. Machida, T. Togashi, S. Watanabe, X.Y. Wang, C.T. Chen, S. Shin, Phys. Rev. Lett. **100**, 017003 (2008)
34. K. Machida, K. Nokura, T. Matsubara, Phys. Rev. B **22**, 2307 (1980)
35. H. Suderow, P. Martinez-Samper, N. Luchier, J.P. Brison, S. Vieira, P.C. Canfield, Phys. Rev. B **64**, 020503(R) (2001)
36. M. Crespo, H. Suderow, S. Vieira, S. Bud'ko, P.C. Canfield, Phys. Rev. Lett. **96**, 027003 (2006)
37. F. Gompf, W. Reichardt, H. Schober, B. Renker, M. Buchgeister, Phys. Rev. B **55**, 9058 (1997)
38. U. Gasser, P. Allenspach, F. Fauth, W. Henggeler, J. Mesot, A. Furrer, S. Rosenkranz, P. Vorderwisch, M. Buchgeister, Z. Phys. B **101**, 345 (1996)
39. D. Wohlleben, E. Paulus, G. Voss, A. Freimuth, Europhys. Lett. **3**, 107 (1987)
40. B. Bergk, V. Petzold, H. Rosner, S.L. Drechsler, M. Bartkowiak, O. Ignatchik, A.D. Bianchi, I. Sheikin, P.C. Canfield, J. Wosnitza, Phys. Rev. Lett. **100**, 257004 (2008)
41. B. Bergk, J. Wosnitza, Low Temp. Phys. **35**, 687 (2009) [Fiz. Nizk. Temp. 35, 872 (2009)]
42. W. Reichardt, R. Heid, K.P. Bohnen, J. Supercond. Incorporating Novel Magnetism **18**, 159 (2005)
43. I.O. Kulik, A.N. Omelyanchouk, R.I. Shekhter, Sov. J. Low Temp. Phys. **3**, 840 (1977) [Fiz. Nizk. Temp. **3**, 1543 (1977)]

Chapter 10
Integrated Submm Wave Receiver: Development and Applications

Valery P. Koshelets, Manfred Birk, Dick Boersma, Johannes Dercksen, Pavel Dmitriev, Andrey B. Ermakov, Lyudmila V. Filippenko, Hans Golstein, Ruud W.M. Hoogeveen, Leo de Jong, Andrey V. Khudchenko, Nickolay V. Kinev, Oleg S. Kiselev, Pavel V. Kudryashov, Bart van Kuik, Arno de Lange, Gert de Lange, Irina L. Lapitsky, Sergey I. Pripolzin, Joris van Rantwijk, Avri M. Selig, Alexander S. Sobolev, Mikhail Yu Torgashin, Vladimir L. Vaks, Ed de Vries, Georg Wagner, and Pavel A. Yagoubov

Abstract A superconducting integrated receiver (SIR) comprises in a single chip a planar antenna combined with a superconductor-insulator-superconductor (SIS) mixer, a superconducting Flux Flow Oscillator (FFO) acting as a Local Oscillator (LO) and a second SIS harmonic mixer (HM) for the FFO phase locking. In

---

V.P. Koshelets (✉) · A.B. Ermakov · A.V. Khudchenko · N.V. Kinev · O.S. Kiselev
Kotel'nikov Institute of Radio Engineering and Electronics, Russian Academy of Science, Mokhovaya st. 11/7, 125009, Moscow, Russia
and
SRON Netherlands Institute for Space Research, 9700 AV, Groningen, The Netherlands
e-mail: valery@hitech.cplire.ru; Khudchenko@hitech.cplire.ru

P. Dmitriev · P.V. Kudryashov · A.S. Sobolev · M. Yu Torgashin
Kotel'nikov Institute of Radio Engineering and Electronics, Russian Academy of Science, Mokhovaya st. 11/7, 125009, Moscow, Russia

D. Boersma · J. Dercksen · L.V. Filippenko · H. Golstein · R.W.M. Hoogeveen · L. de Jong · B. van Kuik · A. de Lange · G. de Lange · I.L. Lapitsky · J. van Rantwijk · A.M. Selig · Ed de Vries
SRON Netherlands Institute for Space Research, 9700 AV, Groningen, The Netherlands

M. Birk · G. Wagner
DLR German Aerospace Centre, Remote Sensing Technology Institute, 82234, Wessling, Germany

S.I. Pripolzin · V.L. Vaks
Institute for Physics of Microstructure, Russian Academy of Science, Ulyanova 46, GSP-105, Nizhny Novgorod, Russia

P.A. Yagoubov
SRON Netherlands Institute for Space Research, 9700 AV, Groningen, The Netherlands
and
European Organization for Astronomical Research in the Southern Hemisphere (ESO), Karl-Schwarzschild-Strasse 2, 85748, Garching bei München, Germany

A. Sidorenko (ed.), *Fundamentals of Superconducting Nanoelectronics*, NanoScience and Technology, DOI 10.1007/978-3-642-20158-5_10,

this report, an overview of the SIR and FFO developments and optimizations is presented. Improving on the fully Nb-based SIR we have developed and studied Nb–AlN–NbN circuits, which exhibit an extended operation frequency range. Continuous tuning of the phase locked frequency has been experimentally demonstrated at any frequency in the range 350–750 GHz. The FFO free-running linewidth has been measured between 1 and 5 MHz, which allows to phase lock up to 97% of the emitted FFO power. The output power of the FFO is sufficient to pump the matched SIS mixer. Therefore, it is concluded that the Nb–AlN–NbN FFOs are mature enough for practical applications.

These achievements enabled the development of a 480–650 GHz integrated receiver for the atmospheric-research instrument TErahertz and submillimeter LImb Sounder (TELIS). This balloon-borne instrument is a three-channel superconducting heterodyne spectrometer for the detection of spectral emission lines of stratospheric trace gases that have their rotational transitions at THz frequencies. One of the channels is based on the SIR technology. We demonstrate for the first time the capabilities of the SIR technology for heterodyne spectroscopy in general, and atmospheric limb sounding in particular. We also show that the application of SIR technology is not limited to laboratory environments, but that it is well suited for remote operation under harsh environmental conditions. Light weight and low power consumption combined with broadband operation and nearly quantum limited sensitivity make the SIR a perfect candidate for future airborne and space-borne missions. The noise temperature of the SIR was measured to be as low as 120 K in double sideband operation, with an intermediate frequency band of 4–8 GHz. The spectral resolution is well below 1 MHz, confirmed by our measurements. Remote control of the SIR under flight conditions has been demonstrated in a successful balloon flight in Kiruna, Sweden.

Capability of the SIR for high-resolution spectroscopy has been successfully proven also in a laboratory environment by gas cell measurements. The possibility to use SIR devices for the medical analysis of exhaled air will be discussed. Many medically relevant gases have spectral lines in the sub-terahertz range and can be detected by an SIR-based spectrometer. The SIR can be considered as an operational device, ready for many applications.

10.1 Introduction

A Superconducting Integrated Receiver (SIR) [1, 2] was proposed more than 10 years ago and has since then been developed up to the point of practical applications [3–5]. Our approach consists in developing a *single chip* heterodyne receiver, which is smaller and less complex than traditional devices. Typically, such a receiver consists of a number of main components (local oscillator (LO), mixer, antenna structure, phase lock circuit, etc.), which are usually built as separate units and are complex (and thus costly). According to our concept (see Fig. 10.1), we have integrated all these components onto one single chip reducing overall system complexity in change for increased on-chip and lithographic fabrication complexity.

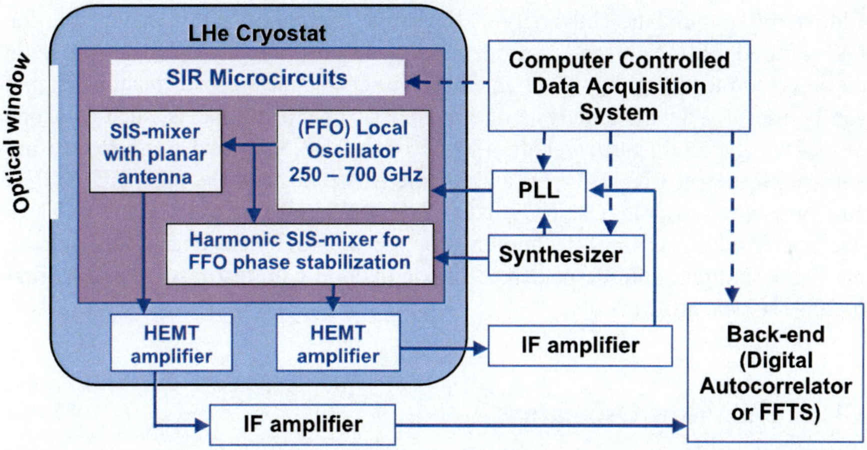

Fig. 10.1 Block-diagram of the superconducting integrated receiver

An SIR comprises on one chip all key elements needed for heterodyne detection: a low-noise superconductor-insulator-superconductor (SIS) mixer with quasi-optical antenna, a flux-flow oscillator (FFO) [6] acting as an LO and a second SIS harmonic mixer (HM) for the FFO phase locking. The concept of the SIR is very attractive for many practical applications because of the compactness and the wide tuning range of the FFO [7]. Presently, the frequency range of most practical heterodyne receivers is limited by the tuning range of the LO, typically 10%–15% for a solid-state multiplier chain [8]. In the SIR, the bandwidth is determined by the SIS mixer tuning structure and the matching circuitry between the SIS and the FFO. A bandwidth up to 30%–40% may be achieved with a twin-junction SIS mixer design. Another potential advantage is the use of arrays of SIR channels within a single cryostat that could operate at the same or different LO frequencies.

One of the important practical application of the SIR is TErahertz and sub-millimeter LImb Sounder (TELIS) [5, 9, 10] – a three-channel balloon-borne heterodyne spectrometer for atmospheric research developed in a collaboration of four institutes: Deutsches Zentrum für Luft- und Raumfahrt (DLR), Germany, Rutherford Appleton Laboratories (RAL), United Kingdom, and SRON – Netherlands Institute for Space Research, the Netherlands (in tight collaboration with Kotel'nikov Institute of Radio Engineering and Electronics, IREE, Moscow). All three receivers utilize state-of-the-art superconducting heterodyne technology and operate at 500 GHz (by RAL), at 480–650 GHz (by SRON + IREE), and at 1.8 THz (by DLR). TELIS is a compact, lightweight instrument capable of providing broad spectral coverage, high spectral resolution and long flight duration. The TELIS instrument serves also as a test bed for many novel cryogenic technologies and as a pathfinder for satellite-based instrumentation.

TELIS is mounted on the same balloon platform as the Fourier transform spectrometer MIPAS-B [11], developed by IMK (Institute of Meteorology and

Climate research of the University of Karlsruhe, Germany) and is operated in the mid-infrared (680–2,400 cm^{-1}). Both instruments observe simultaneously the same air mass, and together they yield an extensive set of stratospheric constituents that can be used for detailed analysis of atmospheric chemical models, such as ozone destruction cycles. In particular, the 480–650 GHz TELIS channel is able to measure vertical profiles of ClO, BrO, O_3 and its rare isotopologues, O_2, HCl, HOCl, H_2O and three rare isotopologues, HO_2, NO, N_2O, NO_2, HNO_3, CH_3Cl, and HCN. In this paper, the design and technology for the 480–650 GHz channel as used in the flight configuration are presented in conjunction with test results and the first preliminary scientific results.

10.2 Flux Flow Oscillators

A Josephson Flux Flow Oscillator (FFO) [6] has proven [4, 5, 7] to be the most developed superconducting LO for integration with an SIS mixer in a single-chip submm-wave SIR [1–5]. The FFO is a long Josephson tunnel junction of the overlap geometry (see Fig. 10.2) in which an applied dc magnetic field and a dc bias current, I_B, drive a unidirectional flow of fluxons, each containing one magnetic flux quantum, $\Phi_0 = h/2e \approx 2 \times 10^{-15}$ Wb. Symbol h is Planck's constant and e is the elementary charge. An integrated control line with current I_{CL} is used to generate the dc magnetic field applied to the FFO. According to the Josephson relation, the junction oscillates with a frequency $f = (1/\Phi_0)V$ (about 483.6 GHz/mV) if it is biased at voltage V. The fluxons repel each other and form a chain that moves along the junction. The velocity and density of the fluxon chain and thus the power and frequency of the submm-wave signal emitted from the exit end of the junction due to the collision with the boundary may be adjusted independently by proper settings of I_B and I_{CL}. The FFO differs from the other members of the Josephson oscillator family by the need for these two control currents, which in turn provides the possibility of independent frequency and power tuning.

We experimentally investigated a large number of the FFO designs. The length, L, and the width, W, of the FFO used in our study are 300–400 μm and 4–28 μm,

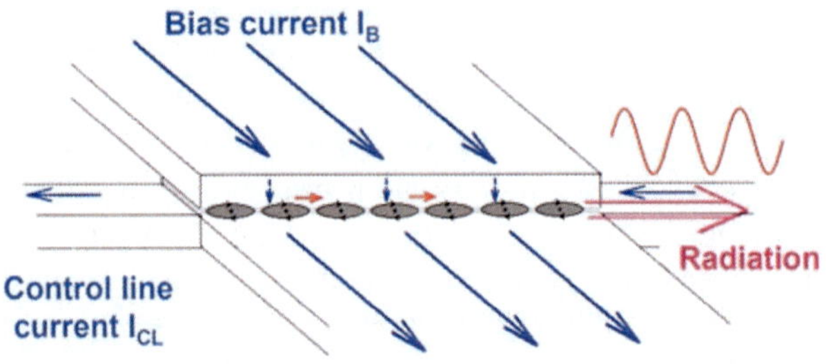

Fig. 10.2 Schematic view of a flux-flow oscillator

respectively. The value of the critical current density, J_C, is in the range 4–8 kA/cm^2 giving a Josephson penetration depth, $\lambda_J \sim$ 6–4 μm. The corresponding value of the specific resistance is $Rn \times L \times W$ *is* $\sim$ 50–25 Ω μm^2. For the numerical calculations, we use a typical value of the London penetration depth, $\lambda_L \approx 90$ nm for all-Nb junctions, and a junction specific capacitance, $Cs \approx 0.08$ pF/μm^2. The active area of the FFO (i.e. the AlO_x or the AlN tunnel barrier) is usually formed as a long window in the relatively thick (200–250 nm) SiO_2 insulation layer sandwiched between the two superconducting films (base and wiring electrodes). The so-called "idle" region consists of the thick SiO_2 layer adjacent to the junction (on both sides of the tunnel region) between the overlapping electrodes. It forms a transmission line parallel to the FFO (not shown in Fig. 10.2). The width of the idle region (W_I = 2–14 μm) is comparable to the junction width. The idle region must be taken into account when designing an FFO with the desired properties. In our design, it is practical to use the flat bottom electrode of the FFO as a control line in which the current I_{CL} produces the magnetic field, which mainly is applied perpendicular to the long side of the junction.

There are a number of important requirements on the FFO properties to make it suitable for application in the phase locked SIR. Obviously, the FFO should emit enough power to pump an SIS mixer, taking into account a specially designed mismatch of about 5–7 dB between the FFO and the SIS mixer, introduced to avoid leakage of the input signal to the LO path. It is a challenge to realize the ultimate performance of the separate superconducting elements after their integration in a single-chip device. Implementation of the improved matching circuits and the submicron junctions for both the SIS and the HM allows delivering optimal FFO power for their operation.

Even for ultra wideband room-temperature PLL systems the effective regulation bandwidth is limited by the length of the cables in the loop (about 10 MHz for typical loop length of two meters). It means that the free-running FFO linewidth (LW) has to be well below 10 MHz to ensure stable FFO phase locking with a reasonably good spectral ratio (SR) – the ratio between the carrier and total power emitted by the FFO [7]. For example, only about 50% of the FFO power can be phase locked by the present PLL system at a free-running FFO LW of 5 MHz. A low spectral ratio results in a considerable error at resolving the complicated spectral line shape [12]. Thus, a sufficiently small free-running FFO LW is vitally important for the realization of the phase locked SIR for the TELIS.

10.2.1 Nb–AlN–NbN *FFO*

Earlier the Nb–AlOx–Nb or Nb–AlN–Nb trilayers were successfully used for the FFO fabrication. Traditional all-Nb circuits are being constantly optimized but there seems to be a limit for LW optimizations at certain boundary frequencies due to Josephson self-coupling (JSC) effect [13] as well as a high frequency limit, imposed by Nb gap frequency (~700 GHz). That is the reason for novel types

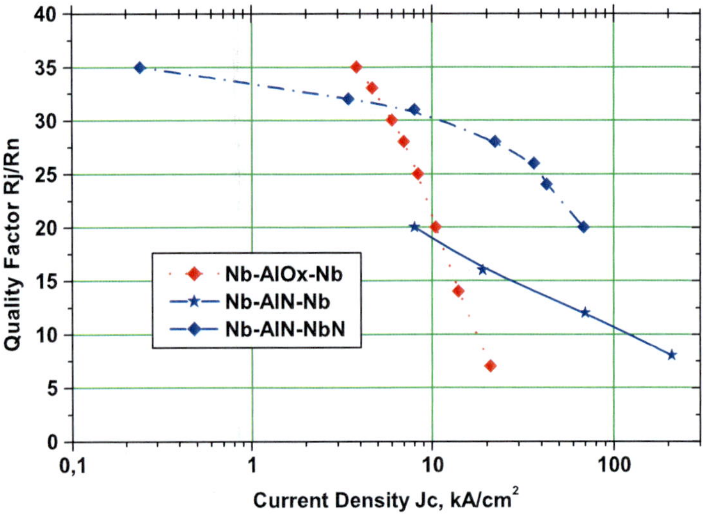

Fig. 10.3 The dependencies of Rj/Rn ratio on critical current density J_c for SIS junctions of different types fabricated at IREE

of junctions based on materials other than Nb to be developed. We reported on development of the high quality Nb–AlN–NbN junction production technology [14]. The implementation of an AlN tunnel barrier in combination with an NbN top superconducting electrode provides a significant improvement in SIS junction quality. The gap voltage of the junction $V_g = 3.7\,\mathrm{mV}$. From this value, and the gap voltage of the Nb film $\Delta_{\mathrm{Nb}}/e = 1.4\,\mathrm{mV}$, we have estimated the gap voltage of our NbN film as $\Delta_{\mathrm{NbN}}/e = 2.3\,\mathrm{mV}$ [15].

The dependency of the ratio of subgap to normal state resistance (Rj/Rn) vs. critical current density (J_c) for different types of the Nb-based junctions fabricated at IREE is presented in Fig. 10.3. One can see that the Nb–AlN–NbN junctions are of very good quality at high current densities, important for implementation in THz mixers. The same technique was further used to produce complicated integrated circuits comprising SIS and FFO in one chip.

The use of Nb for top "wiring" layer is preferable due to lower losses of Nb compared to NbN below 720 GHz; furthermore, the matching structures developed for the all-Nb SIRs can be used directly for the fabrication of receivers with Nb–AlN–NbN junctions. The general behavior of the new devices is similar to the all-Nb ones; even the control currents, necessary to provide magnetic bias for FFO, were nearly the same for the FFOs of similar design.

A family of the Nb–AlN–NbN FFO IVCs measured at different magnetic fields produced by the integrated control line is presented in Fig. 10.4 ($L = 300\,\mu\mathrm{m}$, $W = 14\,\mu\mathrm{m}$, $W_I = 10\,\mu\mathrm{m}$). A single SIS junction with an inductive tuning circuit is employed as a HM for the LW measurements. The tuning and matching circuits were designed to provide "uniform" coupling in the frequency range 400–700 GHz.

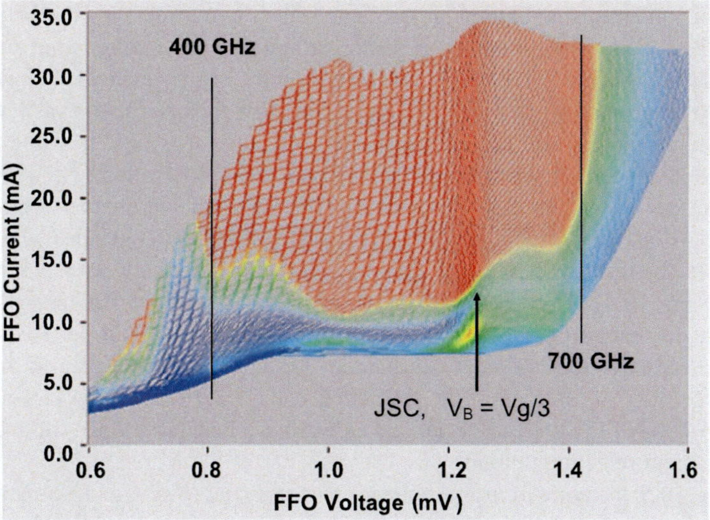

Fig. 10.4 IVCs of the Nb–AlN–NbN FFO measured at different magnetic fields produced by the integrated control line. The *color scale* shows the level of the DC current rise at the HM induced by the FFO. *Red area* marks the region of the FFO parameters where the induced by FFO HM current exceeds 25% of the I_g. This level is well above the optimal value for an SIS-mixer operation

Measured values of the HM current induced by the FFO oscillations (HM pumping) are shown in Fig. 10.4 by the color scale. The HM pumping for each FFO bias point was measured at constant HM bias voltage of 3 mV (pumping is normalized on the current jump at the gap voltage, $I_g = 140\,\mu A$). From Fig. 10.4, one can see that an FFO can provide large enough power over the wide frequency range: limited at higher frequencies only by the Nb superconducting gap in transmission line electrodes (base and wiring layers) and below 400 GHz by design of the matching circuits.

The Nb–AlN–NbN FFOs behave very similar to all-Nb ones. The feature at about 600 GHz where the curves get denser is a Josephson Self-Coupling (JSC) boundary voltage. It was first observed for all-Nb FFOs [13]. The JSC effect is the absorption of the FFO-emitted radiation by the quasi-particles in the cavity of the long junction. It considerably modifies the FFO properties at the voltages $V \approx V_{JSC} = 1/3V_g$ (V_{JSC} corresponds to 620 GHz for the Nb–AlN–NbN FFO). Just above this voltage, the differential resistance increases considerably; that results in an FFO-LW broadening just above this point. This, in turn, makes it difficult or impossible to phase lock the FFO in that region. For a Nb–AlOx–Nb FFO, the transition corresponding *to* $V_{JSC} = V_g/3$ occurs around 450 GHz. So, by using the Nb–AlN–NbN FFOs we can cover the frequency gap from 450 to 550 GHz imposed by the gap value of all-Nb junctions. The feature in Fig. 10.4 around 1 mV is very likely due to a singularity at the difference of the superconducting gaps $\Delta_{NbN} - \Delta_{Nb}$.

Continuous frequency tuning at frequencies below 600 GHz for the Nb–AlN–NbN FFOs of moderate length is possible, although the damping is not sufficient to completely suppress the Fiske resonant structure at frequencies below $V_g/3$. For short junctions with a small α (wave attenuation factor), the distance between the steps in this resonant regime can be as large, that it is only possible to tune the FFO at the certain set of frequencies. For a 300–400 μm long Nb–AlN–NbN junction, this is not the case – the quality factor of the resonator formed by a long Nb–AlN–NbN Josephson junction is not so high at frequencies >350 GHz. Therefore, the resonance steps are slanting and the distance between them is not so big (see Fig. 10.4). This allows us to set any voltage (and any frequency) below V_{JSC}, but for each voltage only a certain set of currents should be used. So, in this case we have the regions of forbidden bias-current values, specific for each voltage below V_{JSC}, instead of the forbidden voltage regions for the Fiske regime in Nb–AlOx–Nb FFO [15]. Special algorithms have been developed for automatic working point selection in flight.

In Fig. 10.5, the typical current-voltage characteristics (IVCs) of a Nb–AlN–NbN SIS junction of an area of about 1 μm^2 is given, both the unpumped IVC (solid line) and the IVC when pumped by a Nb–AlN–NbN FFO at different frequencies (dotted lines). One can see that the FFO provides more than enough power for the mixer pumping. In this experiment, we use the test circuits with low-loss matching circuits tuned between 400 and 700 GHz. Even with the specially introduced 5 dB FFO/SIS mismatch (required for the SIR operation) the FFO delivers enough power for the SIS mixer operation in the TELIS frequency range of 480–650 GHz [7].

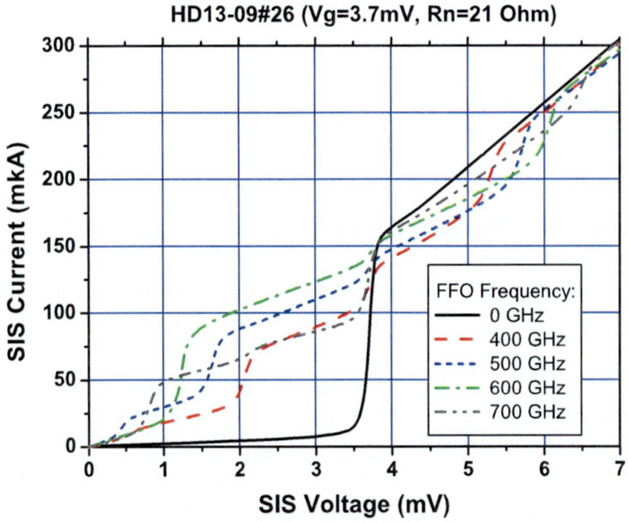

Fig. 10.5 The IVCs of the SIS mixer: unpumped – *solid curve*, pumped at different frequencies – *dashed and dotted lines*

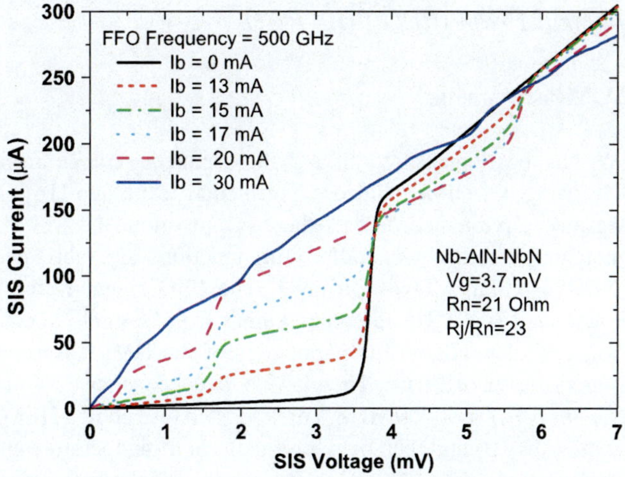

Fig. 10.6 The IVCs of the SIS mixer: unpumped – *black solid curve*, pumped at different FFO bias currents (different powers) – lines with symbols; FFO frequency = 500 GHz

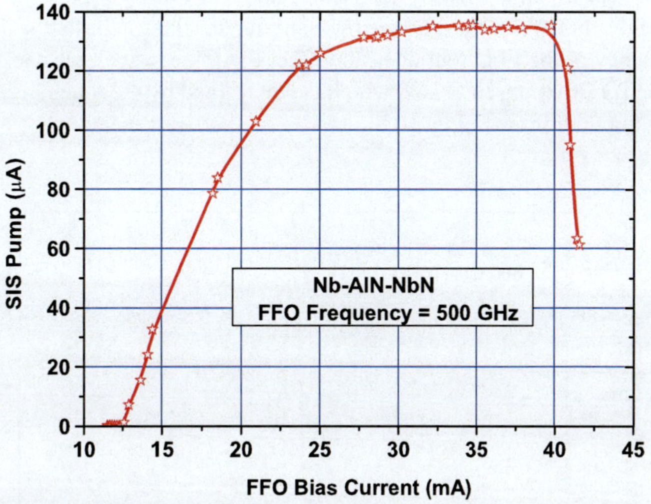

Fig. 10.7 The pump current of the SIS mixer biased at 3 mV as a function of the FFO bias current at the fixed frequency 500 GHz (see Figs. 10.4 and 10.6)

An important issue for the SIR operation is a possibility to tune the FFO power, while keeping the FFO frequency constant. This is demonstrated in Fig. 10.6, where the IVCs of an SIS mixer are shown, while being pumped at different FFO bias currents (different powers). The dependence of the SIS pump current on the FFO bias current is presented in Fig. 10.7, showing that the FFO power can be tuned more than 15 dB, while keeping the same frequency by proper adjustment of the control line current.

10.2.2 Spectral Properties of the FFO

10.2.2.1 LW Measurements

The FFO LW has been measured in a wide frequency range from 300 GHz up to 750 GHz using a well-developed experimental technique [16]. A specially designed integrated circuit incorporates the FFO junction, the SIS HM and the microwave matching circuits. Generally, both junctions are fabricated from the same Nb/AlN/NbN or Nb/AlOx/Nb trilayer. The FFO signal is fed to the SIS HM together with a 17–20 GHz reference signal from a stable synthesizer. The required power level depends on the parameters of the HM; it is about of 1 μW for a typical junction area of 1 μm^2. The intermediate frequency (IF) mixer product ($f_{IF} = \pm(f_{FFO} - n \cdot f_{SYN})$ at ~ *400* MHz is first boosted by a cooled HEMT amplifier ($T_n \sim 5$ K, gain = 30 dB) and then by a high-gain room-temperature amplifier.

To accurately measure the FFO line shape, the IF signal must be time-averaged by the spectrum analyzer. To remove low-frequency drift and interference from the bias supplies, temperature drift, etc., we use a narrow bandwidth (<10 kHz) Frequency Discriminator (FD) system with relatively low loop gain for *frequency locking* of the FFO. With the FD narrow-band feedback system that stabilizes the mean frequency of the FFO (but does not affect FFO line shape), we can accurately measure the free-running FFO LW, which is determined by the much faster internal ("natural") fluctuations (see Fig. 10.8).

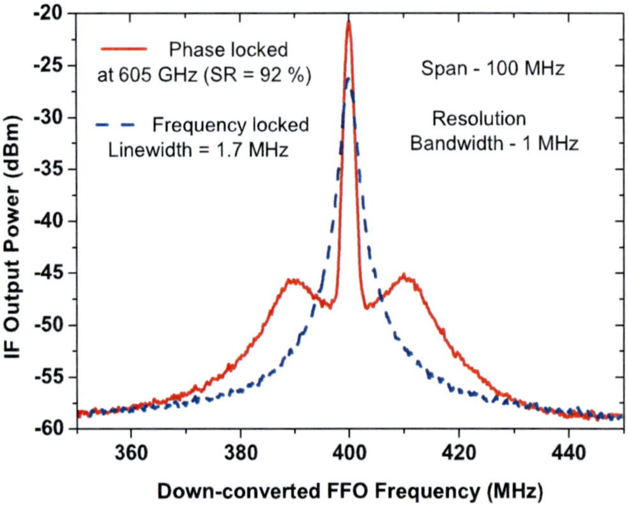

Fig. 10.8 Spectra of the Nb–AlN–NbN FFO operating at 515.2605 GHz (*blue dashed line* – frequency locked by FD; *red solid line* – phase-locked). Linewidth = 1.7 MHz; spectral ratio = 92%

The resulting IF signal is supplied also to the Phase Locking Loop (PLL) system. The phase-difference signal of the PLL is fed to the FFO control line current. Wideband operation of the PLL (10–15 MHz full width) is obtained by minimizing the cable loop length. A part of the IF signal is delivered to the spectrum analyzer through a power splitter (see Fig. 10.8). All instruments are synchronized to harmonics of a common 10 MHz reference oscillator.

The integrated HM may operate in two different regimes, either as a quasi-particle mixer (SIS) or as a Josephson mixer. To exclude the noise from the Josephson super-current fluctuations and thereby realize a pure quasi-particle regime, the super current has to be suppressed by a relatively large magnetic field. This requires a special control line placed near the SIS mixer. The quasi-particle regime of the HM operation can also be realized with sufficient synthesizer power. It has been shown [17] that the FFO LW and signal-to-noise ratio are almost the same for these two regimes, although the phase noise might be somewhat lower in the quasi-particle mode.

10.2.2.2 Dependence of the FFO Linewidth on FFO' Parameters

Detailed measurements of the FFO LW [18, 19] demonstrate a Lorentzian shape of the free-running FFO line in a wide frequency range up to 750 GHz, both at higher voltages on the flux flow step (FFS) and at lower voltages in the resonant regime on the Fiske steps (FSs). This implies that the free-running ("natural") FFO LW in all operational regimes is determined by the wideband thermal fluctuations and the shot noise. This is different from many traditional microwave oscillators, where the "natural" LW is very small and the observed LW can be attributed mainly to external fluctuations. It was found [18, 19] that the free-running FFO LW, δf, exceeds theoretical estimations made for lumped tunnel Josephson junction. The expression for the LW dependency on voltage and differential resistances found for all-Nb FFOs [18, 20] is valid for Nb–AlN–NbN junctions as well:

$$\delta f = \left(2\pi/\Phi_0^2\right)\left(R_\mathrm{d}^\mathrm{B} + K * R_\mathrm{d}^\mathrm{CL}\right)^2 S_\mathrm{i}(0), \tag{10.1}$$

where $S_\mathrm{i}0$ is the power density of low frequency current fluctuations, ${R_\mathrm{d}}^\mathrm{B}$ and ${R_\mathrm{d}}^\mathrm{CL}$ are differential resistances on bias and control line currents, respectively. Note that ratio ${R_\mathrm{d}}^\mathrm{CL}/{R_\mathrm{d}}^\mathrm{B}$ is constant for fixed FFO bias, so $\delta f = A(I_\mathrm{B})\,({R_\mathrm{d}}^\mathrm{B})^2\, S_\mathrm{i}0$.

Earlier, a so-called Super Fine Resonance Structure (SFRS) [21] was observed on the FFO IVCs, resulting in the jumps of the FFO between tiny steps (frequency spacing is of about 10 MHz, see Fig. 10.9). The presence of the SFRS prohibits phase locking at frequencies between the steps. This is unacceptable for practical applications. Recently, we found that the SFRS is related to interference of the acoustic waves created by the FFO (generation of the phonons by Josephson junction, see [22]). A special technological procedure allows us to eliminate this interference and to realize continuous FFO-frequency tuning in the SIR, being

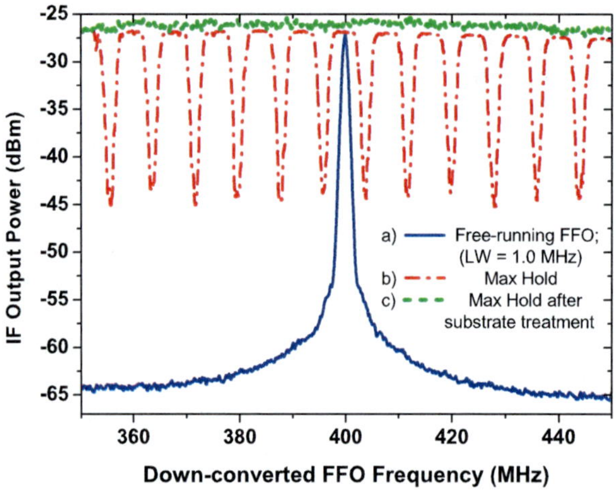

Fig. 10.9 Down-converted spectra of the FFO: (**a**) free-running FFO; (**b**), (**c**) – the lines show the maximum FFO signal level recorded in the MaxHold regime of the Spectrum Analyzer (the top point of curve "a") on the FFO frequency, measured before (**b**) and after (**c**) special Si substrate treatment

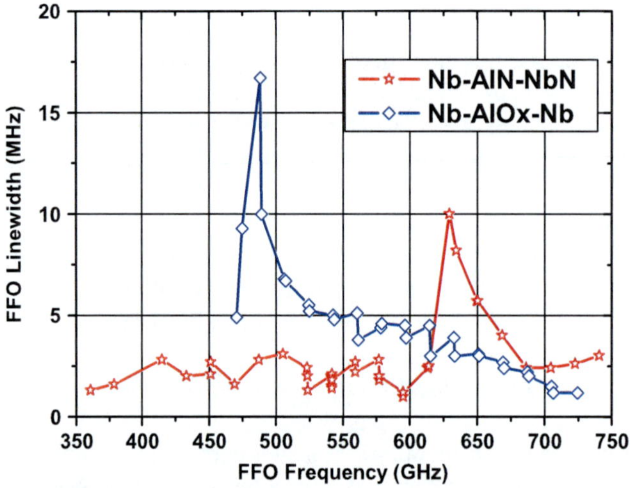

Fig. 10.10 Linewidth dependency on frequency for two types of the FFO

vitally important for TELIS project (see Fig. 10.9). Details of this study will be published elsewhere.

In Fig. 10.10, we present a comparative graph of the free-running FFO LW for two types of the tri-layer. One can see that the LW of Nb–AlN–NbN FFO is twice as small up to 600 GHz. It should be emphasized that due to overlapping FSs continuous tuning is possible and any desirable frequency can be realized. Several "stacked" stars at certain frequencies for the NbN FFO mean that the best LW value can be selected by adjusting FFO bias. Note that the spread in the LW values at a selected frequency is small and all can actually be applied for measurements. Each star corresponds to an "allowed" bias current at an FS (as described above in Sect. 2.1). Although the FFO tuning on an FS is complicated, the benefit in LW (and consequently the spectral ratio) is worth the effort. Linewidths below 3 MHz can be achieved in the whole range between 350 and 610 GHz. An abrupt increase of the FFO LW at some frequencies is caused by the Josephson self-coupling effect. The JSC (absorption of the FFO-emitted radiation by the quasi-particles in the cavity of the long junction, see above) considerably modifies the FFO properties at the voltages $V \approx V_{JSC} = 1/3\, V_g$ [13] (V_{JSC} corresponds to 620 GHz for the Nb–AlN–NbN FFO).

Previous LW measurements have demonstrated [7, 23] the essential dependences of the free-running FFO LW on the FFO voltage, its current density and geometry of the biasing electrodes. In this report, we summarize the results of the FFO study and optimization of the FFO layout for both types of FFOs. Recently, it was shown [4, 7] that the LW decreases considerably with increasing width, W, of the FFO junction. This is valid for all frequencies of interest, and consequently, the spectral ratio of the phase locked FFO for wide junctions is better. We have increased the FFO width up to 28 μm, which is more than five times the Josephson penetration depth λ_J. A number of FFOs with the same electrode layout, but different widths of the FFO junction (W = 4, 8, 12, 16, 20 and 28 μm) are fabricated using the same technological procedure yielding the same junction parameters (normal state resistance × area, $RnS = 30\,\Omega\,\mu m^2$). The results of the LW measurements of these circuits at three frequencies are presented in Fig. 10.11.

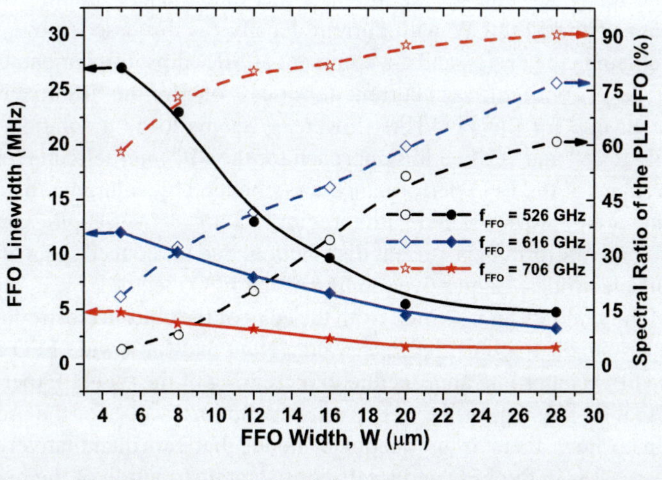

Fig. 10.11 Linewidth of free-running FFOs (*left axis*) and corresponding spectral ratio for the phase-locked FFO (*right axis*) measured at different FFO frequencies as a function of FFO width. All circuits are fabricated by the same technological procedure (RnS = $30\,\Omega\,\mu m^2$)

Even for the largest tested width ($W = 28\,\mu$m), there is no evidence of deterioration in the FFO behaviour. Furthermore, the power delivered to the SIS mixer is getting higher and the LW lower at all frequencies. The decrease of the FFO LW with increasing FFO width is in accordance with existing theoretical models and our expectations. The bias current differential resistance, R_d, decreases approximately inversely proportional to the bias current I_B. Since the FFO LW is proportional to $R_d{}^{2*} I_B$, it scales down linearly with the junction width. Of course, one can expect that the LW decrease will saturate and the FFO performance will deteriorate with further increase of the width (e.g., due to appearance of transversal modes). Without a reliable theory, the optimal value of the FFO width has to be determined experimentally. Note that for a wider FFO the center line of the junction is shifted away from the edge of the control line (the $R_d{}^{CL}$ goes down). This may result in a considerable reduction of extraneous noise from external magnetic fields. Furthermore, a wider FFO presumably will have a more uniform bias current distribution [4]. At the present state, the width of the FFO for TELIS is chosen to be $16\,\mu$m. This is a tradeoff between LW requirements and technical limitation on the maximum bias and control line currents (both should not exceed 70 mA).

In contrast to variation of the FFO LW on the FFO width, previous measurements [7] have demonstrated a considerable increase of the FFO LW with the FFO current density. This contradicts the simplified consideration: the increase of the FFO current density (as it is for increase of the FFO width) should result in the increase of the total FFO bias current, I_B, and reduce the FFO differential resistance on the bias current R_d. Since the FFO LW is proportional to $R_d{}^{2*} I_b$, one should expect the decrease of the measured FFO free-running LW for larger FFO current density. In reality, R_d does not decrease as much as this simple consideration predicts and the LW increases. On the contrary, a high value of the current density ($J_c \geq 8\,\text{kA/cm}^2$) is important for wide-band operation of the SIS-mixer at the submm wave range. The increase of the FFO LW with current density (as discussed above) creates a serious problem in the design and development of SIR chips. Implementation of two separate tri-layers with different current densities – one for the SIS mixer (high J_c) and the other one for the FFO/HM (lower J_c) seems to be a solution. We have successfully tested and verified this approach for the SIR microcircuits for TELIS.

Improvement of the FFO performance was obtained by enlarging the electrodes overlapping area, the so-called "idle region". Larger overlapping presumably provides a more uniform bias-current distribution, due to reduced inductance of the overlapping electrodes. Larger overlapping of the FFO electrodes also implies that the FFO of the same width is shifted from the edge of the bottom electrode, resulting in a considerable decrease of the $R_d{}^{CL}$ value. Note that for a wide FFO also some shift of the FFO center line appears due to increasing of the width. Experimentally, we found that an idle region $W_I = 10\,\mu$m is the optimal value for the present FFO design. Up to now, there is no adequate model that can quantitatively describe both the processes in the FFO and a self-consistent distribution of the bias current. Nevertheless, the presented results are very encouraging and these modifications of the FFO were implemented in the TELIS SIRs.

To further explore this approach, we have developed different designs of the "self-shielded" FFO with a large ground plane in the base electrode. Such FFOs are expected to be less sensitive for variations in the external magnetic field and have to provide more uniform bias current distribution (since all bias leads are laying over superconducting shield and have low inductance). Actually, the low-inductive bias leads provide a possibility of optimal (rather than uniform) current distribution, "regulated" by the FFO itself. The last feature optimizes the emitted FFO power. Indeed, the IVCs of all shielded FFOs are much more reproducible; the power delivered to HM is higher compared to a traditional design. Unfortunately, the free-running LW for all variants of shielded FFOs with separate bias leads is much larger than for FFOs of traditional design. It seems that injection of the bias via separate leads results in some spatial modulation of bias current [23] despite the additional triangular elements added for more uniform current injection. On the contrary, designs that employed three superconducting electrodes provide both perfect pumping and improved LW, details will be published elsewhere.

10.2.2.3 Spectral Ratio, Phase Noise

As it was mentioned above, the free-running FFO LW has to be well below 10 MHz to ensure stable FFO phase locking with a reasonably good spectral ratio (SR, the ratio between the carrier and total FFO power). For example, only about 50% of the FFO power can be phase locked by the present TELIS PLL system at free-running FFO LW of 5 MHz. A low spectral ratio results in a considerable error at resolving of the complicated atmospheric line shapes [12]. For the given PLL system, the value of the SR is fully determined by the free-running FFO LW: these two quantities are unambiguously related (see Fig. 10.12, where data for FFOs of different designs and types are presented). The theoretical curve, calculated in [24], coincides reasonably well with the experimental data. A possibility to considerably increase the SR by application of the ultra-wideband cryogenic PLL system has been recently demonstrated [25].

An important issue for TELIS operations is the possibility to tune the FFO frequency and power independently, while providing the same spectral ratio of PL FFO. The TELIS HM is pumped by a tunable reference frequency in the range of 19–21 GHz from the LO Source Unit (LSU), phase locked to the internal ultra stable 10 MHz Master Oscillator. The HM mixes the FFO signal with the n-th harmonic of the 19–21 GHz reference. The LW and SR of the TELIS FFO are almost constant over a wide range of FFO bias current at fixed FFO frequency (see Fig. 10.13). From this figure, one can see that the SR is about 50% over the range of bias current, I_b, 14–30 mA, while the pumping level varies from 3.5 μA at I_b = 14 mA up to 81 μA at I_b = 30 mA. Furthermore, the SR = 34% can be realized at I_b = 12 mA, where the HM pumping is below 0.5 μA. It means that at proper choice of the HM voltage and LSU power even moderate HM pumping by the FFO is enough for efficient PLL operation (providing sufficient signal-to-noise ratio).

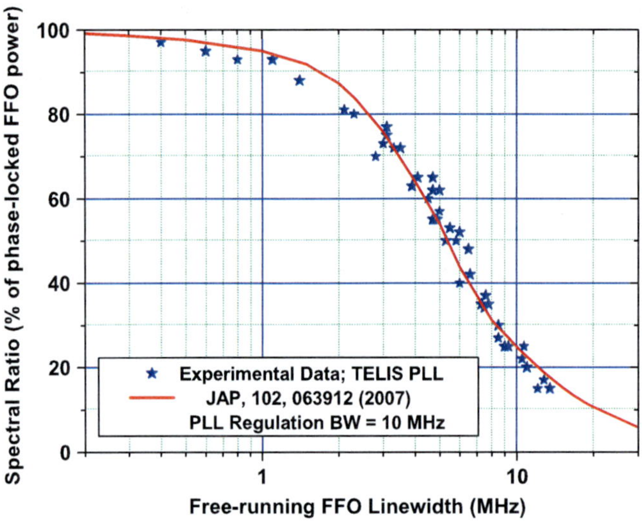

Fig. 10.12 Spectral ratio for the phase-locked FFO of different types and designs as a function of free-running FFO linewidth. *Solid line* – calculated dependence of the SR on FFO LW for PLL bandwidth = 10 MHz

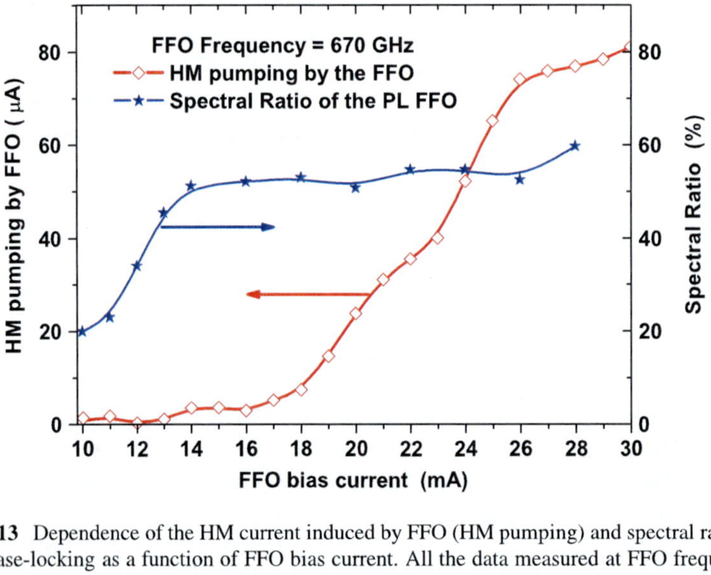

Fig. 10.13 Dependence of the HM current induced by FFO (HM pumping) and spectral ratio after FFO phase-locking as a function of FFO bias current. All the data measured at FFO frequency of 670 GHz

To prove the capabilities for high-resolution spectroscopy, line profiles around 625 GHz of OCS gas have been successfully measured by the SIR operating in the DSB regime [3]. The tests were done in a laboratory gas cell setup at a gas pressure down to 0.2 mBar, corresponding to the FWHM LW < 5 MHz. It was demonstrated that the spectrum recorded by the Digital Auto Correlator (DAC) is a convolution product of the signal (gas emission lines) with the FFO line spectrum; resolution in this experiment is limited by DAC back-end. More detailed spectral measurements data will be presented in the next section.

To investigate the ultimate frequency resolution of the receiver, we have measured the signal of the synthesizer multiplied by a super-lattice structure [26]. The signal recorded in these measurements is a convolution of the narrow-bandwidth (delta-function-like) spectrum of the synthesizer with phase locked spectrum of the FFO with an accuracy of the used resolution bandwidth of the spectrum analyzer (30 kHz). It was confirmed that the frequency resolution of the receiver is better than 100 kHz.

The residual phase noise of the phase locked FFO – measured relative to the reference synthesizer – as a function of the offset from the carrier is plotted in Fig. 10.14. To get the absolute FFO phase noise, one should add the synthesizer noise multiplied by n^2 to the residual phase noise of the FFO. Data for the Rohde&Schwarz®SMF100A Microwave Signal Generator with improved phase noise [27] are also presented in Fig. 10.14, for the case where the FFO, operating at 450 GHz, is locked to the 20th harmonic of the synthesizer, $n^2 = 400$. The total (absolute) FFO phase noise (solid line in Fig. 10.14) is dominated by the synthesizer

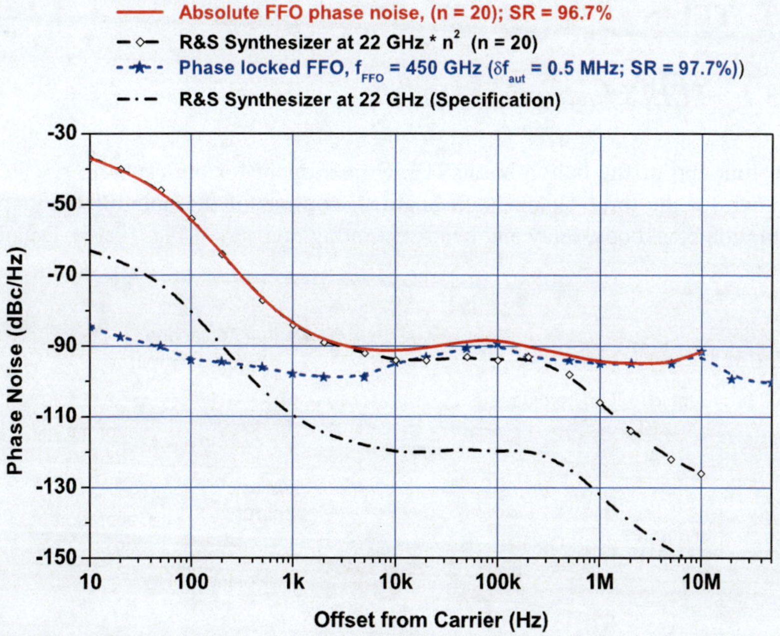

Fig. 10.14 Experimental phase noise of a phase locked FFO at 450 GHz. Since the phase noise of the FFO is measured relative to the 20th harmonic of the synthesizer, the synthesizer noise [27], multiplied by a factor $20^2 = 400$, should be added to the residual FFO noise to get the total (absolute) FFO phase noise – *solid line*

noise for offsets $< 10\,\text{kHz}$. The noise at larger frequency offset is mainly due to PLL system. Note that the FFO phase noise is overestimated since no subtraction of the noise added by the IF amplifier chain was performed; actually at offsets much larger than the PLL regulation bandwidth ($>$20 MHz) the measured phase noise is mainly determined by the IF chain.

This section can be summarized as follows. Continuous tuning of the frequency is possible for Nb–AlN–NbN FFOs due to bending and overlapping of the FSs, so that any desirable frequency can be realized. A possibility to phase lock the Nb–AlN–NbN FFO at any frequency in the range 350–750 GHz has been experimentally demonstrated. An optimized design of the FFO for TELIS has been developed and tested. A free-running LW value from 5 to 1 MHz has been measured in the frequency range 300–750 GHz for a "wide" FFO. As a result, the spectral ratio of the phased locked FFO varies from 50% to 97% correspondingly. The "unlocked" rest of the total FFO power increases the phase noise and the calibration error. To ensure remote operation of the phase locked SIR several procedures for its automatic computer control have been developed and tested. New designs of the FFO intended for further improvement of its parameters are under development, but even at the present state the Nb–AlN–NbN FFOs are mature enough for practical applications.

10.3 TELIS

10.3.1 TELIS Instrument Design

The front-end of the ballon-borne TELIS instrument for atmospheric research is common for the three channels on board. It consists of the pointing telescope, a calibration blackbody, relay and band-separating optics (see Fig. 10.15). Details of

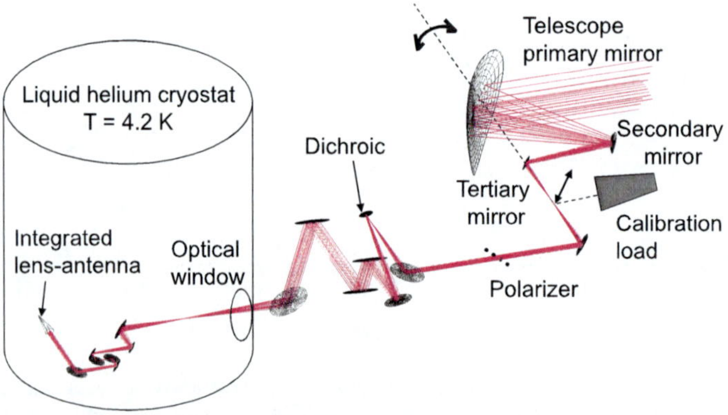

Fig. 10.15 Optical lay-out of the TELIS SIR channel

the optical design can be found in [28–30]. The three mirrors of the dual offset Cassegrain telescope are mounted on a common frame, rotatable around the optical axis of the output beam. Limb scanning is performed between the upper troposphere (8–10 km in the Arctic) to flight altitude (typically 32 km) in 1–2 km steps. At the tangent point of the line of sight, the vertical (elevation) resolution is about 2 km for an observational frequency of 500 GHz, scaling inversely proportional with frequency. In horizontal (azimuth) direction, the spatial resolution is about a factor of 2 less due to the anamorphicity of the telescope. This is allowed as the atmospheric properties within the beam hardly depend on the azimuth.

The radiometric gain of the spectrometers is calibrated once or twice in every Limb scan using a conical blackbody reference source and a measurement of the cold sky. For this, a small flip mirror is included between the telescope and the beam-separating optics. By measuring at two up-looking telescope positions, the impact of the remaining air above the gondola can be assessed.

Simultaneous observation by the receivers is achieved by quasi-optical beam splitting. First, a wire-grid-based polarizing beam splitter is employed to reflect one linear polarization to the 500 GHz channel, the other linear polarization is split by a dichroic filter between the SIR channel and the THz channel. Subsequently, offset mirrors shape and direct the three beams to the cryogenic channels. Inside the custom designed liquid-helium cooled cryostat, each receiver has dedicated cold optics, a superconducting mixing element and IF amplifiers.

The very compact 500 GHz receiver channel consists of a fixed-tuned waveguide SIS mixer, a cryogenic solid-state LO chain and a low-noise IF chain operating at a relatively high IF (IF $=$ 15–19 GHz) [31]. The 1.8 THz channel employs a cryogenic solid-state LO that is loss-less coupled into the mixer via a Martin–Puplett type optical interferometer. The mixer is based on a phonon-cooled NbN HEB (Hot Electron Bolometer) [32]. The 480–650 GHz SIR receiver channel is based on a single-chip SIR, as described in the next section.

The warm optics couples to the SIR channel with a beam that has a waist radius ranging from 2 to 3 mm, located at the cryostat window. The system-pupil is imaged by two additional mirrors on the silicon elliptical lens; on the back surface of this lens, the SIR chip is located. The SIR-channel cold-optics is also frequency independent to fully exploit the wide-band operation of the SIR device.

The amplitude-phase distribution of the near field beam of the SIR cold channel at 600 GHz as measured at the dewar window is shown in Fig. 10.16. The beam waist is measured to be 2.25 mm, which is within 1% of the designed value. The measured Gaussisity of the beam is 92.4%.

The IF processor (located on the main frame of TELIS) converts the amplified IF output signals of the three receivers to the input frequency range of the digital autocorrelator. The digital autocorrelator has a bandwidth of 2×2 GHz with 2,048 spectral channels. Both the IF processor and the digital autocorrelator are developed by Omnisys Instruments AB [33].

The SIR channel is controlled with a battery-operated ultra low-noise biasing system. Since noise on the bias lines of the FFO translates in a wider FFO LW, several precautions, such as decoupling of digital control lines and extensive

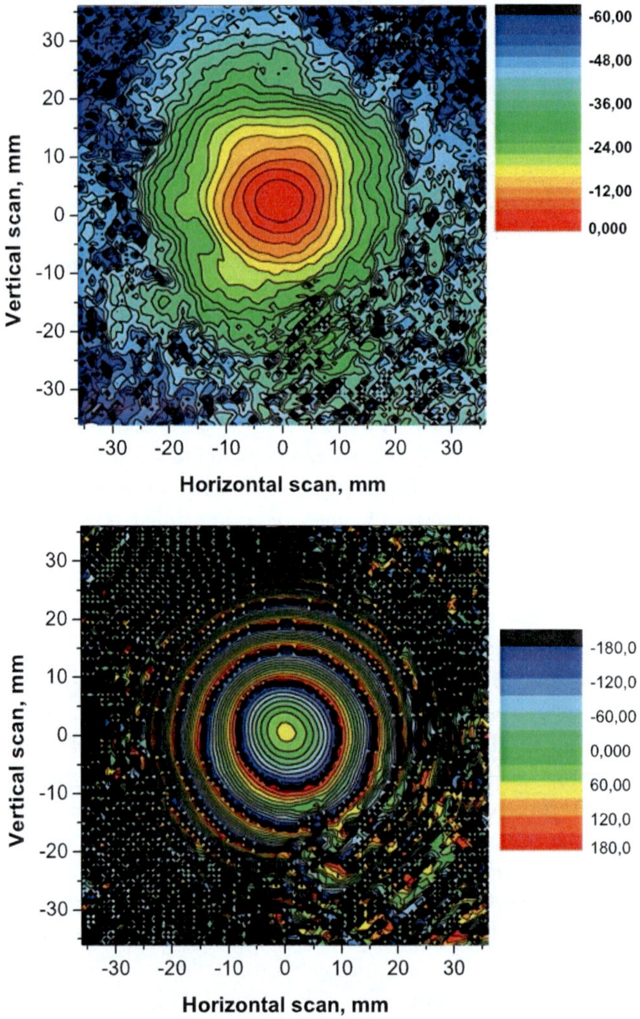

Fig. 10.16 The amplitude (*top figure*) and phase (*lower figure*) distribution of the near field beam of the SIR channel. The amplitude is given in units of dB. The distance from the beam waist is 110 mm and the frequency is 600 GHz

filtering and shielding, are implemented. The SIR bias unit is digitally controlled by the on-board DLR PC-104 computer, that also interfaces with the other channels, the digital autocorrelator, and with the host instrument MIPAS. A radio link provides real-time two-way contact with the ground segment consisting of a server computer with three dedicated client computers, coupled through TCP/IP socket connections. The complete system is dimensioned to have sufficient cooling liquids and battery power for a 24 h flight.

Fig. 10.17 Photo of the SIR microcircuit with double-slot antenna

10.3.2 SIR Channel Design

A key element of the 480–650 GHz channel is the SIR [1–5] that comprises in one chip (size of 4 × 4 × 0.5 mm, see Fig. 10.17) a low-noise SIS mixer with quasioptical antenna, a superconducting FFO [6] acting as an LO and a second SIS HM for FFO phase locking. Since the free-running LW of the FFO can be up to 10 MHz, for spectral applications the FFO has to be locked to an external reference oscillator employing a phase lock loop(PLL) system. The concept of the SIR looks very attractive for TELIS due to a wide tuning range of the FFO. In the SIR, the bandwidth is basically determined by the SIS mixer tuning structure and matching circuitry between the SIS and FFO; bandwidth up to 30–40% may be achieved with a twin-junction SIS mixer design (both for double-slot and double-dipole antennas). To achieve the required instantaneous bandwidth of 480–650 GHz, a twin-SIS mixer with 0.8 μm^2 junctions and new design of the FFO/SIS matching circuitry were implemented. A microscope photograph of the central part of the SIR chip with double-dipole antenna is presented in Fig. 10.18.

The resolution of the TELIS back-end spectrometer is 2.160 MHz, sufficient to resolve the exact shape of atmospheric lines. The FFO line shape and spectral stability should ideally be much better than this. However, the free-running LW of the FFO can be up to 10 MHz and therefore a PLL has been developed to phase lock the FFO to an external reference oscillator [6, 14]. For this, a small fraction of the FFO power is first directed to a so-called HM, placed on the SIR chip. The HM is pumped by an off-chip LSU, which is a tunable reference frequency in the range of 19–21 GHz. The frequency of the LSU is chosen such that the difference frequency

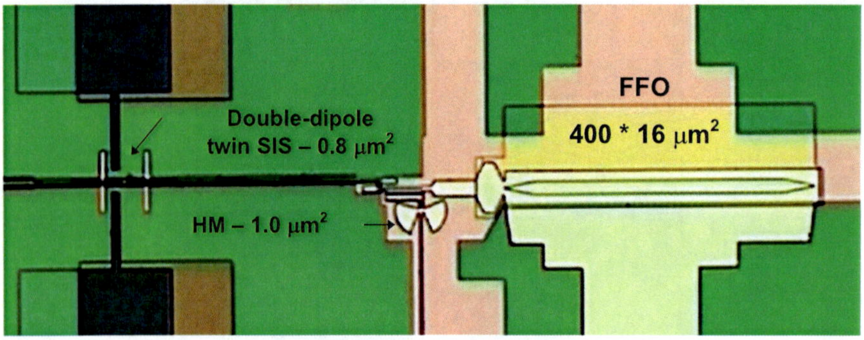

Fig. 10.18 Central part of the SIR chip with double-dipole antenna, twin SIS-mixer, and harmonic mixer for FFO phase-locking

of the nth harmonic of the LSU, generated by the HM, and the FFO is about 4 GHz. This difference signal is then amplified by a cryogenic low-noise HEMT amplifier and down-converted to 400 MHz by using a second reference at 3.6 GHz. Finally, the frequency and phase of this 400 MHz signal is compared against yet another reference frequency of 400 MHz and the resulting error signal is fed back to the FFO. The LSU and the reference signals at 3.6 GHz and at 400 MHz are all phase locked to an internal ultra stable 10 MHz Master Oscillator.

All components of the SIR microcircuits are fabricated in a high quality Nb–AlN/NbN tri-layer on a Si substrate [13]. The receiver chip is placed on the flat back surface of the elliptical silicon lens (forming an integrated lens-antenna) with accuracy 10 μm, determined by the tolerance analysis of the optical system. As the FFO is very sensitive for external electromagnetic interferences, the SIR chip is shielded by two concentric cylinders: the outer cylinder is made of cryoperm and the inner one of copper with a 100 μm coating of superconducting lead. All SIR channel components (including input optical elements) are mounted on a single plate inside a 240 × 180 × 80 mm box cooled by the thermo-straps to the temperature of about 4.2 K.

10.3.3 TELIS-SIR Channel Performance

The TELIS-SIR channel has been characterized in eight micro-windows that have been selected for the flight in (Sweden). These micro-windows have the following LO frequencies:

- 495.04 GHz for $H_2{}^{18}O$
- 496.88 GHz for HDO
- 505.60 GHz for BrO
- 507.27 GHz for ClO
- 515.25 GHz for O_2, pointing, and temperature

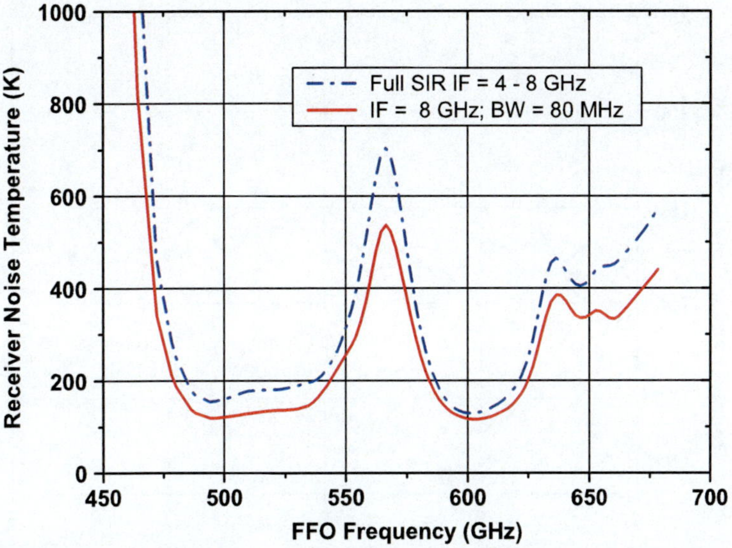

Fig. 10.19 Measured DSB receiver noise temperature of the SIR device selected for flight at 8 GHz IF (*solid line*) and integrated in the 4–8 GHz IF range (*dashed line*)

- 519.25 GHz for BrO and NO_2
- 607.70 GHz for ozone isotopes
- 619.10 GHz for HCl, ClO and HOCl

Initial flight values for the parameters for the FFO, SIS, and HM mixers have been determined for each micro-window. Dedicated algorithms allowing for fast switching between LO frequencies and for in-flight optimization of the SIR have been developed (see below). It takes about 1 min of stabilization and optimization to switch between two LO settings. All experimental results discussed here have been obtained with the SIR flight device.

The measured double sideband (DSB) receiver noise temperature T_R, uncorrected for any loss, is presented in Fig. 10.19 as a function of LO frequency. As can be seen, the noise is well below 200 K at all frequencies of interest, with a minimum of 120 K at 500 and 600 GHz. The noise peak around 540–575 GHz is partially spurious, caused by absorption of water vapor in the path between calibration sources and the cryostat, and partially real – due to properties of the SIS-mixer tuning circuitry. The relatively high noise in this band is of no concern for science observations, since this part of the atmospheric spectrum is obscured by a highly saturated water-vapor line rendering it virtually useless for atmospheric science. The noise as a function of IF is fairly flat in the frequency range 4–8 GHz, as can be seen in Fig. 10.20, where (DSB) receiver noise temperature is plotted as a function of IF. The dependence of the receiver noise temperature on the SIS bias voltage is shown in Fig. 10.21; one can see that for Nb–AlN/NbN circuits there is very wide range of SIS bias voltages where T_R is almost constant.

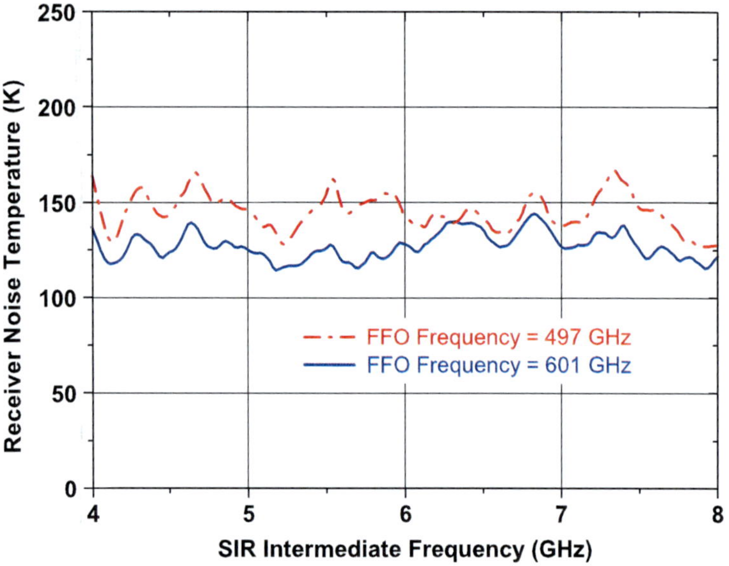

Fig. 10.20 DSB receiver noise temperature as a function of the IF, taken at two FFO frequencies: 497 and 601 GHz

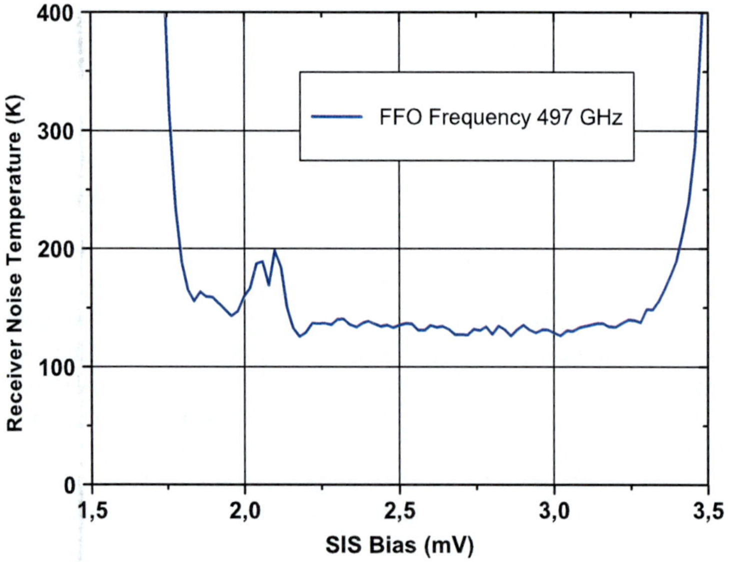

Fig. 10.21 DSB receiver noise temperature as a function of the SIS bias voltage measured at the FFO frequency 497 GHz

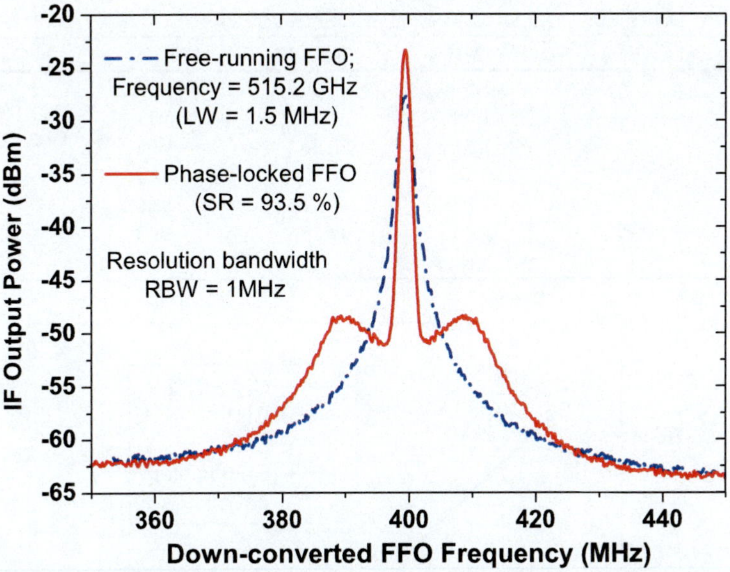

Fig. 10.22 Spectra of the FFO operating at 515.2 GHz (*blue dashed line* – frequency locked; *red solid line* – phase-locked). Linewidth (LW) = 1.5 MHz; signal-to-noise ratio (SNR) = 36 dB; spectral ration (SR) = 93.5%. Spectra measured with RBW = 1 MHz, span = 100 MHz

After optimization of the FFO design, the free-running LW between 7 and 0.5 MHz has been measured in the frequency range 350–750 GHz (see Fig. 10.10), which allows to phase lock from 35% to 95% of the emitted FFO. Example of the free-running (frequency-locked) and phase locked spectra of the FFO measured for flight SIR at one of the frequencies selected for first TELIS flight are presented in Fig. 10.22.

Data for five important TELIS frequencies are summarized in Table 10.1. It should be mentioned that the noise of the digital electronics at frequencies of about 1 MHz slightly increases the measured LW value, while the PLL is able to suppress the interference (that results in larger SR than can be expected from measured LW). Note also the dependence of the SR and LW on the FFO bias current related to variation of the differential resistance along FS.

For the TELIS measurement strategy, it is important to know whether the timing of limb sounding should depend on the stability of the complete receiver chain. The stability determines the optimum achievable measurement time for a single integration, and thus the required frequency of the calibration cycle. The stability of the complete TELIS-SIR system has been determined with a noise-fluctuation bandwidth of 17 MHz, and the results [10] are presented in Fig. 10.23. For the two IF channels that are used to determine the Allan variance, it is found that the Allan stability time is about 13.5 s. When the difference of the two channels is taken to determine the Allan variance (this is the so-called spectroscopic, or differential,

Table 10.1 Data for the flight SIR at selected TELIS frequencies

FFO frequency (GHz)	LW (MHz)	SNR (dB)	SR (%)	FFO Ib (mA)
495.04	1.5	32	90	30.7
496.88	1.5	37	93	31.3
515.25	1.5	36	93.5	29.2
607.7	2.1	32	87.7	30
607.7	1.8	32.6	88.6	34
619.1	5.4	25	63.3	30
619.1	4.6	26.8	70.3	34

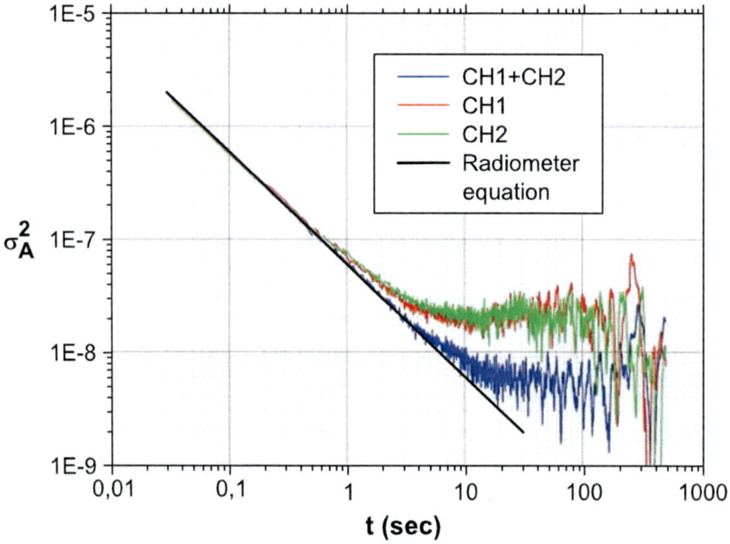

Fig. 10.23 System stability of the SIR channel. FFO is phase locked at 600 GHz. The two lines at the top (*red and green*) represent individual channel variances, the *blue line* is representative of the spectroscopic variance and the *straight black line* corresponds to the radiometer equation

mode), the Allan stability time of 20 s is found. This is comparable to stabilities measured for astronomical receivers.

Within TELIS, a 1.5 s integration time per tangent height is used. This is mainly driven by the required integrated signal levels at the autocorrelator input. The stability of the SIR channel therefore poses no constraints on the observing strategy.

The SIR is a complicated device as it contains multiple interactive superconducting elements: an SIS mixer, an FFO, and an HM for the FFO phase locking. Special algorithms and procedures have been developed and tested to facilitate characterization of the SIR at reasonable timescales and for the SIR control during the flight. These routines include:

- Fast definition of the FFO operational conditions (both on the FS and in the flux-flow regimes).
- Measurements of the free-running FFO LW.

- Optimization of the LSU and HM parameters.
- Optimization of the PLL operation.
- Minimization of the SIR noise temperature.
- Setting all predefined SIR parameters in the exact sequence for control during the flight.
- Continuous monitoring of the main SIR parameters.
- Adjustment (or recovering) of the SIR operational state.

10.3.4 Kiruna Campaigns and Preliminary Science Results

TELIS had two successful scientific campaigns from Kiruna, North-Sweden, in March 2009 and in January 2010. The instrument was launched together with the MIPAS instrument on the MIPAS-B2 gondola (see Fig. 10.24). The launch of both flights took place around midnight. During the ascents, the SIR channel behaved nominally and already after 30 min the first spectra were recorded. In the 2009 flight, the first flight ceiling of 35 km was reached after 3 h and 1 h later the flight continued at 28 km altitude. In the 2010 flight, the ascent took more than 4 h to reach a flight ceiling of 34 km where the balloon stayed for the remainder of the flight. Several night recordings were taken, necessary for background measurements for species with a diurnal cycle and for instrument calibration. The instrument proved to be stable against the strong temperature variations of the atmosphere during ascent (with ambient temperatures as low as −90 °C) and during sunrise. The south eastern wind allowed for long flights of about 12 h over Finland during both

Fig. 10.24 TELIS-MIPAS launch at Esrange, Sweden; March 2009. Balloon size: 400,000 m^3; payload weight: 1,200 kg

campaigns. After sunrise, the diurnal cycle of various species was monitored and in total several hundred limb sequences have been recorded in each flight. The MIPAS-TELIS balloon system performed nominally during the flight and after the parachute landing and recovery, the instruments were found to be undamaged, allowing for post-flight checks and calibration measurements.

The science goals of the campaign from Kiruna, North Sweden, were threefold: investigation of the stratospheric hydrological cycle by measurements of isotopic water, catalytic ozone destruction by chlorine chemistry, and the bromine content of the stratosphere. In addition, measurements were performed for space-borne instruments (ENVISAT satellite and in 2010 also SMILES aboard the International Space Station). Data presented in Fig. 10.25 prove the capabilities of the TELIS-SIR channel for high-resolution spectroscopy. In this case, the FFO frequency is tuned to 505.600 GHz, the telescope is 6 degrees up-looking, and the gondola altitude is 35.780 km. The width of the ozone lines are almost fully determined by atmospheric conditions (Doppler and pressure broadening) and are about 10 MHz, as expected.

Chlorine ozone destruction peaks in the arctic winter and/or spring when the so-called polar vortex breaks up. During this event, the ClO radical, responsible for catalytic ozone destruction, becomes available in huge amounts. However, chlorine is also stored in nonreactive reservoir species of which HCl is an important member. The amount of HCl in the stratosphere is a measure of the total nonactive Cl content and is as such an important species to monitor in ozone chemistry studies.

In Fig. 10.26 measured spectra are shown in which the HCl line at IF = 6.81 GHz is well pronounced for line of sights. A line of sight is determined by the flight

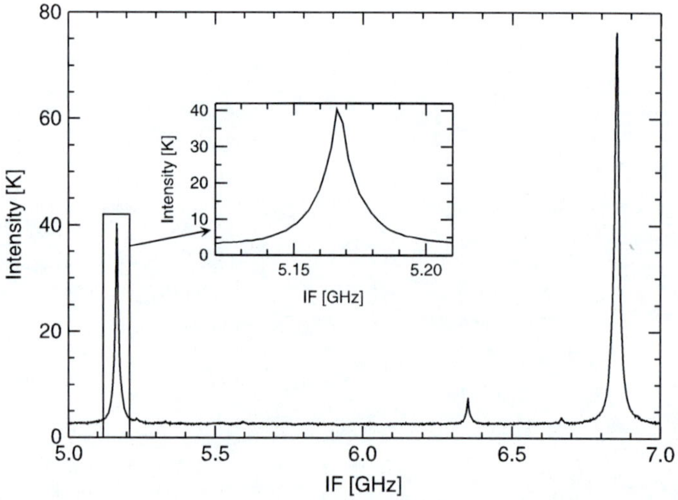

Fig. 10.25 Ozone spectrum recorded during the 2009 campaign with the SIR channel (FFO frequency = 505.6 GHz) from an altitude of 35.780 km and with the telescope pointing 6° upward. The width of the lines is ca. 10 MHz and is fully determined by atmospheric conditions. The intensity of the received signal in Kelvin is plotted as ordinate of the graphs

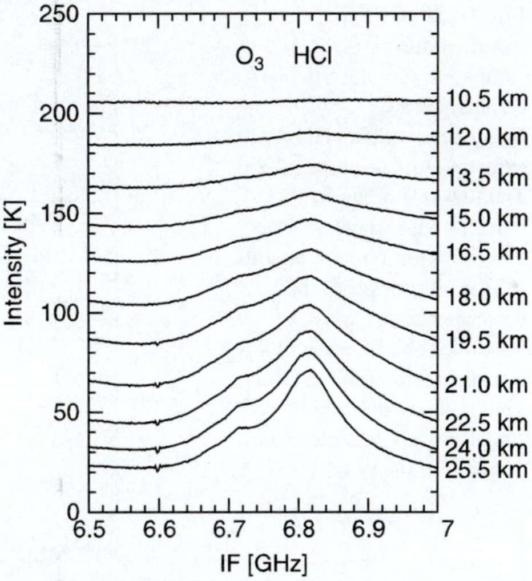

Fig. 10.26 HCl spectra recorded during the 2009 campaign with the SIR channel (FFO frequency = 619.1 GHz) from an altitude of ca. 28 km. The spectra correspond to line of sights with tangent heights in the range of 10.5 (*top*) to 25.5 km (*bottom*), which is also the altitude for which a particular spectrum is most sensitive. The intensity of the received signal in Kelvin is plotted as ordinate of the graphs

altitude of the balloon platform and the tilt of the telescope. Each line of sight results in different altitude sensitivities and effectively probes different parts of the atmosphere. The altitudes mentioned in Fig. 10.26 refer to the lowest probed altitude, the tangent height, for a certain line of sight and generally corresponds to the altitude for which the measurement is mostly sensitive.

Bromine depletes ozone even more aggressively than chlorine on a per molecule basis, but its abundances are much lower. In fact, the total amount of stratospheric bromine is still not settled and is currently one of the main uncertainties in the importance of bromine in ozone depletion.

In 2009, the flight took place in nonvortex conditions whereas in 2010 the flight was in the polar vortex. The diurnal cycle of ClO has been observed in both flights, albeit with a much higher time resolution in the 2010 flight (ca. 1 min). In 2009, BrO has been detected, although barely as the line was superimposed on another spectral feature. However, in 2010 the BrO line, with a level of only ca. 0.3 K, was isolated and clearly detected. The data reduction is on-going but the first spectra for HCl, ClO, and BrO are presented in Figs. 10.26, 10.27, and 10.28, respectively.

10.3.5 SIR for Noninvasive Medical Diagnostics

High sensitivity and spectral resolution of the integrated spectrometer enables the analysis of multicomponent gas mixtures. Exhaled air of human includes about 400 gases, of which some can be indicators of various diseases and pathology. For example, nitric oxide, NO, was detected in the exhaled air of patients suffering from bronchial asthma, pneumonia, and other chronic inflammatory diseases of upper airways. Besides, nitric oxide may have an effect on the reaction of tumors and

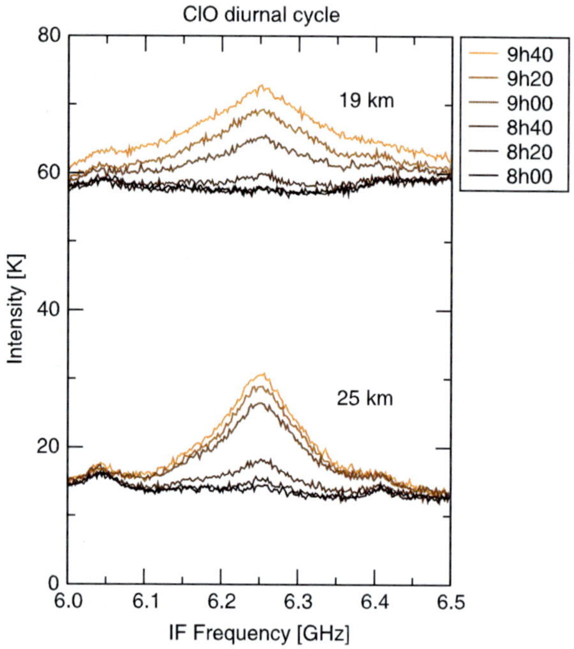

Fig. 10.27 ClO spectra recorded during the 2010 campaign with the SIR channel (FFO frequency = 507.3 GHz) from an altitude of ca. 34 km. The two sets of spectra correspond, respectively, to 25 and 19 km tangent heights, i.e. the altitudes for which these spectra are most sensitive. The increase of ClO over time is clearly visible. The intensity of the received signal in Kelvin is plotted as ordinate of the graphs

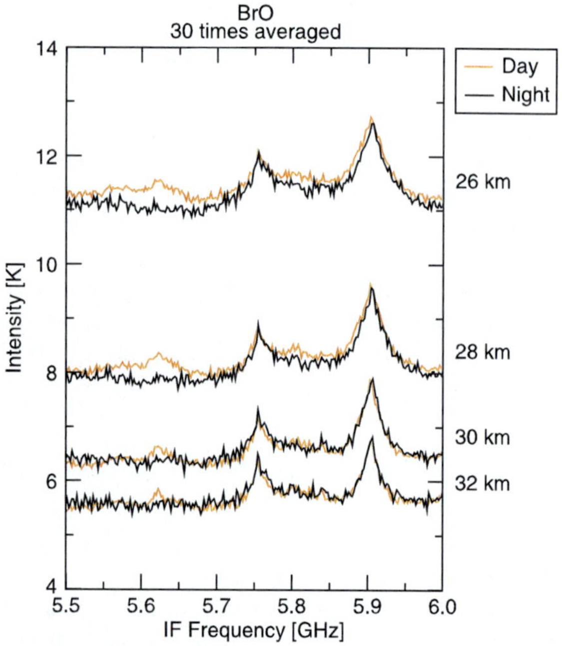

Fig. 10.28 BrO spectra recorded during the 2010 campaign with the SIR channel (FFO frequency = 519.3 GHz) from an altitude of ca. 34 km. In the daytime measurement (*orange*), the BrO line is visible at 5.6 GHz. During the nighttime measurement (*black*), BrO is absent, as expected. The shown tangent heights are from top to bottom 26, 28, 30, and 32 km, i.e. the altitudes for which these spectra are most sensitive. The intensity of the received signal in Kelvin is plotted as ordinate of the graphs

healthy tissues on radiation therapy. Another example may concern the opportunity of noninvasive diagnostics of gastritis or peptic ulcer of the stomach by measuring the concentration of ammonia in exhaled air. Nowadays urease respiratory tests (application of urea with C^{13}) are mainly used to detect the diseases. However, the method is quite expensive and its sensitivity is restricted by natural variations of C^{13} in exhaled air during the procedure. Natural concentrations of ammonia instead are quite low, so the measurement of the ammonia concentration could be a good alternative. Another important application concerns the noninvasive diagnostics of diabetes, where exhaled acetone is an indicator.

A laboratory setup for spectral analysis of the exhalted air has been developed at IREE (input frequency range 480–630 GHz, noise temperature below 200 K over the range, spectral resolution below 1 MHz), based on the integrated spectrometer for atmosphere monitoring. The instrument parameters allow us to measure the spectral lines of the rotational transitions for most of the substances in the exhalted air. The laboratory setup has been developed and demonstrated using the gases OCS and NH_3 in the laboratory gas cell. Clear and well-defined response has been measured at the expected frequencies of spectral lines for pressures down to 10^{-3} mBar. Examples of the NH_3 spectra recorded by the SIR with the Fast Fourier Transform Spectrometer (FFTS) as a back-end, and by the novel technique based on application of the additional oscillator are presented in Fig. 10.29 and 10.30, respectively. The possibility to measure the spectral response in a few seconds has been demonstrated experimentally. This allows to carry on the real-time medical survey. First spectral measurements by the integrated receiver of exhalted air in the sub-THz range have demonstrated good selectivity and speed of the analysis as well as high sensitivity.

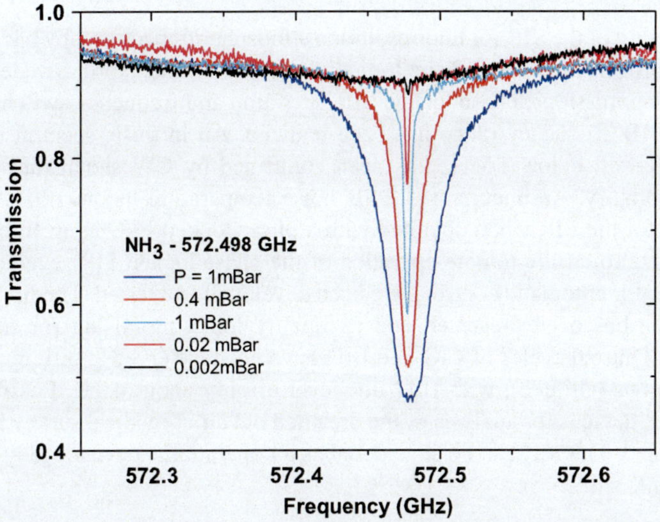

Fig. 10.29 NH_3 spectra measured by the SIR with FFTS back-end at different pressures

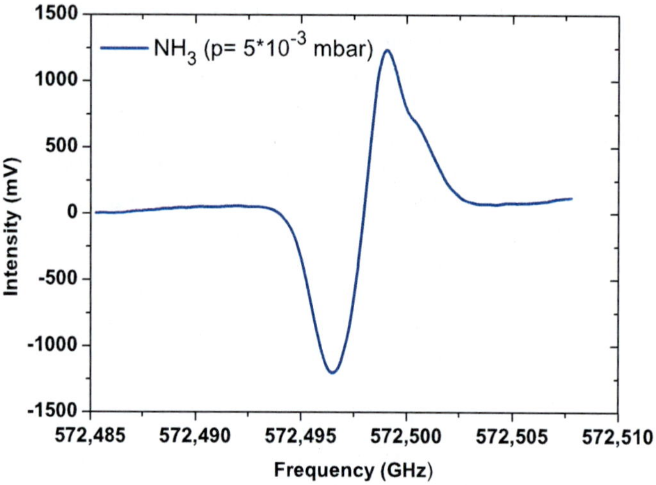

Fig. 10.30 Response (derivative of the spectral line), measured for NH_3 gas by the integrated receiver with implementation of novel technique (details will be published elsewhere)

For example, we have measured ammonia concentration with sensitivity on the order of 10^{-9} (1 ppb).

10.4 Summary

The capability of the SIR for highresolution atmospheric spectroscopy has been successfully proven with scientific balloon flights from Kiruna, North Sweden. During the two 12-h missions, phase locked SIR operation and frequency switching in the 480–650 GHz frequency range has been realized. An intrinsic spectral resolution of the SIR well below 1 MHz has been confirmed by CW signal measurements in the laboratory. An uncorrected DSB noise temperature below 120 K has been measured for the SIR when operated with a phase locked FFO at an IF bandwidth of 4–8 GHz. To ensure remote operation of the phase locked SIR several software procedures for automatic control have been developed and tested. The first tentative HCl profile has been presented and its quality looks promising for future data reduction. Diurnal cycles of ClO and BrO have been observed at different viewing configurations (altitude), with BrO line level of only about 0.5 K. Possibilities to use the SIR devices for analysis of the breathed out air at medical survey have been demonstrated. The SIR can be considered as an operational device, ready for many applications.

Acknowledgements The authors thank colleagues at DLR, IPM, IREE, and SRON for help and assistance in the SIR channel design and characterization: J Barkhof, A Baryshev, J Kooi,

O Koryukin, A Pankratov, D Paveliev O Pylypenko, M Romanini, and S Shitov; as well as T de Graauw and W Wild are acknowledged for their support of this work.

The work was supported in parts by RFBR projects 09–02–00246, 09–02–12172-ofi-m, Grant for Leading Scientific School 5423.2010.2 and State contract No. 02.740.11.0795.

References

1. V.P. Koshelets, S.V. Shitov, L.V. Filippenko, A.M. Baryshev, H. Golstein, T. de Graauw, W. Luinge, H. Schaeffer, H. van de Stadt First implementation of a superconducting integrated receiver at 450 GHz. Appl. Phys. Lett. **68**(9), 1273 (1996)
2. V.P. Koshelets, S.V. Shitov, Integrated superconducting receivers. Supercond. Sci. Technol. **13**, R53 (2000)
3. P. Yagoubov, R. Hoogeveen, M. Torgashin, A. Khudchenko, V. Koshelets, N. Suttiwong, G. Wagner, M. Birk, 550–650 GHz spectrometer development for TELIS. Proc. ISSTT 338 (2006)
4. V.P. Koshelets, A.B. Ermakov, L.V. Filippenko, A.V. Khudchenko, O.S. Kiselev, A.S. Sobolev, M.Y.u. Torgashin, P.A. Yagoubov, R.W.M. Hoogeveen, W. Wild Iintegrated submillimeter receiver for TELIS. IEEE Trans. Appl. Supercond. **17**, 336 (2007)
5. G. de Lange, D. Boersma, J. Dercksen, P. Dmitriev, A. Ermakov, L. Filippenko, H. Golstein, R. Hoogeveen, L. de Jong, A. Khudchenko, N. Kinev, O. Kiselev, B. van Kuik, A. de Lange, J. van Rantwijk, A. Sobolev, M. Torgashin, E. de Vries, P. Yagoubov, V. Koshelets, Development and characterization of the superconducting integrated receiver channel of the TELIS atmospheric sounder. Supercond. Sci. Technol. **23**(4), 045016 (2010)
6. T. Nagatsuma, K. Enpuku, F. Irie, K. Yoshida, Flux-flow type Josephson oscillator for millimeter and submillimeter wave region. J. Appl. Phys. **54**, 3302, (1983), see also Pt. II: J. Appl. Phys. **56**, 3284 (1984); Pt. III, J. Appl. Phys. **58**, 441 (1985); Pt. IV, J. Appl. Phys. **63**, 1130 (1988)
7. V.P. Koshelets, P.N. Dmitriev, A.B. Ermakov, A.S. Sobolev, M.Y.u. Torgashin, V.V. Kurin, A.L. Pankratov, J. Mygind, Optimization of the phase-locked flux-flow oscillator for the submm integrated receiver. IEEE Trans. Appl. Supercond. **15**, 964–967 (2005)
8. I. Mehdi, THz local oscillator technology. Proc. SPIE **5498**, 103 (2004)
9. R.W.M. Hoogeveen, P.A. Yagoubov, A. de Lange, A.M. Selig, V.P. Koshelets, B.N. Ellison, M. Birk, Proc. SPIE **5978**, 440 (2005)
10. R.W.M. Hoogeveen, P.A. Yagoubov, G. de Lange, A. de Lange, V. Koshelets, M. Birk, B. Ellison, Proc. SPIE **6744**, 67441U-1 (2007)
11. F. Friedl-Vallon, G. Maucher, M. Seefeldner, O. Trieschmann, A. Kleinert, A. Lengel, C. Keim, H. Oelhaf, H. Fischer, Appl. Opt. **43**, 3335 (2004)
12. V.P. Koshelets, S.V. Shitov, A.B. Ermakov, O.V. Koryukin, L.V. Filippenko, A.V. Khudchenko, M.Y.u. Torgashin, P. Yagoubov, R. Hoogeveen, O.M. Pylypenko, Superconducting integrated receiver for TELIS. IEEE Trans. Appl. Supercond. **15**, 960–963, 2005
13. V.P. Koshelets, S.V. Shitov, A.V. Shchukin, L.V. Filippenko, J. Mygind, A.V. Ustinov, Self-pumping effects and radiation linewidth of Josephson flux flow oscillators. Phys Rev B **56**, 5572–5577 (1997)
14. P.N. Dmitriev, I.L. Lapitskaya, L.V. Filippenko, A.B. Ermakov, S.V. Shitov, G.V. Prokopenko, S.A. Kovtonyuk, V.P. Koshelets, High quality Nb-based integrated circuits for high frequency and digital applications. IEEE Trans. Appl. Supercond. **13**(2), 107–110 (2003)
15. M.Y.u. Torgashin, V.P. Koshelets, P.N. Dmitriev, A.B. Ermakov, L.V. Filippenko, P.A. Yagoubov, Superconducting integrated receivers based on Nb-AlN-NbN circuits. IEEE Trans. Appl. Supercond. **17**, 379–382 (2007)
16. V.P. Koshelets, S.V. Shitov, A.V. Shchukin, L.V. Filippenko, J. Mygind, Linewidth of submillimeter wave flux-flow oscillators. Appl. Phys. Lett. **69**, 699–701 (1996)

17. V.P. Koshelets, J. Mygind, Flux flow oscillators for superconducting integrated submm wave receivers, in Studies of High Temperature Superconductors, **39**, ed. by A.V. Narlikar (NOVA Science Publishers, New York, 2001) 213–244
18. V.P. Koshelets, A.B. Ermakov, P.N. Dmitriev, A.S. Sobolev, A.M. Baryshev, P.R. Wesselius, J. Mygind, Radiation linewidth of flux flow oscillators. Supercond. Sci. Technol. **14**, 1040–1043 (2001)
19. V.P. Koshelets, S.V. Shitov, P.N. Dmitriev, A.B. Ermakov, L.V. Filippenko, V.V. Khodos, V.L. Vaks, A.M. Baryshev, P.R. Wesselius, J. Mygind, Towards a phase-locked superconducting integrated receiver: prospects and limitations. Phys. C **367**, 249–255 (2002)
20. A.L. Pankratov, Form and width of spectral line of a Josephson flux flow oscillator. Phys. Rev. B. **65**, 054504 (2002)
21. V.P. Koshelets, A.B. Ermakov, S.V. Shitov, P.N. Dmitriev, L.V. Filippenko, A.M. Baryshev, W. Luinge, J. Mygind, V.L. Vaks, D.G. Pavel'ev, Superfine resonant structure on IVC of long Josephson junctions and its influence on flux flow oscillator linewidth. IEEE Trans. Appl. Supercond. **11**, 1211–1214 (2001)
22. P. Berberich, R. Buemann, H. Kinder, Monochromatic phonon generation by the Josephson effect. Phys. Rev. Lett. **49**(20), 1500–1503 (1982)
23. V.P. Koshelets, S.V. Shitov, L.V. Filippenko, P.N. Dmitriev, A.B. Ermakov, A.S. Sobolev, M.Y.u. Torgashin, A.L. Pankratov, V.V. Kurin, P. Yagoubov, R. Hoogeveen Superconducting phase-locked local oscillator for a submm integrated receiver. Supercond. Sci. Technol. **17**, $127–$131 (2004)
24. A.L. Pankratov, V.L. Vaks, V.P. Koshelets, Spectral properties of phase locked flux flow oscillator. J. Appl. Phys. **102**, 0629 (2007)
25. A.V. Khudchenko, V.P. Koshelets, P.N. Dmitriev, A.B. Ermakov, P.A. Yagoubov, O.M. Pylypenko, Cryogenic phase detector for superconducting integrated receiver. IEEE Trans. Appl. Supercond. **17**, 606–608 (2007)
26. E. Schomburg, R. Scheuerer, S. Brandl, K.F. Renk, D.G. Paveliev, Y.u. Koschurinov, V. Ustinov, A. Zhukov, A. Kovsh, P.S. Kopev, Electron. Lett. **35**(17) (1999)
27. http://www2.rohde-schwarz.com/product/smf100a.html
28. P. Yagoubov, H. van de Stadt, R. Hoogeveen, V. Koshelets, M. Birk, A. Murk, in *Proceedings of the 28th ESA Antenna Workshop on Space Antenna Systems and Technologies*, Noordwijk, 2, 763 (2005)
29. P.A. Yagoubov, W.J. Vreeling, H. van de Stadt, R.W.M. Hoogeveen, O.V. Koryukin, V.P. Koshelets, O.M. Pylypenko, A. Murk, in *Proceedings of the 16th Intern. Conf. on Space Terahertz Technology, Gothenburg*, 438 (2005)
30. A. Murk, P. Yagoubov, U. Mair, M. Birk, G. Wagner, H. van de Stadt, R. Hoogeveen, N. Kämpfer, *Proc. of the 28th ESA Antenna Workshop on Space Antenna Systems and Technologies, Noordwijk*, 757 (2005)
31. B.N. Ellison, B.P. Moyna, D.N. Matheson, A. Jones, S.M.X. Claude, C. Mann, B.J. Kerridge, R. Siddans, R. Munro, W.J. Reburn, in *Proceedings of 2nd ESA Workshop on Millimetre Wave Technology and Applications, Espoo*, (1998)
32. S. Cherednichenko, V. Drakinskiy, T. Berg, P. Khosropanah, E. Kollberg, Rev. Sci. Instrum. **79**, 034501 (2008)
33. A. Emrich, S. Andersson, M. Knis *Proceedings of the joint 31st International Conference on Infrared Millimeter Waves and 14th International Conference on Teraherz Electronics, Shanghai*, 314 (2006)

Chapter 11
Cryogenic Phase-Locking Loop System Based on SIS Tunnel Junction

A.V. Khudchenko, V.P. Koshelets, and K.V. Kalashnikov

Abstract An ultra-wideband cryogenic phase-locking loop (CPLL) system is a new cryogenic device. The CPLL is intended for phase-locking of a Flux-Flow Oscillator (FFO) in a Superconducting Integrated Receiver (SIR) but can be used for any cryogenic terahertz oscillator. The key element of the CPLL is Cryogenic Phase Detector (CPD), a recently proposed new superconducting element. The CPD is an innovative implementation of superconductor–insulator–superconductor (SIS) tunnel junction. All components of the CPLL reside inside a cryostat at 4.2 K, with the loop length of about 50 cm and the total loop delay 5.5 ns. Such a small delay results in CPLL synchronization bandwidth as wide as 40 MHz and allows phase-locking of more than 60% of the power emitted by the FFO even for FFO linewidth of about 10 MHz. This percentage of phase-locked power three times exceeds that achieved with conventional room-temperature PLLs. Such an improvement enables reducing the FFO phase noise and extending the SIR operation range.

Another new approach to the FFO phase-locking has been proposed and experimentally verified. The FFO has been synchronized by a cryogenic harmonic phase detector (CHPD) based on the SIS junction. The CHPD operates simultaneously as the harmonic mixer (HM) and phase detector. We have studied the HM based on the SIS junction theoretically; in particular we calculated 3D dependences of the HM output signal power versus the bias voltage and the LO power. Results of the calculations have been compared with experimental measurements. Good qualitative

A.V. Khudchenko (✉)
Kotel'nikov Institute of Radio Engineering and Electronics, Russian Academy of Science, Mokhovaya st. 11/7, 125009, Moscow, Russia
and
SRON Netherlands Institute for Space Research, 9700 AV Groningen, The Netherlands
e-mail: Khudchenko@hitech.cplire.ru

V.P. Koshelets · K.V. Kalashnikov
Kotel'nikov Institute of Radio Engineering and Electronics, Russian Academy of Science, Mokhovaya st. 11/7, 125009, Moscow, Russia

A. Sidorenko (ed.), *Fundamentals of Superconducting Nanoelectronics*, NanoScience and Technology, DOI 10.1007/978-3-642-20158-5_11,

and quantitative correspondence has been achieved. The FFO phase-locking by the CHPD has been demonstrated. Such a PLL system is expected to be extra wideband. This concept is very promising for building of the multi-pixel SIR array.

11.1 Introduction

A cryogenic phase-locking loop (CPLL) system is primarily intended for phase stabilization of a Local Oscillator (LO) of a Superconducting Integrated Receiver (SIR) [1, 2]. The SIR circuit embodied on a chip (size of $4 \times 4 \times 0.5$ mm) comprises a low-noise superconductor–insulator–superconductor (SIS) mixer with quasioptical planar antenna, a superconducting Flux-Flow Oscillator (FFO) acting as a tunable LO in the frequency range 400–700 GHz and a second SIS harmonic mixer (HM) to phase-lock the FFO.

The FFO emission spectrum was found to be the Lorentz line [3,4] at frequencies up to 750 GHz. Such line shape indicates that the free-running ("natural") FFO linewidth is determined by the wideband noise, that is thermal fluctuations and the shot noise. This is contrary to many traditional microwave oscillators (e.g., backward-wave and Gunn oscillators), where the "natural" linewidth is rather narrow and is broadened primarily by external fluctuations. In fact, the FFO linewidth may be up to 10 MHz. Therefore, for higher spectral resolution, FFO frequency should be stabilized with a wideband phase-locking loop (PLL) system. The wider is PLL synchronization bandwidth (BW), the larger part of the FFO power can be phase-locked (PL), resulting in lower phase noise.

The ratio of the phase-locked power to the total power emitted by the oscillator is called a "spectral ratio" (SR). Actually, the SR is determined from a measured LO spectrum as a ratio of the phase-locked power of the carrier to the total emitted power comprising the carrier and the intrinsic phase noise power generated by LO [4]. So, the SR would be 100% for an ideal LO with a delta-shaped spectrum and zero phase noise level. Usually, the noise power is embedded in the bandwidth 100 MHz. It should be noted that measured phase noise includes a contribution of the IF amplifier, and we do not subtract this level in SR calculations (so the actual SR is even a little higher). The error introduced due to IF amplifier noise is about 1–3%. A specially designed semiconductor room temperature PLL (RT PLL) with the BW of about 12 MHz is used for the FFO phase locking. The RT PLL provides SR value of about 50% for the 5 MHz-wide FFO emission line.

There are several reasons to extend BW of the PLL system much beyond the present 12 MHz. The first one is that the FFO linewidth exceeds 10 MHz at the voltages exceeding one-third of the FFO gap voltage, where the Josephson self-coupling effect drastically modifies FFO *I–V* curves increasing differential resistance and internal damping [5, 6]. In case of such a wide line, an essential part of the emitted FFO power cannot be phase-locked with the RT PLL (the SR is as low as 20–25%). The second reason is that FFOs based on NbN or NbTiN films are the most attractive for future SIR applications at frequencies of about 1 THz. The linewidth of such FFOs can considerably exceed 10 MHz due to higher surface losses. In this situation, the PLL's bandwidth has to be as large as 50 MHz (to reach at least 50% of the SR value required for most applications). The third reason is that using SIR for interferometery applications requires an LO with extremely high phase stability. For example, for the ALMA project (interferometer in Chile with baseline up to 15 km) the LO rms phase noise should be considerably less than 75 fs (the value of rms atmospheric fluctuation at interferometer site) [7]. The required value of the SR for this rms phase noise level at frequency of about 600 GHz is more than 95%. To provide such a high SR value even for the FFO with relatively narrow linewidth of 2 MHz, the PLL should have BW of about 50 MHz.

The BW is determined by the group delay τ in the loop [8, 9]. It should be emphasized that the noise of RT PLL elements is low enough as compared to the noise of IF amplifier, and it does not affect phase-locking efficiency. For RT PLL, the BW is limited to 12 MHz due to the total delay of about 15 ns (including 5 ns contribution from the PLL filters and semiconductor electronics and 10 ns from the 2 m-long cables connecting the RT PLL electronics with cryogenic FFO). The minimal cabling length is restricted by the size of a cryostat and cannot be essentially reduced without increasing the heat flux into the cryogenic space. Note that the traditional PLL is a semiconductor-based device designed for 300 K, and it cannot be placed directly into the cryostat.

To improve spectral characteristics of the FFO and to overcome the limitation of the RT PLL, we propose the CPLL system based on a Cryogenic Phase Detector (CPD) [8–10]. All elements of the CPLL can be located very close to the FFO inside the cryostat to avoid any temperature gradients between FFO and phase detector and to minimize the loop length. Negligible CPD delay and small group delays in the short loop enable an ultra-wide BW.

11.2 CPD Properties

The CPD is a key element of the CPLL. It is new cryo-electronic device based on micron-sized SIS tunnel junction with a tunnel barrier of about 1–2 nm. The junction has current density 5–10 kA/cm^2, and the gap voltage V_g is 2.8 MV for Nb–AlO$_x$–Nb circuits.

Synchronization systems theory shows that any mixer can operate as a phase detector [11]. The SIS junction is a well-known mixer element due to its nonlinear properties [12, 13] and can be utilized as a CPD for the CPLL.

There are several advantages of the SIS implementation for the FFO phase-locking. First of all, the SIS junction has a power consumption of about 10^{-6} W that is much smaller than 10^{-3} W for semiconductor diode phase detectors, and second, both the SIS and the FFO operate at temperature 4.2 K. These peculiarities allow place the CPD and the FFO close to each other, and even integrate them on the same chip. Such a PLL system will be very compact and extra wideband.

The output signal of the phase detector with a sinusoidal response is:

$$\varepsilon(t) = mA_1A_2 \sin\varphi(t) = K_{\mathrm{PD}} \sin\varphi(t),$$

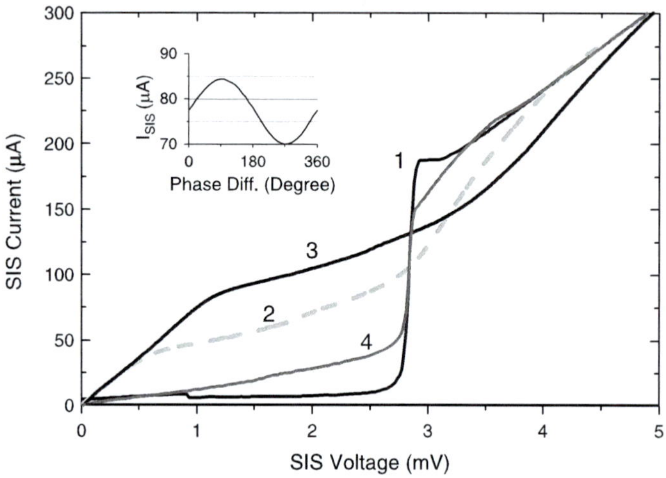

Fig. 11.1 *I–V* curves of the SIS junction measured at various settings of the microwave signals (frequency 5 GHz): curve "1" – autonomous; "2" – pumped by one microwave signal; "3" – pumped by two in-phase microwave signals; "4" – pumped by two anti-phase microwave signals; PSynth1 = 0.3 μW, PSynt2 = 0.1 μW. *Inset* shows a sinusoidal dependence of the SIS current vs phase difference between the signals. Junction aria is about 2 μ^2 with $R_n S$ product about 30 $\Omega\mu^2$

where $\varphi(t)$ is a phase difference between input signals, K_{PD} – a phase detector gain factor, which is equal to the maximum amplitude of the output signal; m – a mixer efficiency, A_1 and A_2 – input signals amplitudes. The phase characteristic of an SIS-mixer (it can be nonsinusoidal), frequency and amplitude properties of output signal and optimal amplitudes and frequencies of input signals must be determined. In this case, such a junction will be specified as a phase detector.

A principle of CPD operation can be demonstrated by the tunnel junction *I–V* curves. A typical autonomous *I–V* curve of the SIS junction is shown in Fig. 11.1 (curve 1); the curve 2 corresponds to the *I–V* curves of the SIS pumped by a microwave signal [8]. Two microwave signals applied in phase give a higher power and a higher pumping level (curve 3), while antiphase results in the lower pumping level (curve 4). The difference between curve 3 and curve 4 corresponds to the phase response amplitude. It is important that this difference is rather large in the wide range of the CPD bias voltages.

The CPD is a unique phase detector due to nonlinearity of the SIS junction [12]. Although the frequency of the synthesizer for curves 2–4 is a few gighertz (corresponding voltage is rather low compared to the smearing of the junction superconducting gap), the shape of the pumped CPD *I–V* curve looks like a result of irradiation by a high frequency signal. Apparently, a certain number of higher order harmonics of the applied signal are excited in the SIS junction due to its nonlinearity; these harmonics effectively pump the tunnel junction. We have compared experimental *I–V* curves and dependence of the CPD pump current as a function of power of the applied signal with the simulated ones. The calculations were performed on the basis of the Tien–Gordon formula for photon-assisted tunneling (see, e.g. [12]). The best fit to the experimental results was found for frequency of applied signal in the range of 80–100 GHz; at that point, the shape of experimental curves qualitatively resembles the theoretical ones for synthesizer frequency in the range of 0.2–20 GHz [9].

11.2.1 Phase Characteristics

It was shown [8] that dependence of the CPD output versus phase difference between the two input signals can be sinusoidal for the CPD under the proper experimental conditions (see inset in Fig. 11.1). The phase response shape is close to sinusoidal if its amplitude is less than 0.1 I_g (I_g is a current jump at the gap voltage). If the amplitude becomes larger, the response shape changes. Small shape deviations are not critical for operation of the PLL system because it works on the slope part of the phase characteristic, i.e. at $\varphi(t) << 1$. For the nonsinusoidal response, the main parameters are the derivative $d\varepsilon(t)/d\varphi(t)$ (that gives the slope K_{PD}) and the amplitude of the linear section of the phase response (these two parameters are equal to the K_{PD} for sinusoidal signal). However when the phase response amplitude is very high and even approaches to I_g, the response shape is usually deformed so strongly that the PLL system works unstable.

11.2.2 Frequency Characteristics

The capacitance of the used SIS junctions is about 0.1 pF. Due to such a small capacitance, the imaginary part of the junction impedance is very large at low frequencies until the value of order 1 GHz. Therefore, the CPD response amplitude remains flat in this range that was confirmed experimentally for frequencies up to 750 MHz [8]. This magnitude essentially exceeds 100 MHz required for the CPLL operation.

The CPD input signal frequency can be ranged from 0.2 MHz to 20 GHz due to nonlinear SIS properties described above. In this range, the operation frequency of the CPLL can be chosen. Actually, the CPLL operation frequency has the upper limit that is restricted only by working frequency of cryogenic HEMT amplifier (value of order 10 GHz). And the lower limit about 0.4 GHz is determined as a magnitude that should be considerably larger than the CPLL output bandwidth BW = 30–50 MHz.

11.2.3 Amplitude Properties

The CPD amplitude properties can be described by the dependence of the output signal versus input signal amplitudes and bias voltage of the SIS junction. From such a dependence, the maximum output signal and optimal input signal amplitudes can be found.

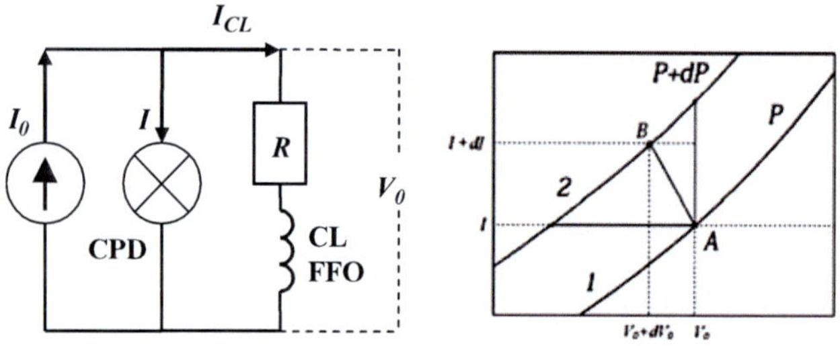

Fig. 11.2 Diagram of the CPD – FFO connection through the resistor R (*left figure*). Part of the CPD *I–V* curves (*right figure*) in large scale at the voltage $V_0 < V_g$

The CPD of the CPLL is connected to the FFO control line (CL FFO) to tune the FFO frequency [4]. The simplified diagram of the CPD–FFO interface is shown in Fig. 11.2 (left). This equivalent circuit is certainly valid at frequencies of interest 0–100 MHz. The current source generates I_0 and a part of it I splits to the CPD. The CPD voltage V_0 can be expressed in terms of load R and current I by Ohm's law $I_0 = I + V_0/R$.

The tunneling current of the junction changes with increasing of the applied microwave power P (Fig. 11.2 right curves 1 and 2). For the power variation $\mathrm{d}P$, the CPD biasing point goes along the load line $(1/R)$ from point A to point B. The changing of the CPD voltage $\mathrm{d}V_0$ is given by the relation:

$$\mathrm{d}V_0 = -\frac{r_\mathrm{d} R}{r_\mathrm{d} + R}\frac{\partial I}{\partial P}\mathrm{d}P,$$

where r_d is a CPD differential resistance, $\partial I/\partial P$ is a partial derivative (can be calculated from the dependence *I(P)* measured at the fixed CPD voltage). The maximal available value of the derivative $\partial I/\partial P$ rises with enhancing of the current jump at gap voltage I_g, that is with increasing of the tunnel junction area.

The combined power of two microwave signals of power P_1 and P_2 applied to the junction with a phase difference φ is: $P(\varphi) = P_1 + P_2 + 2\sqrt{P_1 P_2}\cos\varphi$. In PLL system, one of these two signals is generated by local oscillator (LO) and the second one by a reference oscillator. Then the LO is phase-locked the phase difference φ is close to $\pi/2$ and the deviation is small $(\varphi(t) - \pi/2) << 1$, for the CPD we have:

$$\mathrm{d}V_0 = -2\frac{r_\mathrm{d} R}{r_\mathrm{d} + R}\frac{\partial I}{\partial P}\sqrt{P_1 P_2}\mathrm{d}\varphi,$$

here $P = P_1 + P_2$. The r_d and $\partial I/\partial P$ are experimentally measured; the optimal R is of about $10\,\Omega$. The derivative $\partial I/\partial P$ is a function of powers P_1 and P_2 and bias voltage V_0. The detailed analysis of this formula and experimental results show that output signal amplitude can achieve the value of about $0.2\,V_g$ at the bias voltage $0.55\,V_g$. For the Nb–AlO_x–Nb SIS junction, which *I–V* curves are shown in the Fig. 11.1, the maximum output signal riches −50 dBm. The dependence of the CPD maximum output versus V_g is rather flat and stay in 3 dB range at the interval 0.25–0.8 V_g. For the optimal value of the input signals, the induced pumping current of the SIS junction is of about 0.3 I_g.

The contribution to the current I_{CL} from the CPD is determined by the dV_0/R value. Thereby, the phase deviation $d\varphi$ leads to the variation of the FFO frequency df_{FFO}:

$$df_{FFO} = kdV_{FFO} = k\frac{dV_0}{R}Rd_{CLFFO} = -kRd_{CL\_FFO}\frac{r_d}{r_d + R}\frac{\partial I}{\partial P}\sqrt{P_1 P_2}d\varphi,$$

here $k = 483.6\,\text{MHz}/\mu\text{V}$ is the Josephson constant and $Rd_{CL\ FFO} = dV_{FFO}/I_{CL}$ is the FFO differential resistance by I_{CL} (typically, about 0.01 Ω).

This formula describes the efficiency of the CPD–FFO coupling; it has been experimentally verified with a good accuracy [9, 14], and used for design of the CPLL. An important result is that the value of df_{FFO} is found to be linearly proportional to the derivative $\partial I/\partial P$, magnitude $Rd_{CL\ FFO}$, amplitudes of the microwave signals, and is determined by the r_d and the R.

The coefficient $K = df_{FFO}/d\varphi$ is the open loop gain in the CPLL system [11,15]. It can be adjusted by variation of the powers P_1 and P_2 to achieve the optimal feedback in the loop providing the maximum SR in the FFO spectra. For the Nb–AlO_x–Nb SIS junction presented in Fig. 11.1, the coefficient K achieves the value 250 MHz/rad at $R = 10\,\Omega$ and $Rd_{CL\ FFO} = 0.01\,\Omega$.

The K also shows the maximum frequency deviation of the FFO, which can be tuned by CPD, that is the holding range of the CPLL system that is an upper limit of the BW. It means that the BW of the PLL system based on the CPD connected directly to the FFO (without any DC amplifier between) can reach the value of about 250 MHz.

11.3 CPLL System: Description and Experimental Results

Several experimental embodiments of the CPLL systems with various operation frequencies and various loop lengths have been developed and tested [8–10, 16]. A flowchart of the latest design is shown in Fig. 11.3. The FFO emission at the frequency of the order of 600 GHz is downconverted by the HM to the frequency 4 GHz and amplified by two HEMT-amplifiers, HEMT 1 and HEMT 2. This signal is compared with the external reference signal by the CPD and the resulting output error signal (proportional to the phase difference) is applied to the FFO to control its frequency. A certain fraction of the microwave power after HEMT 1 is tapped by the directional coupler to the spectrum analyzer. A filter between the CPD and the FFO comprises a rejecter filter to suppress signals close to the reference frequency and an integrating loop filter for locking and holding of the FFO frequency. The integrating loop filter significantly increases stability of the CPLL system. It also considerable decreases a phase noise of the phase-locked FFO near the carrier at frequencies

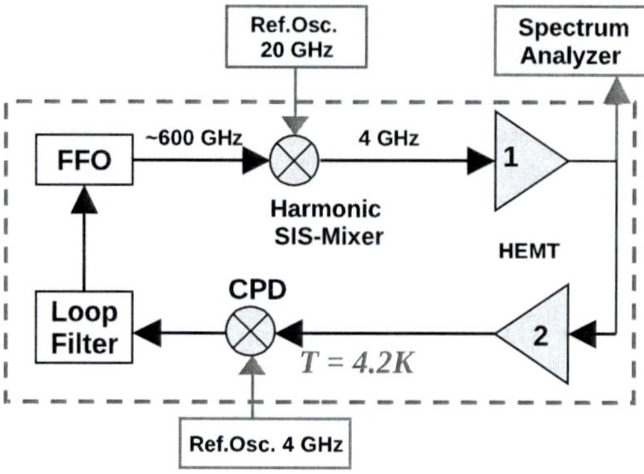

Fig. 11.3 Flow chart of the CPLL for FFO phase locking

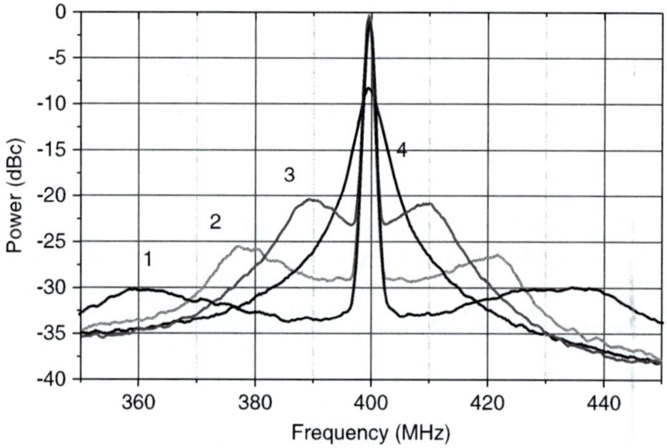

Fig. 11.4 Downconverted spectra of the FFO operating at 600 GHz: curve "1" – Free running, linewidth 2 MHz; "2" – phase-locked with CPLL, BW 40 MHz (SR = 93.5%); "3" – phase-locked with CPLL, BW 25 MHz (SR = 90.5%); "4" – phase-locked with RT PLL, BW 12 MHz (SR = 82%)

offset less 100 kHz. The essential moment is that the filters are based on a passive elements operating at 4.2 K and can be placed close to the FFO and the CPD [16].

In the latest design, the loop length has been reduced down to 50 cm, which corresponds to delay 2.5 ns. The operating frequency for this system was chosen 4 GHz. For such a frequency, a band-stop filter (between CPD and FFO) with delay less than 0.5 ns has been developed. The reduction of the loop delay due to the described modifications results in increasing of the BW up to 40 MHz (see Fig. 11.4, curve 2).

In Fig. 11.4, downconverted spectra of the FFO phase-locked by different PLL systems are presented (spectrum of the free-running FFO is also shown for comparison). It should be noted that all phase-locked spectra have typical shape consisting of a central peak containing locked power, and phase noise shoulders on

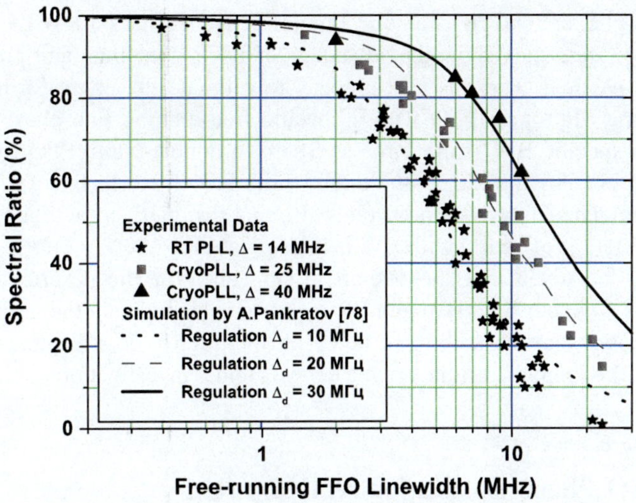

Fig. 11.5 SR vs FFO linewidth for PLL systems with various synchronization bandwidths

sides of the peak. For example, the tops of shoulders for the CPLL with BW 40 MHz are located at offsets 35–40 MHz from the carrier (curve 2 on the Fig. 11.4). At these frequencies, phase of the return signal is shifted by π due to the delay in the loop. So instead of noise suppression, the increasing of the phase noise occurs in the PLL system; position of these shoulders allows estimation of the PLL synchronization BW. These shoulders rise and become sharp with increasing of the PLL loop gain. The optimal PLL gain for any given PLL system is achieved for the shoulders about 3dB higher than phase noise near the carrier. In this case, the maximum SR and signal-to-noise ratio is reached.

From Fig. 11.4 one can see that increasing of the PLL BW leads to expansion of the frequency range, where the phase noise of the PL FFO is lower than the level for the free-running FFO. The reason is that the wider BW of the CPLL enables us using higher loop gain bringing essential reduction of phase noise in the entire regulation bandwidth (frequency of unity gain is shifted to the right) without considerable increase of the phase noise in the shoulders. For the FFO linewidth of 2 MHz, the CPLL is capable of phase-locking of 93.5% power, as compared to 82% for the RT PLL (Fig. 11.4). For the FFO linewidth of 7 MHz, the new CPLL yields SR = 81% instead of 41% for the RT PLL. Even higher gain was obtained at a broader linewidth of 11 MHz for a free-running FFO: SR = 63% in case of CPLL against 23% for RT PLL.

Summary of the obtained results is presented in Fig. 11.5. Experimental data for the RT PLL with the BW of 12 MHz are shown by asterisks, for the CPLL with BW 25 MHz – by squares, and triangles show results for the described above CPLL with BW of 40 MHz. Lines show results of numerical simulations [17] to fit the experiment. The "regulation BW" of 10 MHz in simulations roughly corresponds to the synchronization bandwidth "BW" = 12 MHz in experiment. Data for the described CPLL are shown as a solid line (effective regulation BW of 30 MHz in simulations); these data show the advantage of BW broadening and prospects for future improvement. There is an easy way to estimate the effect form PLL BW widening: the SR for an FFO with a specific free-running line phase-locked by PLL with a specific BW is the same as SR for a double-width FFO free-running line phase-locked by the PLL with double-width BW (it works like a scale effect). Experimental data analysis demonstrates also that the relative fraction of the phase noise (1 − SR) falls linearly with BW increasing at SR > 50%.

It should be noted that the overall group delay for the described CPLL is estimated as 3 ns, but both BW and SR measurements indicate the delay of about 5.5 ns. It means that the additional delay of about 2.5 ns is still present in some elements of the loop; the origin for this delay is under investigation.

11.4 FFO Phase-Locking Directly by HM

The radical improvement of the described above CPLL system may be achieved by implementation of a single SIS junction as the HM and the phase detector simultaneously. This cryogenic harmonic phase detector (CHPD) substitutes the HM, the HEMT-amplifier, and the CPD in the CPLL (Fig. 11.3). The CHPD and all loop elements can be placed on the same chip with the FFO, which leads to further loop group delay reduction. Such a PLL system will be ultra-wideband. A block diagram of the new PLL system is shown in Fig. 11.6. The signals from the FFO and the LO#1 are applied to the CHPD. The frequency of harmonic of the LO#1 signal is equal to the FFO frequency (of the order of 600 GHz). The CHPD generates an output signal proportional to the phase difference between the FFO and the appropriate harmonic of the LO#1. This error signal is applied directly to the FFO control line through a low-pass filter. For demonstration of the CHPD operation, the additional SIS Mixer is used. This mixer operates as a HM (see Fig. 11.6). It is utilized for observation of the FFO radiation line spectrum and monitoring of the phase-locking effect. The LO#2 frequency is chosen to obtain the intermediate frequency (IF) of this mixer of about 6 GHz. Such an IF is determined by the operating range 4–8 GHz of the HEMT-amplifiers in IF chain of the SIS mixer.

11.4.1 On the Theory of HM

It is crucial to achieve a high-power output signal of the HM based on SIS to reach the effective FFO synchronization. The experimental data shows that power of the IF signal depends in a complicated way on the HM bias voltage, frequencies and powers of the LO and FFO signals. To study the HM properties, it has been theoretically analyzed and the 3D dependences of its output signal power versus the bias voltage of SIS junction and the LO power have been calculated. The

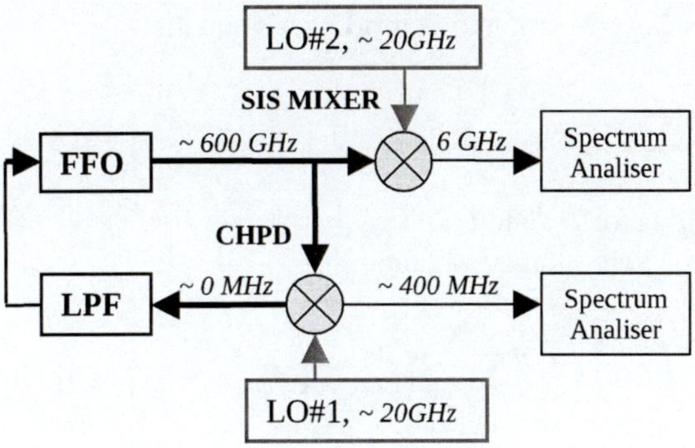

Fig. 11.6 Block diagram of the system for the FFO synchronization by the CHPD

detailed theoretical description of the SIS junction under action of microwave frequency signals is given in [12, 18, 19]. In [12], a simplifying assumption about the small amplitude of the input signal and shunting of the highest harmonics of the LO by junction capacitance was used. Papers [18, 19] involved the method of the SIS junction description in general case, that is for any powers of input signals. Nowadays the theory [18, 19] is the most complete; it takes into account existence of the harmonics generated by SIS junction and influence of the external electromagnetic environment. The calculations of some HM characteristics on the SIS based on this theory are given in [19].

We present the simplified model, which gives us the opportunity to reduce the time of HM characteristics calculation, but good qualitative and quantitative agreement with the experimental data still can be obtained. Let us consider the model of the weakly interacting quasi-particles under the influence of the periodic electric field without taking into account spin effects as in [20].

Wave function of a quasi-particle with energy E without applying high-frequency electric field is $\Psi = f(x, y, z)\exp(-\mathrm{i}Et/\hbar)$, where $f(x, y, z)$ – certain function of coordinates, i – imaginary unit, t – time, $\hbar$ – Planck's constant. This wave function is the eigen function of a nonexcited system Hamiltonian H_0. The voltage across the junction is $V_{\omega 1}\cos(\omega_1 t) + V_{\omega 2}\cos(\omega_2 t)$ then two periodic signals of frequencies ω_1, ω_2 and of amplitudes $V_{\omega 1}$, $V_{\omega 2}$ are applied to junction electrodes. The Hamiltonian of quasi-particles system is then:

$$H = H_0 + \mathrm{eV}_{\omega 1}\cos(\omega_1 t) + \mathrm{eV}_{\omega 2}\cos(\omega_1 t),$$

where H_0 – nonexcited system Hamiltonian, H – Hamiltonian of system influenced by two harmonic signals, e – charge of electron. The new wave function is:

$$\Psi = f(x, y, z)\exp(-\mathrm{i}Et/\hbar)\left(\sum_n B_n \exp(-in\omega_1 t)\right)\left(\sum_m C_m \exp(-im\omega_2 t)\right),$$

where B_n and C_m – unknown functions.

Applying this wave function to Schrödinger's equation

$$i\hbar\frac{\partial\psi}{\partial t} = \widehat{H}\psi,$$

we obtain equations for B_n and C_m.

Solution of this Schrödinger's equation is:

$$\Psi = f(x)\exp(-\mathrm{i}Et/\hbar)\left(\sum_n\sum_m J_n\left(\frac{\mathrm{eV}_{\omega 1}}{\hbar\omega_1}\right)J_m\left(\frac{\mathrm{eV}_{\omega 2}}{\hbar\omega_2}\right)\exp\left[-\mathrm{i}(n\omega_1+m\omega_2)t\right]\right),$$

where $J_n(\alpha)$ Bessel function nth order

One can see that quasiparticles energy levels are split into levels described by wave functions Ψ_{nm} with energies $E + n\hbar\omega_1 + m\hbar\omega_2$; $n, m = 0, \pm 1, \pm 2 \ldots$. Probability of occupation of such levels is proportional to $J_n\left(\frac{\mathrm{eV}_{\omega 1}}{\hbar\omega_1}\right)J_m\left(\frac{\mathrm{eV}_{\omega 2}}{\hbar\omega_2}\right)$.

Quasi-particle tunnel current is provided by quasiparticles transport between SIS junction electrodes. This current is described as complex function of current response $j(V)$, here V is a DC voltage applied to junction. The function $j(V)$ is calculated in [21]: $j(V) = \mathrm{i}I_{\mathrm{dc}}(V) + I_{\mathrm{KK}}(V)$. Here, $I_{\mathrm{dc}}(V)$ is unpumped I–V curve of the SIS, and the $I_{\mathrm{KK}}(V)$ relates to the $I_{\mathrm{dc}}(V)$ as Kramers–Kronig transform.

It should be noted that the experimentally measured $j(V)$ function was used in the calculations. The I–V curve of the HM contains all the information about the gap voltage, its smearing, the gap current (current step at the gap voltage), and leakage current below the gap.

The quasiparticle increases its energy by $\hbar\omega$ when the radiation quantum is absorbed. One can describe this process by application of the voltage $\hbar\omega/e$ to the junction. Therefore, the tunnel current is defined by function $j(V + \hbar\omega/e)$. As far as the quasiparticle is able to absorb several photons of the energy $\hbar\omega_1$ and $\hbar\omega_2$, in order to find the total tunnel current we should sum the current response functions $j_{\mathrm{nm}} = j\left(V + \frac{n\hbar\omega_1}{e} + \frac{m\hbar\omega_2}{e}\right)$ subject to the probability of the quasiparticle tunneling. The quasiparticle transmission probability of state Ψ_{nm} to state Ψ_{lk} is defined by the matrix element $\langle\Psi_{\mathrm{lk}}|\Psi_{\mathrm{nm}}\rangle$, where

$$|\Psi_{\mathrm{nm}}\rangle = f(x)\exp(-\mathrm{i}Et/\hbar)\left(J_n\left(\frac{\mathrm{eV}_{\omega 1}}{\hbar\omega_1}\right)J_m\left(\frac{\mathrm{eV}_{\omega 2}}{\hbar\omega_2}\right)\exp\left[-\mathrm{i}(n\omega_1+m\omega_2)t\right]\right),$$

$$\langle\Psi_{\mathrm{lk}}| = g(x)\exp(-\mathrm{i}Et/\hbar)\left(J_l\left(\frac{\mathrm{eV}_{\omega 1}}{\hbar\omega_1}\right)J_k\left(\frac{\mathrm{eV}_{\omega 2}}{\hbar\omega_2}\right)\exp\left[\mathrm{i}(l\omega_1+k\omega_2)t\right]\right).$$

The changing of the summation variable leads to:

$$I(V,t) = \mathrm{Im}\sum_{n,m,l,k} J_n(\alpha_1)J_{n+l}(\alpha_1)J_m(\alpha_2)J_{m+k}(\alpha_2)^* * \exp\left[-\mathrm{i}(l\omega_1+k\omega_2)t\right]j\left(V + \frac{n\hbar\omega_1}{e} + \frac{m\hbar\omega_2}{e}\right),$$

where $\alpha_i = \frac{\mathrm{eV}_{\omega i}}{\hbar\omega_i}$.

It should be noted that we can overwrite the current function in the form:

$$I(V,t) = a_0 + \sum_{l=1}^{\infty}\sum_{k=1}^{\infty}(2a_{lk}\cos((l\omega_1 + k\omega_2)t) + 2b_{lk}\sin((l\omega_1 + k\omega_2)t)).$$

This formula shows that the signals of the frequencies described as $l\omega_1 + k\omega_2$ (l and k are integers) are generated on the SIS junction. For the practical application of the HM, the first signal frequency is close to the frequency of second signal harmonic k, i.e. $l = 1$ and $\omega_1 - k\omega_2 << \omega_2$. Let us put that frequency of RF signal is $\omega_1 \equiv 2\pi f_{\mathrm{RF}}$, and for LO signal is $\omega_2 \equiv 2\pi f_{\mathrm{LO}}$. Then current amplitude of IF $f_{\mathrm{IF}} = f_{\mathrm{RF}} - nf_{\mathrm{LO}}$ is given by

$I_{\mathrm{IF}} = \sqrt{{a_{1k}}^2 + {b_{1k}}^2}$, here

$$a_{1k}(V) = \sum_{n,m} J_n(\alpha_1)J_m(\alpha_2)\left[J_{n+1}(\alpha_1)J_{m-k}(\alpha_2) + J_{n-1}(\alpha_1)J_{m+k}(\alpha_2)\right] \times I_{\mathrm{dc}}\left(V + \frac{n\hbar\omega_1}{e} + \frac{m\hbar\omega_2}{e}\right),$$

$$b_{1k}(V) = \sum_{n,m} J_n(\alpha_1)J_m(\alpha_2)\left[J_{n+1}(\alpha_1)J_{m-k}(\alpha_2) - J_{n-1}(\alpha_1)J_{m+k}(\alpha_2)\right] \times I_{\mathrm{kk}}\left(V + \frac{n\hbar\omega_1}{e} + \frac{m\hbar\omega_2}{e}\right).$$

The dependences of the IF signal power versus the input signals parameters and the junction bias voltage have been calculated by the presented formula. The result of such calculations is shown in Fig. 11.7 (top part).

The experimental study of such dependences is also performed. In Fig. 11.7 (bottom part), there is a dependence of the IF power versus LO power and bias voltage for the RF signal frequency 636 GHz and the LO frequency 18 GHz. Good qualitative and quantitative correspondence between theory and experiment is achieved (Fig. 11.7). The Nb–AlOx–Nb SIS junction with area of 1 μ^2 and gap current 90 μA was used in the experiment. The maximum output signal power of about −90 dBm has been obtained for described frequencies. However, the calculations show that 5 dB larger value would be achieved at more careful adjustment.

11.4.2 Experimental Demonstration

A test circuit presented by the diagram in Fig. 11.6 has been experimentally realized. The feedback loop between the HM and the FFO was implemented in two ways: by lumped elements on a contact plate and by the microstrip lines directly on the chip.

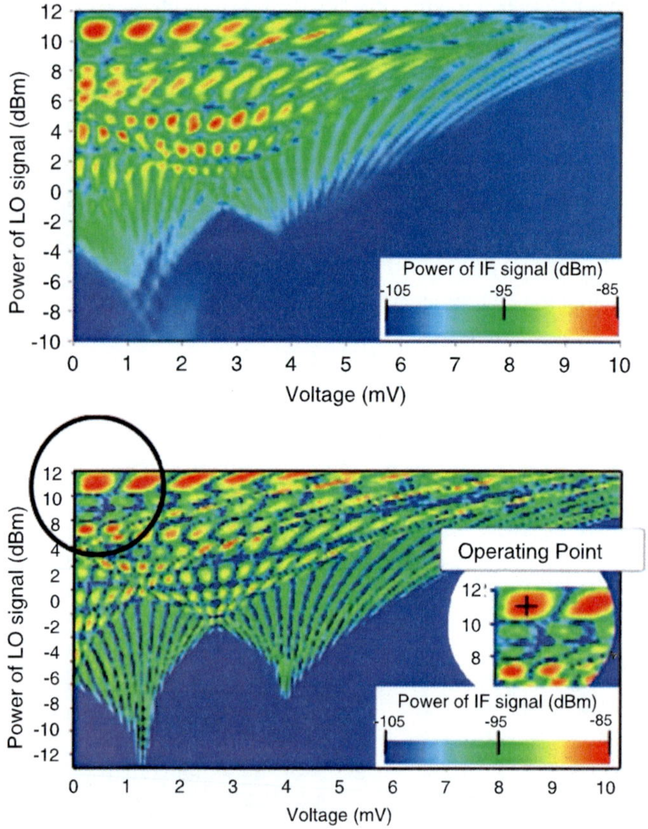

Fig. 11.7 Experimental (*up*) and theoretical (*bottom*) dependence of IF signal's power versus LO signal's power and bias voltage across the HM. The operating point with the maximum power of IF signal (−85 dBm) is shown on the theoretical dependence. The FFO frequency is 636 GHz, LO frequency – 18 GHz (the 35th harmonic of LO is used) as result frequency of IF signal is equal 6 GHz. The critical current of the SIS junction was suppressed by the magnetic field

In the experiment, the FFO signal is split into two channels to pump both mixers. Each part of the signal is downconverted, so that output signals frequencies are 400 MHz for the CHPD and of about 6 GHz for the SIS (Fig. 11.6). At the first stage of the experiment, the CHPD output signal is observed at 400 MHz by the spectrum analyzer and is maximized by determining of the optimal operating point (see Fig. 11.7). After the optimum is found, the LO#1 frequency is tuned so that the IF of the CHPD becomes "0" instead of 400 MHz and the FFO becomes phase-locked. At the same time, the IF of the HM #2 is also changed but the FFO radiation line is still presented on spectrum analyzer screen and synchronization effect is observed.

The result of the FFO synchronization by the CHPD is shown in the Fig. 11.8. This spectrum demonstrates the validity and the potential of the concept. The group

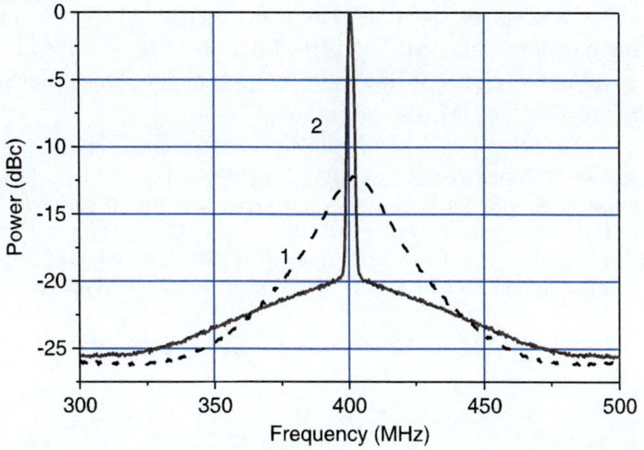

Fig. 11.8 Downconverted spectrum of the FFO: curve "1" – free running line, LW about 20 MHz; curve "2" – FFO is phase-locked by the CHPD

delay of the loop is less than 2 ns, and the bandwidth of such a PLL system is expected to be about 100 MHz. For the data presented in Fig. 11.8 (curve "1"), free running FFO linewidth is about 20 MHz and the PLL system based on CHPD can phase-lock as much as 70% of the emitted FFO power (curve "2"). According to the scale effect, described for Fig. 11.5, the BW of such a PLL system should be also about 100 MHz. However, the HM output signal is limited and the open loop gain is not large enough to for the optimal synchronization. As a result, the phase noise shoulders like in the Fig. 11.4 demonstrating BW are not observed and the maximum SR is not achieved.

11.5 Conclusions

The concept of the ultra-wideband CPLL system based on the CPD has been developed and experimentally proven. The CPD based on the SIS junction has been studied in detail. Synchronization bandwidth as large as 40 MHz has been realized for cryogenic PL; that considerably exceeds the value of about 12 MHz obtained for regular room-temperature PLL. The innovative CPLL system can phase-lock more than 50% of the FFO spectral line if the free-running FFO is about 12 MHz. Practical implementation of the CPLL looks especially promising for phase-locking of new superconducting LOs utilizing NbN/NbTiN films, as well as for SIR applications in the interferometry where the extremely low LO phase noise is required.

The novel application of the SIS junction – the CHPD has been proposed. The theoretical and experimental studies of the HM are performed. Comparison of the theoretical and experimental data demonstrates a good qualitative and quantitative agreement. The concept of the CHPD is experimentally realized and the FFO phase-locking has been obtained. The part of the phase-locked FFO power is not ultimate because output signal of HM was not large enough. This problem would be overcome by utilizing the HM with larger area.

Acknowledgements The authors thank colleagues at IREE and SRON for support and assistance in experiments and fruitful discussions: A.M. Baryshev, P.N. Dmitriev, A.B. Ermakov, P.A. Yagoubov, H. Golstein, and M. van den Bemt.

The work was supported in parts by RFBR projects 09-02-00246, 09-02-12172-ofi-m, Grant for Leading Scientific School 5423.2010.2 and State contract No. 02.740.11.0795.

References

1. V.P. Koshelets, S.V. Shitov, L.V. Filippenko, A.M. Baryshev, H. Golstein, T. de Graauw, W. Luinge, H. Schaeffer, H. van de Stadt, First implementation of a superconducting integrated receiver at 450 GHz. Appl. Phys. Lett. **68**(9), 1273 (1996)
2. V.P. Koshelets, S.V. Shitov, A.B. Ermakov, O.V. Koryukin, L.V. Filippenko, A.V. Khudchenko, M.Y.u. Torgashin, P. Yagoubov, R. Hoogeveen, O.M. Pylypenko, Superconducting integrated receiver for TELIS. IEEE Trans. Appl. Supercond. **15**, 960–963, 2005
3. V.P. Koshelets, S.V. Shitov, A.V. Shchukin, L.V. Filippenko, J. Mygind, Linewidth of submillimeter wave flux-flow oscillators. Appl. Phys. Lett. **69**, 699–701 (1996)
4. V.P. Koshelets, S.V. Shitov, P.N. Dmitriev, A.B. Ermakov, L.V. Filippenko, V.V. Khodos, V.L. Vaks, A.M. Baryshev, P.R. Wesselius, J. Mygind, Towards a phase-locked superconducting integrated receiver: prospects and limitations. Phys. C **367**, 249–255 (2002)
5. V.P. Koshelets, S.V. Shitov, A.V. Shchukin, L.V. Filippenko, J. Mygind, A.V. Ustinov, Self-pumping effects and radiation linewidth of josephson flux-flow oscillators. Phys. Rev. B **56**, 5572–5577 (1997)
6. V.P. Koshelets, P.N. Dmitriev, A.B. Ermakov, A.S. Sobolev, M.Yu. Torgashin, V.V. Kurin, A.L. Pankratov, J. Mygind, Optimization of the phase-locked flux-flow oscillator for the submm integrated receiver. IEEE Trans. Appl. Supercond. **15**, 964–967 (2005)
7. S. AlBanna, R. Brito, B. Shillue, ALMA 1st LO photonic reference: status of phase drift measurements. NRAO website. Available: http://www.tuc.nrao.edu/~bshillue/E2E_Phase_Drift_Status_RevB.doc (2005)
8. A.V. Khudchenko, V.P. Koshelets, P.N. Dmitriev, A.B. Ermakov, P.A. Yagoubov, O.M. Pylypenko, Cryogenic phase detector for superconducting integrated receiver. IEEE Trans. Appl. Supercond. **17**, 606–608 (2007)
9. A.V. Khudchenko, V.P. Koshelets, P.N. Dmitriev, A.B. Ermakov, O.M. Pylypenko, P.A. Yagoubov, Cryogenic phase locking loop system for flux-flow oscillator, in *Proceedings of the 19th International Symposium on Space Terahertz Technology (ISSTT-08)* **2**, 511–515 (2009)
10. A.V. Khudchenko, V.P. Koshelets, P.N. Dmitriev, A.B. Ermakov, P.A. Yagoubov, O.M. Pylypenko, Cryogenic phase locking loop system for superconducting integrated receiver, Supercond. Sci. Technol. **22**(8), (2009)
11. V.C. Lindsey, *Synchronization Systems in Communication and Control* (Prentice-Hall, New Jersey, 1972)
12. J.R. Tucker, M.J. Feldman, Rev. Mod. Phys. **57**(4), 1055 (1985).
13. K.K. Likharev, *Dynamics of Josephson junctions and circuits* (Gordon and Breach, 1986)
14. A.V. Khudchenko, V.P. Koshelets, P.N. Dmitriev, Ermakov, A cryogenic phase detector for a cooled wideband phase-lock loop system. J. Commun. Tech. Electron. **53**(5) 594–599 (2008)
15. F.M. Gardner, *Phaselock Techniques* (Wiley, 1979)

16. A.V. Khudchenko, V.P. Koshelets, P.N. Dmitriev, A.B. Ermakov, O.M. Pylypenko, Cryogenic phase locking loop system for flux-flow oscillators, *Extended Abstracts of the International Superconductive Conference ISEC'2009* HF-P17 Japan (2009)
17. A.L. Pankratov, V.L. Vaks, V.P. Koshelets, Spectral properties of phase locked flux-flow oscillator. J. Appl. Phys. **102**, 0629 (2007)
18. S. Withington, P. Kittara, G. Yassin, Multitone quantum simulations of saturating tunnel junction mixers. J. Appl. Phys. **93**, 9812–9822 (2003)
19. P. Kittara, S. Withington, G. Yassin, Theoretical and numerical analysis of very high harmonic superconducting tunnel junction mixers. J. Appl. Phys. **101**, 024508 (2007)
20. P.K. Tien, J.P. Gordon, Multiphoton process observed in the interaction of microwave fields with the tunneling between superconductor filmsPhys. Rev. **129**, 647–653 (1963)
21. N.R. Werthamer, Nonlinear self-coupling of josephson radiation in superconducting tunnel junction Phys. Rev. **147**, 255 (1966)

Conclusion

As already mentioned in preface, it is impossible in one book to highlight all the aspects of a rapid developing Superconducting Electronics. To extend the knowledge about the problems, presented in 11 chapters of the book, the reader will be helped with some relevant publications that we can recommend in the last chapter, Bibliography.

A. Sidorenko (ed.), *Fundamentals of Superconducting Nanoelectronics*, NanoScience and Technology, DOI 10.1007/978-3-642-20158-5,

Bibliography (Some relevant books and reviews)

1. A.I. Larkin, A.A. Varlamov, *Theory of Fluctuations in Superconductors*, 2nd edn. (Oxford University Press, New York, 2009)
2. K.Yu. Arutyunov, D.S. Golubev, A.D. Zaikin, Superconductivity in one dimension. Phys. Rep. **464**(N 1–2), 1–70 (2008)
3. M. Sigrist, K. Ueda, Phenomenological theory of unconventional superconductivity. Rev. Mod. Phys. **63**, 239 (1991)
4. J. Clarke, *Nonequilibrium Superconductivity*, ed. by D.N. Langenberg, A.I. Larkin (North-Holland, Amsterdam, 1986)
5. R. Tidecks, *Current-Induced Nonequilibrium Phenomena in Quasi-One-Dimensional Superconductors* (Springer, New York, 1990)
6. P. Cadden-Zimansky, J. Wei, V. Chandrasekhar, Cooper-pair-mediated coherence between two normal metals. Nat. Phys. **5**, 393 (2009)
7. A. Levy Yeyati, F.S. Bergeret, A. Martin-Rodero, T.M. Klapwijk, Entangled Andreev pairs and collective excitations in nanoscale superconductors. Nat. Phys. **3**, 455 (2007)
8. A.I. Buzdin, Proximity effects in superconductor-ferromagnet heterostructures. Rev. Mod. Phys. **77**, 935–976 (2005)
9. H. Hahn, A. Sidorenko, I. Tiginyanu (eds.) *Nanoscale Phenomena – Fundamentals and Applications* (Springer, New York, 2009)
10. Yu.G. Naidyuk, I.K. Yanson, *Point-Contact Spectroscopy, Springer Series in Solid-State Sciences*, vol. 145 (Springer Science + Business Media, Inc, New York, 2005)
11. K.H. Müller, V. Narozhnyi (eds.) *Rare Earth Transition Metal Borocarbides (Nitrides): Superconducting, Magnetic and Normal State Properties*, vol. 14 (Kluwer Academic Publishers, Dordrecht, 2001)
12. J.W.A. Robinson, J.D.S. Witt, M.G. Blamire, Controlled injection of spin-triplet supercurrents into a strong ferromagnet. Science **329**, 59 (2010)
13. T. Klapwijk, Magnetic nanostructures: supercurrents in ferromagnets. Nat. Phys. **6**, 329 (2010)

A. Sidorenko (ed.), *Fundamentals of Superconducting Nanoelectronics*, NanoScience and Technology, DOI 10.1007/978-3-642-20158-5,

14. V.V. Ryazanov, Josephson superconductor—ferromagnet—superconductor π-contact as an element of a quantum bit (experiment). Usp. Fiz. Nauk **169**, 920 (1999) [Phys.-Usp. **42**, 825 (1999)]
15. A.K. Feofanov, V.A. Oboznov, V.V. Bol'ginov, J. Lisenfeld, S. Poletto, V.V. Ryazanov, A.N. Rossolenko, M. Khabipov, D. Balashov, A.B. Zorin, P.N. Dmitriev, V.P. Koshelets, A.V. Ustinov, Implementation of superconductor/ferromagnet/ superconductor π-shifters in superconducting digital and quantum circuits. Nat. Phys. **6**, 593 (2010)
16. V.P. Koshelets, S.V. Shitov, Integrated superconducting receivers. Supercond. Sci. Technol. **13**, R53 (2000)
17. V.P. Koshelets, J. Mygind, Flux flow oscillators for superconducting integrated submm wave receivers, in *Studies of High Temperature Superconductors*, ed. by A.V. Narlikar (NOVA Science Publishers, New York, 2001)
18. T. Van Duzer, C.W. Turner (eds.), *Principles of Superconductive Devices and Circuits*, 2nd edn. (Prentice Hall, Upper Saddle River, 1998)
19. V.I. Zdravkov, J. Kehrle, G. Obermeier, S. Gsell, C. Müller, H.-A. Krug von Nidda, J. Lindner, J. Moosburger-Will, E. Nold, R. Morari, V.V. Ryazanov, A.S. Sidorenko, S. Horn, R. Tidecks, L.R. Tagirov. Reentrant superconductivity in superconductor/ferromagnetic-alloy bilayers. Phys. Rev. B **82**, 054517 (2010)

About the Editor

Prof. Dr. Anatolie Sidorenko

Director of the Institute of Electronic Engineering and Nanotechnologies ASM, Kishinev, Moldova

Thirty-five years experience in the field of superconductivity, Ph.D. (1979) in the Department of Superconducting Electronics of Institute for Low Temperature Physics, Kharkov, Ukraine; Doctor of Science (1991) in the Cryogenic Laboratory of Institute of Applied Physics, Kishinev, Moldova, 1993–2003 Humboldt-fellow and researcher in Karlsruhe University-Forschungszentrum Karlsruhe (now "KIT" – Karlsruhe Institute of Technology).

Research activity – *superconductivity of low-dimensional and layered superconductors*:

- Critical fluctuations (change of dimensionality, 2D–0D crossover for superconducting thin films in strong magnetic field)
- Double-dimensional crossover 3D–2D–3D of layered superconductor in magnetic field

A. Sidorenko (ed.), *Fundamentals of Superconducting Nanoelectronics*, NanoScience and Technology, DOI 10.1007/978-3-642-20158-5,

- Superconducting fractals – peculiarities of artificially layered superconductors with fractal topology
- Problem of "positive curvature" of the upper critical magnetic field in layered superconductors
- Giant oscillations of interlayer coupling strength in layered superconductor
- Nonergodic behavior of high-temperature superconductors (long-term relaxations of thermoremanent magnetization)
- Multiband superconductivity: two-dimensional character of superconductivity nucleation in MgB_2 and anomalous thermally activated flux flow (TAFF)
- Superconductor/ferromagnet-layered systems investigation. Detection of the double re-entrant superconductivity in Nb/CuNi nanostructures

The author has to his credit 210 publications and 26 patents on superconducting sensors and devices.

Index

A. Sidorenko (ed.), *Fundamentals of Superconducting Nanoelectronics*, NanoScience and Technology, DOI 10.1007/978-3-642-20158-5,